S0-BLA-436

ECOLOGY OF INLAND WATERS AND ESTUARIES

REINHOLD BOOKS IN THE BIOLOGICAL SCIENCES

Consulting Editor:
PROFESSOR PETER GRAY

Department of Biological Sciences
University of Pittsburgh
Pittsburgh, Pennsylvania

CONSULTING EDITOR'S STATEMENT

I AM HAPPY that the author has called his book "Inland Waters and Estuaries" rather than "Limnology." This is not to suggest that this addition to the REINHOLD BOOKS IN THE BIOLOGICAL SCIENCES slights the biogeochemistry of lakes. It is only to emphasize that the Reinhold Series is once more able to add a new dimension to an established field. Dr. Reid's long experience as an investigator and teacher particularly qualifies him to discuss the ecology of those long-neglected waters which bridge the gap between the lakes, the rivers, and the oceans.

This book follows the integrative approach. Topics of broad scope, such as the oxygen content of waters, are first discussed in their theoretical aspects, and the facts brought out in this discussion are then applied specifically to lakes, streams, and estuaries. This technique results in a book from which the beginner can derive a broad basis of understanding but in which the specialist can also check his facts.

PETER GRAY

Pittsburgh, Pennsylvania
January, 1961

ECOLOGY OF INLAND WATERS AND ESTUARIES

GEORGE K. REID
Professor of Biology
Florida Presbyterian College
St. Petersburg, Florida

REINHOLD PUBLISHING CORPORATION, New York
Chapman & Hall, Ltd., London

Library of Congress Catalog Card Number: 61-7436
Printed in the United States of America

TO

Pauline and Eugenie Reid

AND TO THE MEMORY OF

Professor W. C. Allee

PREFACE

THIS BOOK is intended as an introduction to the elemental factors and processes that operate in lakes, streams, and estuaries as dynamic systems. In seeking to fulfill this intention the aim is primarily two-fold. First, the book attempts to bring into summary some of the major aspects of the knowledge that has been amassed from the study of inland waters and estuaries by many investigators. Such a study, drawing as it does from geology, hydrology, physics, chemistry, and biology, is obviously vast and complex. In order to appreciate so broad a field it is necessary to view a considerable body of basic factual information.

Knowledge of basic facts in any science serves a further purpose in providing background for the understanding of important principles. Therefore, in its second aim, the book attempts to place related principles in proper perspective with respect to the given facts. The author has tried to emphasize principles without becoming enmeshed in a web of detailed information and terminology.

As is the case with most textbooks in aquatic ecology, this one is essentially derived from the teaching of college courses. Experience gained from student reaction and consultation with numerous teachers of the subject has led to the belief that the most logical approach to the study of natural communities is through early concern with the "elements" of the subject followed by consideration of these fundamentals as they pertain to principles. It is the author's belief that to understand the whole, one must have knowledge of the origin and nature of the parts. Such is the organization of this book. It develops from the particular to the general, while incorporating themes of fitness of the environment, energy traffic, and adaptations of organisms.

To the student taking his first taste of a subject of this nature, the approach used here logically brings the complex of interrelating factors into more real perspective. There is also reason to believe that the student feels more secure in his background when he reaches "limiting factors," "energy relationships," "community dynamics," and the like. For the person who may be more interested in species ecology than in community ecology, or in a particular factor such as heat or dissolved solids, the elemental approach serves as a starting point.

The book is organized into five parts. Part I is concerned with the origins of lake basins, and stream and estuary channels. Water is then added, and the morphology of what may be termed lakes, streams, and estuaries is considered. Stream dynamics and hydrology are emphasized for the first time in an introductory textbook in ecology. This emphasis is based on the importance of these aspects in the development and maintenance of stream communities. Part II briefly presents something of the nature of water as a substance with many unique properties. These properties account for various environmental features and processes in natural waters. Part III describes physical and chemical characteristics of natural water and emphasizes the dynamic interrelationships of these factors. Part IV presents the plants and animals which have become adapted to lakes, streams, and estuaries. This part simply acquaints the student with the general nature of various kinds of organisms (remember that in many instances students may have had only general biology and chances are great that they learned only those plants and animals typical to such courses). In Part V the "particulars" are considered as they relate to the over-all systems: that is, population and community organization principles are discussed.

Any textbook represents a synthesis of information contributed to knowledge through publication of research results in scientific journals. This book is no exception. The author has drawn freely from the original works of others. Citations of these articles and papers have not been made within the text in belief that doing so detracts from the readability, particularly at the introductory level. All sources are, however, given in the Bibliography.

For assistance in the preparation of this book I am indebted to many persons, only a few of whom can be acknowledged here. To Dr. Robert W. Pennak, University of Colorado, who read the entire manuscript, I am greatly indebted for invaluable criticisms and suggestions. I am especially grateful to Mrs. Aline Hansens, North Brunswick, New Jersey, for the illustrations. Gratitude is expressed to Dr. H. T. Odum for comments on the original outline.

Dr. Peter Gray, University of Pittsburgh, Consulting Editor to the publisher, and, indeed, the entire editorial staff of the publisher have been most helpful and encouraging in the task of getting out the book.

Special thanks are accorded the following for reading and offering constructive criticisms of certain chapters: Drs. D. E. Fairbrothers and E. T. Moul, Department of Botany; W. B. Foster and P. G. Pearson, Department of Zoology; P. E. Wolff, Department of Geology; and R. A. Barnes, Department of Chemistry, Rutgers University. At the University of Pittsburgh, Department of Biological Sciences: Drs. R. Dugdale, R. T. Hartman, and C. A. Tryon, Jr. At Florida Presbyterian College, St. Petersburg: Dr. D. E. Anderson.

For specific information on certain phases of the material I should like to express gratitude to Drs. L. Berner and M. J. Westfall, University of Florida; H. H. Hobbs, University of Virginia; W. J. Woods, University of Pittsburgh; and to Mr. R. W. Pride, U. S. Geological Survey, Ocala, Florida.

Dr. Paul G. Pearson, Rutgers University, contributed the section on "Population Density" in Chapter 13. For this, and much worthwhile discussion on the approach and contents of the book, I am indeed grateful.

Appreciation is expressed to individuals and agencies for furnishing photographs and other illustrative materials. Specific credit is given in the captions. Especial mention is made of the cooperation of Mr. Seldin Tinsley, U. S. Soil Conservation Service, New Brunswick, New Jersey.

Mrs. Arlene Hahn and Miss Marion Connolly, New Brunswick, and Mrs. Cynthia Lankford, St. Petersburg, typed the manuscript; to them thanks are given.

The author wishes to take this opportunity to express his deep appreciation to his wife, Eugenie, for her patience and cooperation during the two years in which much of his time was directed toward work on the manuscript. Her material contributions toward indexing and in reading manuscript and proof have aided immeasurably to the undertaking.

Although great effort has been exerted toward presenting carefully edited and corrected information, some inaccuracies are certain to occur. The author will greatly appreciate receiving notice of any faults in the present book.

George K. Reid

St. Petersburg, Florida
January, 1961

CONTENTS

Part II. Water

Part III. Natural Waters as Environment

Part V. Relationships of Organisms and Environment

Chapter 1

INTRODUCTION

In Section 1 of Book 1 of his *Historia Animalium* Aristotle recognized that all inland waters are not alike, at least insofar as the animal inhabitants are concerned. For in this writing he erected a system of classification which included, among others, such categories as "river dwellers," "lake dwellers," and "marsh dwellers." Theophrastus, a student of Aristotle's, gave cognizance to plants in a somewhat similar sense in distinguishing among "plants of deep fresh water," "plants of shallow lake shores," "plants of wet banks of streams," and others.

Although the ancestry of freshwater study is varied and old, its place as a circumscribed and directed area of scientific endeavor and achievement was not established until 1869. That year witnessed the publication of a paper on the bottom fauna of Lac Léman (now Lake Geneva), Switzerland, by F. A. Forel, and, according to many historians, marked the beginning of the scientific study of lakes. To this study, the name *limnology* (from the Greek *limnē*, meaning "marsh") was applied. The term was first used in another report by Forel, this one entitled *Le Léman, monographie limnologique*. The first two volumes of this monograph described the geology, chemistry, and physics of the Swiss lake, and were published in 1892 and 1895. The biological aspects of the lake appeared in 1904. Altogether, this monumental treatise contains some two thousand pages, and it, together with an impressive list of other publications, has earned for Forel, a professor at the University of Lausanne, Switzerland, the title, "Father of Limnology."

From the content of the volumes of Forel's monograph can be seen the disciplines which form the core of the study of inland waters, namely: geology, chemistry, physics, and biology. To these can be added the more modern concept of ecology, which interrelates the several aspects into an integrated, functional system. Forel's major contribution was in the consideration of the water, or what might be termed the *environment* of the organisms inhabiting the lake; his treatment of the plants and animals of Léman consisted only of a partial list of the biota.

It remained for an American, Professor E. A. Birge, of the University of Wisconsin, to wed biology to the Forelian physicochemical limnology.

This Birge did through his studies of the microscopic, floating assemblage, the *plankton*, of Lake Mendota, Wisconsin. These investigations very early led Professor Birge to appreciate the relationships between the plankton and the physical and chemical aspects of lakes. Beginning in 1873, and continuing to 1941, Birge's researches form a major support in the structure of limnology. His importance rests not alone in his studious investigations of the plankton; his work took him into the realm of physical limnology, and there he discovered much concerning light penetration, thermal properties, and currents in lake waters. In 1909, there appeared a paper written by Birge and Juday. With that paper, Chancey Juday took his place beside Birge to form a team that made notable gains in American limnology for nearly thirty years.

By 1900, limnology was becoming more ecological in approach, being stimulated to a great extent in this direction by Professor S. A. Forbes of the University of Illinois–this institution, itself, gaining considerable stature in the field of hydrobiology. A paper by Forbes in 1897, entitled "The Lake as a Microcosm," described the lake as a "small world" composed of environmental features and living organisms bound and organized by interdependences and interrelationships. It is basically this concept that permeates much of modern limnology.

Today, limnologists may be specialized as physical limnologists, or chemical limnologists, or biological limnologists. But whatever their interests, the ultimate aim of limnology is in the direction of recognizing the components of aquatic systems and understanding their functions in the dynamics of the whole. In point of fact, this is aquatic ecology.

The term *Oekologie* was first used in a scientific sense by a German biologist, Ernst Haeckel, in a paper published in 1870. The word derives from the Greek noun for "home" (*oikos*), and, in its present spelling and meaning, *ecology* is the study of the relationships between an organism and its environment. By "organism" we mean any living thing; and since a living thing is a representative of a species, our fundamental unit is usually (although not necessarily always) the species, or "species population." The term *environment* in the definition refers in a broad, but by no means rigidly delineated, sense to the physical, chemical, and biological features of the habitat, or locality, in which the organism lives.

Since a given environment normally harbors more than a single animal or plant species, or species population, we come to recognize another level of relationships, that of an assemblage of populations, or the *community* (see definitions in Chapter 14). Thus, several inroads are open for undertaking the study of "relationships." The study of interrelationships between a single individual or species population and the environment is *population ecology* ("autecology" in some literature). For example, the investigation of the distribution and activities of the perch

FIGURE 1·1. The physical, chemical, and biological features of natural waters may differ greatly. The upper photograph shows a rock-strewn, Pennsylvania stream characterized by variable flow attributes and temperature. The lower photograph depicts a sand-bottom Florida stream having generally uniform flow features and temperature. The organisms inhabiting each stream are those which have become adapted to the respective environmental characteristics.

(*Perca flavescens*) in Lake Doe is population ecology. An aggregation of several species populations constitutes a community and the study of such an assemblage is termed *community ecology* ("synecology"). To return to the aforementioned hypothetical lake, an investigation of the general, or over-all, ecology of the shallow shore-zone, or the bottom, or the entire lake, would be called community ecology. Community ecology demands knowledge of the individual kinds of organisms present

FIGURE 1·2. Examination of bottom deposits reveals information on chemical processes, and on the biology and ecology of the bottom community. In the background hangs a Petersen dredge, used for obtaining bottom samples. (Photograph courtesy of University of Michigan News Bureau.)

and leans heavily upon taxonomy. The members of a community react with the physical and chemical features of the environment to form an *ecosystem*. Ecosystem ecology is a form of community ecology which places more emphasis on mass of organisms and less upon the individual species; it is concerned with functional aspects of biogeochemical cycles and energy transfer. This approach is similar to the physiological study of an organism.

Thus, in its more comprehensive and modern concept, limnology is an area of science concerned with the characteristics of inland waters, the forces and processes which mold and maintain the integrity of these waters, the interrelationships between water and basin, and the community of living organisms inhabiting the environment. Limnology, therefore, embraces certain aspects of related sciences. Geology supplies principles and applications necessary for the appreciation of the origin of lake basins and drainage systems, and the processes involved in basin modification. Geochemistry relates the composition of the substrate to the chemical nature of the waters of lakes. Limnology depends upon physics for concepts and understanding of such important aspects as light penetra-

FIGURE 1·3. Most natural waters support a community of microscopic "drifting" organisms, the "plankton." Highly important in the metabolism of a body of water, these plants and animals are obtained by straining through a fine-meshed net of silk. The net shown in the photograph is a Clarke-Bumpus model.

tion, heat dynamics, and water movements. From inorganic and organic chemistry, limnology takes techniques and principles necessary for describing and measuring a great array of processes involving gases, liquids, and solids. Biology contributes to the knowledge of plants and animals, their metabolism, life histories, and living habits. It also contributes to studies of biological processes concerned with the synthesis of energy-containing protoplasm, the transfer of energy through the ecosystem, and the decomposition of organic matter with the release of basic nutrients.

The study of streams has, in general, lagged behind that of lakes. Consequently our knowledge of stream dynamics and ecological relationships is not extensive. Notable pioneering efforts toward stream investigations

FIGURE 1·4. The rubble of stream floors is typically inhabited by various kinds of plants and animals. The larger individuals can be taken by loosing them from the substrate and allowing the current to carry the organisms into a net placed immediately downstream. A Surber-type net is being used by the student in the photograph.

in North America were begun about 1875 in Illinois, under the direction of S. A. Forbes mentioned previously. The work of C. A. Kofoid on the plankton of the Illinois River, from 1894 to 1899, has long stood as classic in this field. Although the term *limnology* was originally applied solely to the study of lakes, stream investigation is, in the minds of some, also in the realm of limnology. As will be seen throughout our text, these two types of waters differ in many respects, and consequently demand, in many instances, different methods of study. It may well be that a different term is needed to distinguish lake and stream studies. Indeed, only recently the first Institute of Potamology (*potamos*, "river") in North America was founded at the University of Louisville, in Kentucky, for the study of rivers.

Estuaries represent the regions of entrance of coastal streams into the seas. Because of our attention to streams in this text, it seems desirable and altogether proper to include some consideration of the zone where sea and river meet. Most of the researches in North American estuaries have been carried on by institutions and individuals with oceanographic backgrounds and interests. On the Atlantic coast, however, several institutions located on Chesapeake Bay are engaged in nearly full-time investigations of that estuary.

The approach to the study of lakes, streams, and estuaries which we wish to take is essentially stated in our earlier definition of limnology. It is important from the very beginning to keep in mind the concept of a body of water as a dynamic system in which many interrelated and often interdependent factors and processes are operative. In order to appreciate these factors, however, it seems logical to view them separately in some reasonable sequence and arrangement even though the categories are sometimes artificial. After considering the fundamental characteristics and processes in natural waters, we can relate them to the functional whole. Our organization and plan is, therefore, as follows:

(1) The nature and method of formation of basins and valleys which are occupied by water to form lakes, streams, and estuaries.
(2) The source of water and maintenance of water in lakes, streams, and estuaries in the basins and valleys.
(3) The physical features and processes in these bodies of water.
(4) The chemical characteristics and processes in inland waters and estuaries.
(5) The animal and plant inhabitants of lakes, streams, and estuaries.
(6) The relationships between the inhabitants and the physicochemical aspects of the environment as shown in population and community dynamics.

These approaches can be reduced to three "working principles" which we have endeavored to weave as themes throughout this introduction to inland waters and estuaries. These principles pertain to the nature of the environment, the nature of the inhabitants, and the relationships of the two, as follows:

(1) "Fitness of the environment." This phrase is borrowed from the very stimulating book by L. J. Henderson (1913). While our use is in a somewhat different sense from the original, the thought here pertains

to ways in which physical and chemical attributes are variously combined to present environmental conditions which determine, to a high degree, the composition of biotic communities. Chapters 2 through 10 should be studied with this idea in mind.

FIGURE 1·5. Chemical analyses for dissolved solids and gases are highly important in the study of natural waters. (Photograph courtesy St. Petersburg, Florida, Publicity Dept.)

(2) Evolutionary adaptations of the living components. The plants and animals which come to successfully occupy the various environments found in natural waters do so by virtue of adaptations derived through the evolutionary history of each population. For example, habitation of stream rapids typically demands streamlined bodies or holdfasts, or other devices to prevent the animal from being swept downstream; animals of this environment characteristically possess such structures. These modifications are often in sharp contrast to those of lake dwellers. Chapters 11 and 12 are devoted to the kinds of organisms in inland waters and estuaries.

(3) Traffic in energy. Where animals and plants have become adapted to the "fitness" complex of an environment, an ecosystem is developed. Within this system, the plants capture solar energy and produce energy-containing substance. This substance serves as the basis of "food webs" of varying complexities in which herbivores graze upon the plants, small carnivores prey upon the herbivores, and larger carnivores upon the smaller ones. Thus, energy is moved through the system by consumption of organic material at various levels. These concepts are given attention in Chapters 13 and 14.

PART I

The Origin and Features of Basins and Channels

Chapter 2

LAKE BASINS AND LAKES

STANDING WATERS ARE usually named lakes, ponds, swamps, and marshes. Definition of two of these is not especially difficult; swamps are wet lowlands supporting large trees and shrubs, while marshes are broad, wet areas on which abundant grasses and sedges grow. Lakes and ponds each defy precise definition, chiefly because of the diversity of ways in which they originate, because they change with age, and because numerous features of them differ latitudinally. There are also other reasons for the difficulty in definition, not the least of which is the use of the terms among peoples of various regions. What is called a lake in one part of the country may be a pond in another. In fact, lakes in the southeastern United States may not always be standing waters; for example, several wide places in the slow-flowing St. John's River of Florida are called lakes.

It becomes evident that the problem of definition condenses to one of degree, with large, open-water bodies at one extreme and small bodies, often thick with plant growth, at the other. Since, to laymen and limnologists alike, the term *pond* generally connotes a small, quiet body of standing water with rooted plants growing across it (or usually capable of supporting plants all the way across), there is little reason to make a more complex definition. Similarly, any large sheet of standing water occupying a basin is qualified to be called a lake. The fact that the water is standing and occupying a basin suggests that some process (or combination of processes) has been at work to form the basin. These processes will now be considered.

ORIGIN OF LAKE BASINS

The phenomena concerned in the origin of lake and pond basins are varied but may be reduced to a dozen major processes. These modes of origin offer a system of classification, for generally the lakes and ponds in a given geographic area have originated, if naturally, as a result of one particular force. For example, most lakes in central North America were formed by the actions of continental glaciers, while a majority of Florida lakes occupy basins dissolved from a soluble, porous limestone; basins in the Rift Valley of Africa and regions to the north are the aftermath of considerable geologic catastrophe. Nearly all geologic disturbances are capable of producing some sort of lake basin.

Tectonic Basins

Tectonic basins are those depressions formed as the result of some movement of the earth's crust. Lake Okeechobee, in southern Florida, gives evidences of being a remnant depression in the Pliocene sea floor which retained its form when uplifted to become land. This lake covers approximately 1880 km^2 and in the wider area is nearly 40 km across. Except for Lake Michigan, Lake Okeechobee is the largest freshwater body completely within the United States. During periods of drought the depth of this lake may be no more than 4.5 m. The only tributary of any import to Lake Okeechobee is the Kissimmee River, which drains some 13,000 km^2 and empties into the lake on the north. Several lakes (Apopka and Weir) of central Florida may have formed in the same manner, part of the evidence in the latter cases being the presence of a killifish (*Cyprinodon hubbsi*) whose nearest and closest relative is found in salt and mildly saline waters along the present coasts.

Earthquakes have been responsible for the formation of lake basins such as certain ones in Arkansas and Missouri, and Reelfoot Lake in Tennessee. These were formed by the New Madrid earthquakes of 1811-1813. The basin of Reelfoot is about 32 km long and approximately 6 m deep. Geologic faulting, the mass shifting of great blocks of rock, forms lake basins by tilting of the land surface and by the formation of a basin in which water collects. In Oregon, Lake Abert occupies a basin formed by tilting and faulting. The basin of Lake Tahoe, California, was formed by the accumulation of water in a *graben*, or trough, resulting from the displacement of crustal masses to form steep walls. The deepest known lakes occupy grabens; Lake Baikal, in Siberia, covers three basins, the deepest of which is 1741 m deep. Lake Tanganyika, in Africa, has a maximum depth of 1435 m. Tanganyika is one of many lakes in a belt of rifts, or displacements of crustal blocks along fractures, extending some 4800 km from Rhodesia to Israel. In these rift valleys are found such famous waters as the River Jordan, the Sea of Galilee, and the Dead Sea.

Less spectacular crustal deformations account for Great Salt Lake in Utah. Here mountains have been thrown up to form a basin from which there is no drainage. Inflowing waters carry dissolved minerals, and rapid evaporation concentrates the salts.

Subtle warping of the earth's crust over which a stream courses may have two effects. If the surface bends with the direction of stream flow, the velocity of the stream is increased as a result of steeper gradient; if the bending is against the direction of flow, the stream gives way to a dammed lake. Lake Victoria, in Africa, is 418 km long and resulted from the natural impoundment of the Katonga River through upwarping of the land.

Basins of Volcanic Origin

Various volcanic activities may form lake basins. A most notable example is Crater Lake, Oregon. This lake occupies a *caldera,* or volcanic basin, near the summit of the volcano, produced by explosion. The surface of the lake lies nearly 610 m below the caldera rim, and the depth of the lake itself is also about 610 m. No streams enter Crater Lake, the water being derived from rain and snow melt.

Lava issuing from an active volcano may flow across a stream valley and dam it. Various lakes in the volcano regions of Oregon and Washington were so produced. Sheet lava may, upon hardening, contain depressions into which water drains, thereby forming a lake.

Basins Resulting from Glacial Action

Glaciation has been and is a conspicuous force in the formation of lake basins in certain regions of the earth. The greatest concentrations of lakes in North America and Europe are found within regions formerly covered by Pleistocene ice sheets and where valley glaciers have scoured and dammed basins in mountainous country. Not only are basins formed by erosional scouring but also by moraines (deposition of debris and drift by glaciers).

Valley glaciers in Alaska and much of the Rocky Mountains, for example, contribute to the formation of lakes in several ways. Cirque lakes are common in the Cordilleran ranges from Alaska to Cape Horn. These lakes occupy amphitheater-shaped basins known as *cirques* eroded at the heads of glaciated valleys. Steep walls surround the basins except at the outlet (Figure 2·1). Glacier National Park contains many classical examples of lakes in cirques. Often the valley draining the cirque will be marked by a chain of lakes occupying basins in stair-step fashion along the axis of the valley; these are called *paternoster* lakes.

Many small lakes throughout much of Canada and northern United States have resulted from some force associated with continental glaciation. These may range from lakes and ponds in basins scoured by ice to depressions dammed by materials carried in or pushed by the ice. The Finger Lakes of New York occupy basins formed partly by the erosion of original valleys by ice, and partly through damming by glacier moraines. Several well-known lakes of the Alps, Lakes Lucerne, Como, and Constance, for example, were formed in glaciated valleys closed by end moraines.

Scattered throughout the Coastal Plain of New Jersey are hundreds of shallow basins (less than 3 m deep) ranging in size from a few hundred square meters to over one and a half square kilometers. These basins are

FIGURE 2·1. Mountain Lakes of British Columbia. The elongate lake behind the peak to the right of center occupies a cirque. Drainage from this lake enters the lower lake. These valley lakes are called paternoster lakes. A braided stream pattern, to be described in Chapter 3, is seen in the broad valley to the left. (Photograph courtesy United States Air Force.)

considered to be *periglacial frost-thaw basins*, formed by subsidence of the ground surface. This surface depression resulted from alternate thawing and freezing during Pleistocene times. These basins have been likened to thaw basins being formed presently in arctic regions. In areas once covered by continental glaciers, numerous small lakes and ponds occupy basins called *kettles*. These basins were formed by the melting of masses of buried ice.

The Great Lakes are excellent examples of basins formed by glacial action, and their history has been recorded rather accurately. The original basins, quite different from those of today, were formed during the late Pleistocene by the scouring of previously formed stream valleys. These basins were later shaped by the depression of the land under the weight of the continental ice sheets or, as some authorities believe, by continued erosion of easily fractured minerals.

The history of the Great Lakes begins with the recession of the last ice sheet of the glacial period. During one of its several advances and recessions this ice sheet reached as far south as the junction of the Ohio and Missouri Rivers in the central United States. As the ice retreated, old Lake Chicago formed between the ice and the higher elevations southward. Lake Whittlesey formed from earlier and smaller Lake Maumee in the present Lake Erie region. The beginning of Lake Huron and Lake Saginaw occupied an area northward and between the other early lakes. The extent of the glacier and the original lakes are shown in Figure 2·2A. New York's Finger Lakes to the east of the Great Lakes region drained temporarily into the Susquehanna River. Drainage from Lake Whittlesey was through Lake Saginaw into Lake Chicago, eventually to empty into the Mississippi by way of the Illinois River.

As the glacier melted and withdrew farther, a single lake, Lake Warren, was formed by the joining of Saginaw and Whittlesey. As shown in Figure 2·2B, the Finger Lakes probably drained into Lake Warren to a greater extent than into the Susquehanna.

Later still, with the continued retreat of the glacier front, Lake Duluth was formed to the northwest in the region of present-day Lake Superior (Figure 2·2C). Lake Lundy formed from the glacier and joined the waters of Whittlesey and the Finger Lakes. Lake Lundy was large, extending from Lake Saginaw into the basin of Lake Huron and north of Lake Ontario. The drainage pattern from Lake Lundy was through the Mohawk River to the Hudson. Lake Chicago continued to increase in size and at that time was nearly as large as Lake Michigan is today.

Further recession of the glacier left a large lake, Lake Algonquin, occupying the regions now taken by Huron, Michigan, and Superior (Figure 2·2D); Lake Erie and Lake Ontario possessed somewhat the same morphology as today. The time of Lake Algonquin was also the time of another spectacular event. The St. Lawrence Valley was exposed by the retreating glacier, and water entered this valley from the sea. This large body of water has been called the Champlain Sea. One arm of the sea reached into Lake Ontario, and the basin proper extended from New York Bay up the St. Lawrence. Traces of the shorelines of this sea are seen today near Montreal.

The final stage preceding the present form of the Great Lakes has been called the Nipissing Great Lakes Stage. At that time Lake Erie and Lake Ontario were separated from the other three lakes. Later the ice sheet withdrew far to the north, a connection was established between Erie and Ontario and the Nipissing Great Lakes, the lakes decreased in size, and the present configuration was established. The St. Lawrence system was

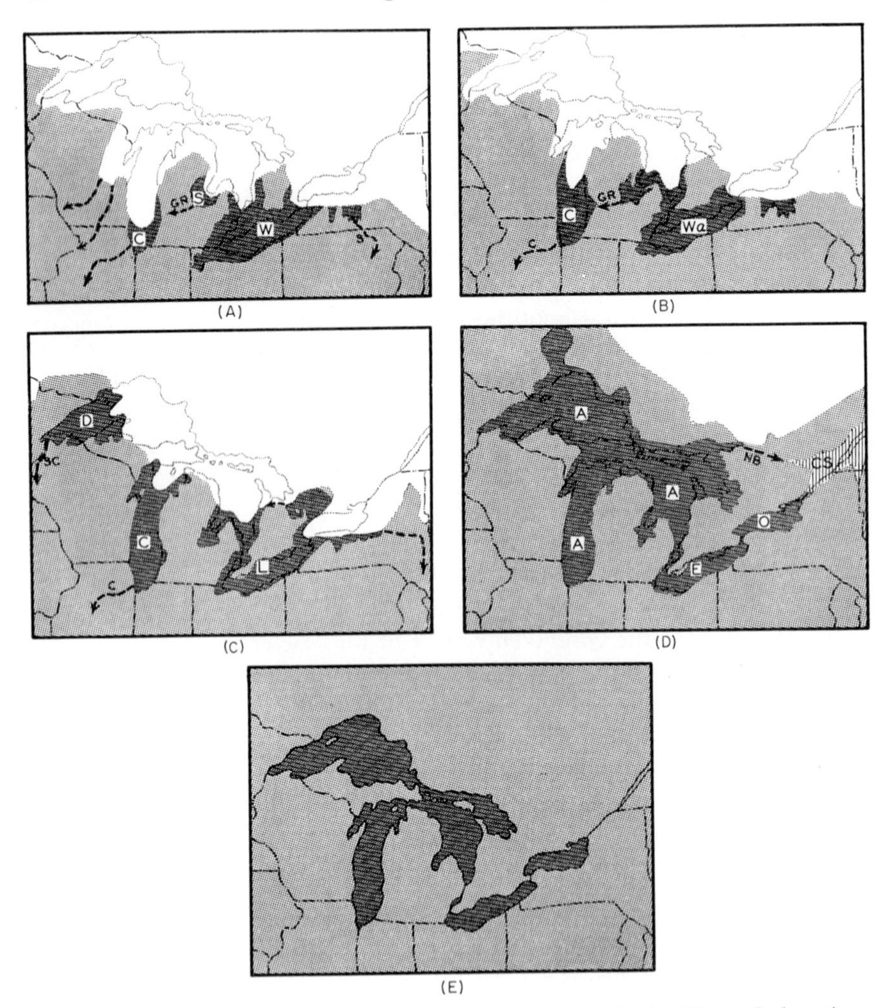

FIGURE 2·2. Developmental Stages in the Formation of the Great Lakes Accompanying Recession of the Pleistocene Continental Glaciation. Processes are described in the text.

A–Lake Algonquin
C–Lake Chicago
c–Chicago River
CS–Champlain Sea
D–Lake Duluth
E–Lake Erie
GR–Grand River
L–Lake Lundy
NB–North Bay Outlet
O–Lake Ontario
S–Lake Saginaw
s–Susquehanna River
sc–Saint Croix River
W–Lake Whittlesey
Wa–Lake Warren

(After Thwaites, F. T., 1959. "Outline of Glacial Geology," publ. by the author at Madison, Wis.; and after Flint, R. F., 1957. "Glacial and Pleistocene Geology," John Wiley & Sons, Inc., New York, N.Y., from various authors.)

developed from drainage of the lakes eastward rather than into the Mississippi as previously.

For about 1000 years the greatest of all glacier lakes existed over much of North Dakota, Minnesota, Ontario, Manitoba, and Saskatchewan. This was Lake Agassiz, and it covered an area 1120 km long and 400 km wide at its maximum. It was over 210 m deep. The ice dam which held the waters melted, the land became elevated, and most of the great lake drained. Today Lake Winnipeg, Lake Manitoba, Lake of the Woods, Lake Winnipegosis, and a few smaller lakes occupy basins on the old floor of Lake Agassiz.

Solution Basins

Various sized lakes in sinkholes occur in widespread parts of the world, but notably in the Balkan Peninsula, the Calcareous Alps, Yucatan Peninsula, and in Indiana, Kentucky, and Florida in the United States. These are extensive limestone regions. The funnel-shaped sinks, or Slavonic *dolines*, which form the basins have developed from the solution and erosion of some soluble substratum.

The rolling, karst topography of central Florida contains thousands of solution lakes formed in light-colored, usually highly fossiliferous limestone, of Tertiary times (Figure 2·3). These solution depressions appear wherever the porous limestone has been dissolved by percolating surface water or flowing ground water. In the percolation process surface waters may begin dissolving and eroding the limestone at points of fracture or other weakness. The inflowing water is carried away, and with continued flow the doline develops. The second mechanism involves the solution of the limestone from below. This occurs when water moving in a subsurface aquifer weakens the roof of an underground chamber, thus permitting the roof of the chamber to cave in, forming a fairly regular cone-like doline.

Not all sinkholes serve as lake basins, the determining factor being the proximity of the bottom of the sink to the existing water table. If the bottom of the doline is situated above the water table, the sink is essentially dry. Where the lower limit of the sink extends within the range of fluctuation of the water table a lake appears during the rainy season but disappears with lowering of the table. A permanent lake exists if the sink reaches below the water table. Occasionally a sink may serve to drain a lake not wholly of solution origin.

Basins of Wind Origin

In arid regions the movement of fine, light materials such as clay and sand may result in the development of lake basins. The movement of the soil may be of such nature as to block an existing stream, thus giving rise to a dammed lake. Moses Lake, in Washington, was formed when wind action shifted sand dunes across Crab Creek. Removal of loose materials from an area may also form what has been termed *deflation basins,* and many lakes in Nebraska, New Mexico, and other regions east of the Rocky Mountains lie in basins scoured and shaped by winds. Obviously, sufficient rainfall and proper substratum are necessary for the development of permanent lakes in deflation basins. However, since these conditions do not always exist in these regions, many of the basins do not hold water. On the other hand, lakes do occur in some arid localities and it has been suggested that the basins were formed under previously existing

FIGURE 2·3. Sinkhole Lakes in the Karst Topography of Marion County, Florida. The rather extensive shore zones evident in the photograph result from lowered water level during a drought at the time the photograph was taken. Note the nearly dry basins scattered throughout the area. (Photograph courtesy Commodity Stabilization Service, U.S.D.A.)

arid conditions, to become filled at a later time following changes in climate. In some areas, depressions created by shifting dunes and hollowing of troughs between the dunes contain small ponds and shallow swamps. Cherry County, Nebraska, contains such waters.

Basins Formed by Stream Action

The action of waves and currents of streams may form lake basins by deposition of materials in suspension and by erosion. Streams with steep gradient and relatively high velocity lose their carrying power and drop their loads when joining with larger streams. Lake Pepin, on the Minnesota-Wisconsin state line, was formed by the deposition of alluvium carried by the Chippewa River. Here the Chippewa, a tributary of the Mississippi, dropped its load across the main stream, thereby creating a *fluviatile* dam on the Mississippi above which Lake Pepin formed. Similarly, Tulare Lake, California, had its origin in an alluvial fan built by the King's River across the valley of the San Joaquin.

Often the reverse process occurs. A large stream builds its banks with sediments at a faster rate than a tributary. The result is a lake impounded by the action of the main river. These are termed *lateral* lakes, and examples are found along the Red and the Sacramento Rivers.

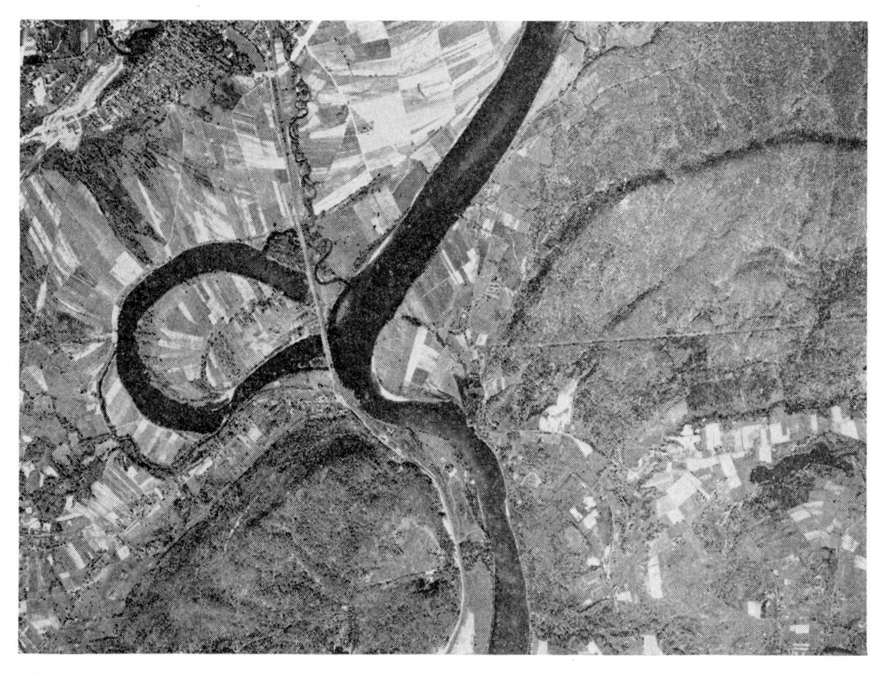

Figure 2·4. Oxbow Lake Formed from a Loop in the Connecticut River. (Photograph courtesy Soil Conservation Service, U.S.D.A.)

As a meandering stream erodes the outside shores of its broad bends, the loops become cut off, forming one of the more impressive types of lakes originating by stream action. These shallow, crescent-shaped lakes are called *oxbow lakes* (Figure 2·4). The mature flood plain of the Mississippi and some of its tributaries in Louisiana and neighboring states contain hundreds of oxbow lakes now occupying abandoned river channels. The lower ends of many of the oxbows are blocked by sediments deposited by the parent stream as the meanders were cut off.

The plunging of a stream over a cliff to form a waterfall will, of course, excavate a pool at the foot of the fall. This pool may become enlarged and the original stream may eventually become extinct, leaving a lake. The lakes at Dry Falls in the region of the Grand Coulee, Washington, represent former *plunge basins*.

Basins Resulting from Landslides

Large masses of rock, soil, mud, and the like may fall away from an adjacent slope and block a stream valley. Lakes in basins formed by landslides are found in the Alps, western North America, and other regions. In 1925, a landslide on the Gros Ventre Range in Wyoming dammed the Gros Ventre River and formed a lake nearly three miles long. Two years later the water level reached the top of the dam, overflowed, and created a disastrous flood. The dam was not completely destroyed, however, and a lake remains today. Since the materials in the slide are not sufficiently consolidated, eventual erosion and destruction of the dam are rather to be expected.

If the slide material is compacted and the impounded stream small and slow to fill the basin, long-lived lakes may result. Deep Lake, Wyoming, is the result of a landslide of early geologic time which dammed Clark's Fork Branch, a tributary of the Yellowstone River, behind some 800 feet of rubble. A mudflow was responsible for the formation of Lake San Cristobal in Colorado.

Pluvial Lakes

During the Pleistocene, glaciations influenced the climate to a great extent, and in some regions conditions were quite unlike those of today. In the western United States, for example, the environment was periodically much more moist than at present. During those times, the action of glaciers produced *pluvial* periods, or times of greater rainfall and lowered evaporation. Many lakes were formed. Only a few of these lakes remain today, and with recent increased aridity, evaporation, and lack of outlets, most of the lakes have become saline. Lake Bonneville was, during the Pleistocene, a large lake covering some 51,000 km^2 and having a maximum depth of over 300 m. Its drainage was northward to the Pacific through the Snake River and the Columbia River. Today all that remains of Lake Bonneville are three saline lakes, one of which is Great Salt Lake. The Bonneville Salt Flats are the remains of most of the original lake bottom and shore zones. Thus, although the original basin of Lake Bonneville was of tectonic origin, the basins occupied by the present saline lakes resulted essentially from the drying of the earlier lake. These basins might conceivably be called "fossil basins."

Lakes Developed from Shoreline Activities

In large bodies of water, currents moving masses of water in which sediments are carried may be responsible for the formation of lakes. If the

FIGURE 2·5. "Ponds" Along the New England Coast. These bodies of water have been formed by deposition of materials across the mouths of estuaries. The materials have been derived from erosion of headlands (see also Figure 4·1), and distributed by wave action and longshore currents. (Lowry Aerial Photo Service.)

sediments are deposited opposite a small embayment, a lake may develop. Along sea coasts, sand spits formed by currents moving parallel to shore may effectively block a lagoon or bay, thereby creating an enclosed body. Some of the most impressive shore-line lakes, or "ponds," are found along the New England coast, especially in the vicinity of Martha's Vineyard and Nantucket (Figure 2·5). Similar processes operating in an inland lake may divide the parent lake by isolating an embayment. In Minnesota, Buck Lake was cut off from larger Cass Lake by the eventual closure of a spit.

Lakes of Organic Origin

Lakes and lake basins resulting from the activities of living organisms and blocking of valleys with organic materials may be considered briefly.

The role of beavers in creating ponds and lakes is well established and familiar to all. The construction of dams with pieces of logs and sticks is apparently a typical characteristic of the American beaver (*Castor canadensis*) and its ponds occur throughout the range of the animal over most of North America. Some of the dams may reach considerable size, there being a report of one nearly 600 m long in Montana. Mostly, however, the dams are considerably shorter and seldom more than 2 m high.

Through impoundment of streams and excavation of the landscape, man, himself, is a major force in the formation of lakes. In recent years vast bodies of water have been created for purposes of hydroelectric power, flood control, water storage, and for other reasons. The farm pond movement has resulted in the creation of thousands of small bodies of water in various regions of the world. Because of the diverse use and manipulation of these artificial lakes and ponds, the limnological features and processes often differ from those of natural waters. The limnology of impoundments promises to be a rich and enlightening field of study.

Plant material, both living and dead, may contribute to the formation of a lake by blocking the normal outflow. Dense growth of aquatic and marsh plants serves rather effectively to impound waters, especially in the more southern regions where there are long and vigorous growing periods. Lake Okeechobee, in Florida, is impounded to some extent by vegetation. In Louisiana, dense mats of alligator weed (*Alternantheria philoxeroides*) may choke water courses (in that part of the United States known as the "bayous") resulting in the formation of ponds. Log jams and accumulation of debris and detritus may also block a stream, thereby creating a pond.

Basins of Meteoritic Origin

Although insignificant as a basin-forming process, the impact of meteors against the earth's surface occasionally creates depressions which later may be occupied by lakes. Several such lakes are known, Chubb Lake in Quebec being a classical example.

Lakes of Unknown Origin

For all of our knowledge of lake-producing processes, there remain some basins whose origins are not firmly established. Such is the case for a number of lakes situated in the Atlantic coastal plain (Figure 2·6). Of this group, a cluster of small, shallow lakes in eastern North Carolina, known as the Carolina Bays, have probably received more attention than others. These bays (so called by the local inhabitants but, of course, not bays in any technical sense) occupy basins of elliptical form and are oriented in the same direction, essentially northwest-southeast. The lakes generally have sandy shores and bottoms, and the water is relatively clear. Recorded depths range to about 6 m. Suggestions as to the origin of the

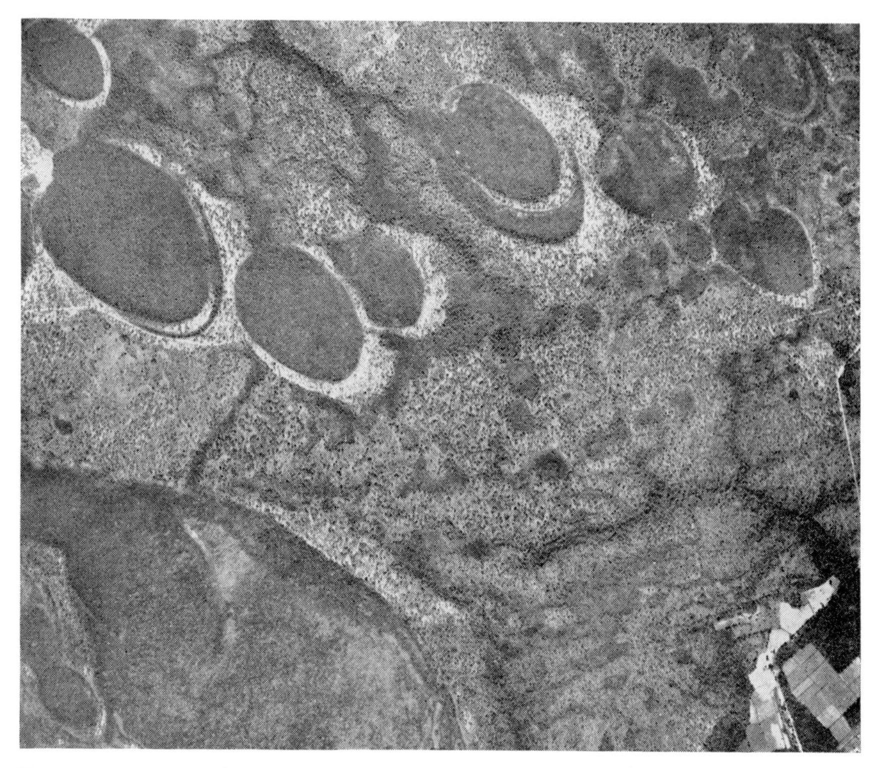

FIGURE 2·6. Carolina Bays, Horry County, South Carolina. (Photograph courtesy Commodity Stabilization Service, U.S.D.A.)

lakes have included meteoritic impact, wind action resulting in deflation basins, erosion resulting from currents set into motion by schooling fishes, burning of peat accumulations, and others. To date no single explanation is satisfactory and acceptable to all biologists and geologists who have studied the bays.

SURFACE FORM OF LAKE BASINS

We have all observed that lakes vary greatly in form and size. In the preceding section various forces which produce lake basins were discussed, and, in a sense, a classification of lake basins based upon origin was developed. In a general way, the form of a lake basin is a function of its origin. This is to say that often the shape or morphology of the basin will, barring extensive post-origin modification, reflect the nature of the mechanism responsible for the basin. A few examples will illustrate this principle. The terminology is that of Hutchinson (1957).

Geometrically Shaped Basins

(1) *Circular Basins:* As suggested previously, typical solution sinks, or doline lakes, are primarily circular in shape. However, the lakes of Florida often overflow a portion of the land to one side of the sink, resulting in a shallow marsh, but these usually show gently rounded outlines. Basins of meteoritic origin, as well as crater and caldera lakes of volcanic beginnings, are generally circular.

(2) *Subcircular Basins:* Cirque lakes (tarns) in mountainous areas of valley glaciers are mostly of subcircular form.

(3) *Elliptical Basins:* The Carolina Bays, those lakes of unknown origin, are examples of elliptically shaped basins.

(4) *Subrectangular Basins:* Most of the lakes of tectonic origin situated in grabens show an elongate, angular pattern in the form of a rough rectangle.

(5) *Dendritic Basins:* When a highly branched stream valley is ponded by the development of a dam, the subsequent flooding of the main channel and tributaries results in a lake with many arms, or embayments.

(6) *Lunate Basins:* Oxbow lakes in mature river valleys typically occupy crescentric, or lunate basins.

(7) *Triangular Basins:* Coastal lakes formed by the development of a barrier beach or sand spit across the mouth of a stream valley are usually triangular.

Irregular Basins

Lakes formed by the scouring process of continental glaciers often fuse and join several basins to give an irregular outline. Deposition of glacial debris to form dams, and differential scouring may also form irregular basins.

Not all lakes encountered will fit into one or another of the above categories. This is due to the fact that intrinsic processes such as currents, sedimentation, and freezing, to name a few, are continually operating to change the original form. External factors, man for example, can also modify the morphology of a lake basin through dam and water-way construction and by changing original drainage patterns.

Islands

Islands are often conspicuous features of inland waters and may bear heavily and directly upon physical attributes such as currents and shore-building processes and upon biological characteristics. Because most natural islands are directly related to the origin of the basin, a few generalities concerning lacustrine islands should be considered here.

The work of long-shore currents in moving sediments has been discussed above in relation to the origin of lakes. It was pointed out that sand spits projecting at various angles from a shore could be formed through the deposition of loose materials carried by the currents. The near-shore portion of a spit may later be eroded by changes in currents and/or wind direction and intensity, leaving an island developed in the lake. Long Point, on the Canadian side of Lake Erie, has apparently developed in such a manner. Wizard Island in famous Crater Lake, Oregon, represents the remnant of a secondary volcanic cone formed in the earlier caldera now occupied by the lake. Lakes in regions subjected to Pleistocene continental glaciers typically contain islands formed of materials resistant to the scouring action of the glacier or debris dropped or pushed by the moving glacier.

Islands that bear no relationship to the basin are common in certain lakes in various parts of the world. These are floating islands. They have been described for lakes in Germany, the English Lake District, Chile, and other localities. In North America, floating islands have been reported for Minnesota bog lakes, and the upper St. Johns River and Orange Lake in Florida. In Orange Lake, the islands are freely floating structures composed of a substratum of densely compacted peat-like material supporting shrubs and small trees. The islands vary in size up to several acres.

LAKE WATERS

Having considered the major processes responsible for the formation of lake basins and the most conspicuous forms developed by these processes, it is well to consider the sources of water and mechanisms involved in filling the basins and maintenance of water levels. The broad, universal system of evaporation of water from the earth's surface, movement as vapor through the atmosphere, and precipitation back to the surface is termed the *hydrologic cycle* (Figure 2·7).

It has been calculated that there is a mass of about $13{,}967 \times 10^{20}$ g of water on the accessible areas of the earth's surface as liquid and ice and in the atmosphere as vapor. Approximately 99 per cent of the total is in the oceans and seas, and most of the remainder is locked in glaciers, snow, and

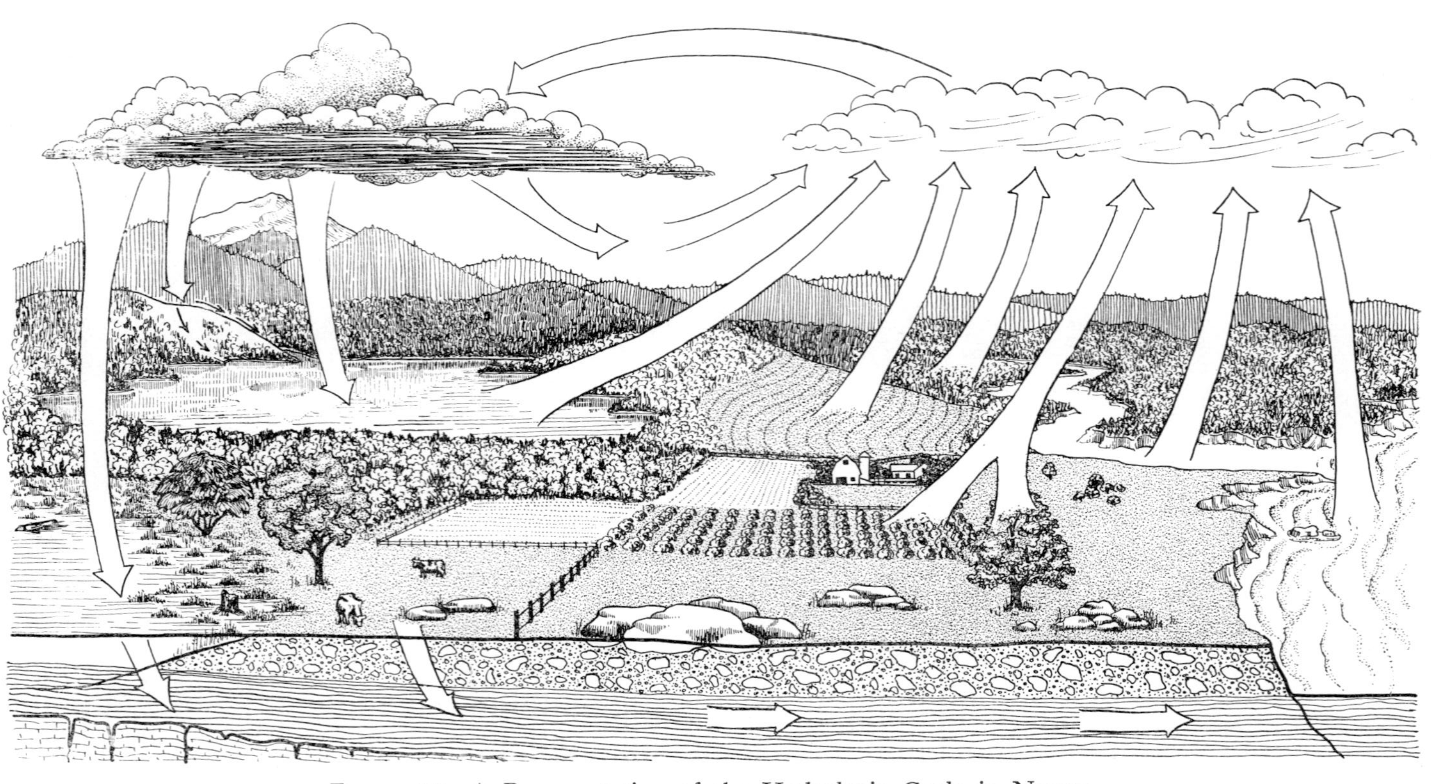

FIGURE 2·7. A Representation of the Hydrologic Cycle in Nature.

ice. The mass of inland waters constitutes only about 0.25×10^{20} g. Water vapor in the atmosphere amounts to only a minor fraction of 1 per cent of the total. The greatest mass of water, estimated at $250{,}000 \times 10^{20}$ g, occurs in the rock mantle of the earth.

The cycle is complex insofar as precipitation (snow, rain, etc.), the path of water upon the earth, and the mode of loss of surface water back to the atmosphere are concerned. Studies by the United States Geological Survey in various regions of North America have shown that in the general region from Connecticut to North Dakota, evaporation and plant transpiration return from about 50 to 97 per cent of the experienced precipitation back into the atmosphere; from 2 to 27 per cent of the precipitation finds its way to streams and the sea; infiltration into the ground accounts for 1 to 20 per cent of the water falling upon the surface.

Lake waters, then, may be derived from one or more sources. Although lakes do receive water directly from precipitation, land surface drainage in some form contributes most of the water. Water of infiltration arriving in the basin through seepage or springs is another source. In some instances all three sources may contribute. Since surface runoff and infiltration depend ultimately upon precipitation, the distribution of lakes is naturally associated with rainfall drainage patterns as well as with geologic lake districts.

Lakes have been described as "temporary stopping places for water on its way to sea." This suggests that lakes and lake districts are most numerous in those regions of the world where streams arise and find their way to ocean basins. There are other regions in which streams are common, but end in dry valleys or lakes without outlets; lakes are less common in these regions. In desert and semiarid places streams are rare, and as would be expected, so are lakes.

We have seen that lakes may be classified on the basis of the presence or absence of outlets, or effluents. Most lakes possess some form of outlet and are termed *open* lakes. The effluent may be a stream which drains the lake, or drainage may be effected by seepage in the form of ground water. Whatever the influent and effluent of open lakes, there is a relationship between the two systems that determines the water level of the lake. Open lakes serve as "settling basins" for sediments introduced by inflowing streams. As we shall see presently, the dropping of materials into the lake as the tributary loses its carrying capacity at the entrance increases the rate at which the lake is eventually filled. The water in the effluent stream usually is clearer than that in the influent, attesting to the settling action of the lake.

Lake Geneva (Lac Léman of Forel) is an open lake in the western Alps. Its major tributary is the Rhone River. The river annually contributes great amounts of sediments to the basin of the lake. Although originally about 75 km long, the lake is now nearer 64 km as a result of delta formation in its upper region. It is about 300 m deep in places, and the delta is quite thick. Continued sedimentation brings about the encroachment of the delta toward the lower part of the lake and the eventual filling of the basin. At the same time, the effluent is reducing the water level by erosion of the stream channel. It has been calculated that the Rhone waters take over 11 years to travel the length of the lake.

In open lakes, substances carried in solution (various salts, for example) by the influent generally move through the lake. Some materials, however, may enter into biochemical cycles and be transformed. Fluctuations in the quantity of dissolved materials introduced into the lake usually average out in time.

In addition to discrete streams as effluents, or drainages, of open lakes, overflow water may be carried away by other means. Among some open lakes, water may be lost through shallow seepage—no outlet being apparent. Lakes in karst regions often drain into subsurface channels through openings in the lake bottom. Indeed, an inherent characteristic of most sinkhole lake in limestone areas is the extreme lowering of water level and even complete emptying. Either of these phenomena may result from lowering of the water table, solution of a portion of the basin permitting drainage into underground channels, or the opening of a previously clogged channel to subsurface drainage.

It has been pointed out that in some regions of the world streams occur which do not find their way to the sea, and that these streams end in dry beds or lakes without effluent. The regions in question are usually those arid or semiarid areas where evaporation takes place at a faster rate than precipitation. The lake never fills to the point of overflowing; its basin may even dry completely during certain seasons. These lakes are termed *closed* lakes and are typically saline. Many closed lakes are found in glaciated areas; these waters are often called *seepage lakes.*

All fresh waters contain some amount of dissolved salts of various kinds depending upon the nature of the soil and the chemical composition of underlying geologic formations which may contribute to the content of the stream water through seepage. In some regions, precipitation may influence the composition of stream waters. The tributary streams of a closed lake introduce into it small to large amounts of dissolved salts, which, in turn, become further concentrated in the lake through evaporation of the water. Thus the nature of the salt or salts in a lake is dependent upon the geochemistry of the drainage basin. Great Salt Lake, in Utah, derives its salinity mainly from sodium chloride and sodium sulfate. The Dead Sea salt is almost wholly chlorides. Several lakes in Oregon are characterized by high concentrations of carbonates, chiefly of sodium. *Bitter* lakes, in the arid lands between the Sierra Nevada and the Rockies, are high in sulfates. The chemical nature of lake waters will be discussed further in Chapters 9 and 10.

SOME IMPORTANT PARAMETERS OF LAKES

Having considered the nature of lake basins, the sources of water to form the lake, and the maintenance of the water in the basin, we can now give attention to certain physical features relating to the form of the lake. We have seen the great diversity in form exhibited by lakes. Similarly, they vary in size. Such aspects as depth, mean depth, length, breadth,

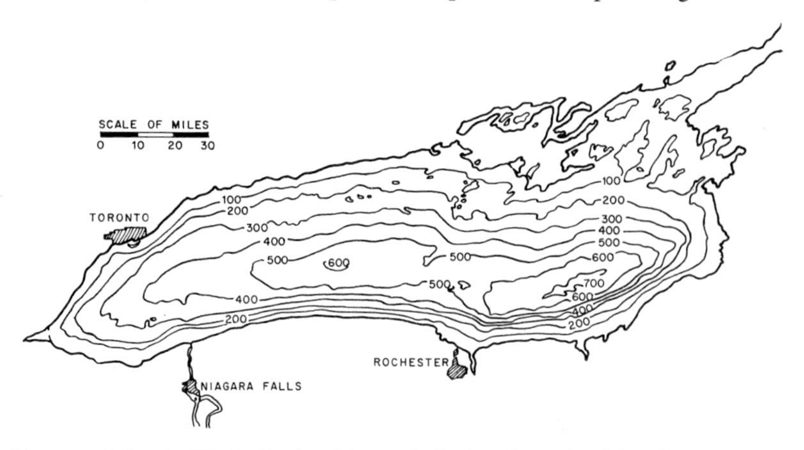

FIGURE 2·8. A Hydrologic Map of Lake Ontario Showing Bottom Topography. (From Hough, 1958. "Geology of the Great Lakes," Univ. of Illinois Press, Urbana, Ill.)

area, volume, extent and development of shoreline, water level, and elevation above sea level are basic to limnological investigations of lakes. These characteristics are also primary agents in the dynamics of the lake. Since these features can be expressed quantitatively, they become especially useful as parameters of morphology in comparative limnology. One of the early phases in a limnological study is the construction of a bathymetric map of the area investigated. Figure 2·8 illustrates such a map and the features are discussed below. For details and methods in obtaining the morphometric data and map construction, the student is referred to Welch, "Limnological Methods," Blakiston, 1948.

MAXIMUM DEPTH

This parameter is best determined by sounding the lake in the field, or from previously constructed maps. It must be remembered, however, that sedimentation, erosion, and water level fluctuation can rapidly alter the depths in a lake. Thus the use of old data must be approached with caution.

Depths of lakes may range from a few meters to nearly 1800 m. Lakes of great depths are not necessarily large in area. Lake Baikal, in Siberia, is the deepest lake known. While its maximum depth is 1741 m, its expanse of 31,500 km^2 places it quite low on the list of the lakes of great area. Lake Tanganyika, in Africa, is the second deepest lake on record, the maximum recorded depth being 1470 m. The Caspian Sea is 946 m deep, but it will be recalled that the waters are saline. Lake Nyasa, also in Africa, has a maximum depth of 706 m. In the U.S.S.R., Lake Issyk-kul is 702 m deep. All of these spectacularly deep lakes are of tectonic origin and, with the exception of the Caspian Sea, all occupy basins in grabens. In addition to

TABLE 2·1. MORPHOMETRIC DATA FOR THE GREAT LAKES

(Data from Hutchinson, G. E., 1957. "A Treatise on Limnology, Vol. I, Geography, Physics, and Chemistry," John Wiley & Sons, Inc., New York, N. Y.)

Lake	Area (km^2)	Maximum Depth (m)	Mean Depth (m)	Volume (km^3)	Shoreline Length (km)	Shoreline Development
Superior	83,300	307	145	12,000	3000	2.93
Huron	59,510	223	76	4,600	2700	3.1
Michigan	57,850	265	99	5,760	2210	2.6
Erie	25,820	60	21	540	1200	2.1
Ontario	18,760	225	91	1,720	1380	2.8

these, Hutchinson (1957) has listed 16 lakes with depths greater than 400 m. North American deep lakes include Crater Lake, Oregon, which has a maximum depth of 608 m (recall that Crater Lake occupies a volcanic basin); Lake Tahoe, a lake in a graben, between California and Nevada, which is about 500 m deep; and Lake Chelan, formed by glacial activity in Washington, which has a maximum depth of 458 m. Data for the Great Lakes are given in Table 2·1.

MEAN DEPTH

The mean depth ($\bar{d}$) is a relationship between volume (V) and area (A) of a lake. The formula is $\bar{d} = \frac{V}{A}$. This parameter is of particular value because it renders a more accurate description of depth-area proportions than does maximum depth. For example, the Caspian Sea with a maximum depth of 946 m has a mean depth of only 182 m when its area of 436,400 km^2 is considered in relation to the volume of 79,316 km^3. On the other hand, Lake Baikal, with a volume of 23,000 km^3 and covering an area of 31,500 km^2 to a maximum depth of 1741 m, has a mean depth of 730 m.

Length

This is the distance between the farthest points on the shore of a lake.

Breadth

Breadth is the distance from shore to shore measured at right angles to the longitudinal axis. In view of the often irregular nature of lake shorelines, this parameter may be quite variable. *Mean breadth* ($\bar{b}$) is of somewhat greater value than breadth and is defined as the area (A) divided by the length (l), or $\bar{b} = \frac{A}{l}$.

Area

Area is the extent of surface of a lake. As a result of sporadic flooding or extreme drainage, this feature may also fluctuate considerably. In artificial impoundments "drawdown" can significantly reduce the area as well as other aspects of a lake. Area is determined most accurately and easily by the use of a planimeter operated on a well-drawn map or aerial photograph of known scale.

Of the world's most extensive lakes, the Caspian Sea has the greatest area, covering 436,400 km^2. Lake Superior is the second largest lake in the world (Table 2·1). Lake Victoria, in Africa, has an area of 68,800 km^2. The Sea of Aral, U.S.S.R., is fourth; it covers 62,000 km^2. The two last-mentioned lakes are of tectonic origin. Lakes Huron and Michigan, glacial lakes of North America, are also among the largest of the world.

Volume

Determination of volume of a lake involves deriving the volume contained in each of the strata bounded by depth contours. Here a bathymetric map (Figure 2·8) is needed. Because of the slope of the bottom, it is necessary to consider the area of the upper and lower surface of each contour stratum. The volume of the stratum may be calculated from various formulas, one of which is:

$$V = \frac{1}{3}(A_1 + A_2 + \sqrt{A_1 A_2})h$$

where A_1 is the area of the upper surface of a contour stratum and A_2 is the area of the lower surface of the same stratum. The height of the stratum is shown by h. The volume for each stratum is computed from the formula, the sum of the volumes being the total lake volume.

The Caspian Sea has the greatest known volume, 79,319 km^3. As we have seen, this volume is derived mainly from the great area of the lake. Lake Baikal, by virtue of its considerable depth, has a volume of 23,000 km^3, and is the second most "massive" lake in the world. Lake Tanganyika contains 18,940 km^3 of water. Volumes for the Great Lakes are shown in Table 2·1.

Extent and Development of Shoreline

The extent of shoreline is a simple measurement of the length of the shore. This parameter can be obtained from a finely detailed map by means of a mechanical map measurer. These devices are variously called rotometers or chartometers. Shoreline measurement is valuable in calculating area of the lake, extent of shallow shore zone, and, as below, in deriving an important index called *shore development,* or *development of shoreline.*

Shore development is the quantitative expression describing the configuration of a shoreline. This index is derived as the ratio of the shoreline length to the length of the circumference of a circle of the same area as the lake. Shoreline development may be calculated from the formula:

$$SD = \frac{S}{2\sqrt{\pi A}}$$

in which S is the length of the shore and A is the lake area. Since the ratio is related to a circle, it is seen that a perfectly round basin would result in an index of 1. Increasing irregularity of shoreline development in the form of embayments and projections of the shore is shown by departures from the value of 1.

Small doline lakes of Florida are often nearly circular and consequently give shoreline development indexes near 1. Fjord lakes with numerous embayments in valleys, and lakes formed by drowned river mouths, would be expected to yield high values. Lake Salsvatn, a fjord lake in Norway, has a shore development of 5.5.

Mean Slope and Area of Shoal

The extent of shallow water in a lake is important in respect to the total biological activity. Where sunlight penetrates to the bottom, photosynthesis and the development of bottom organisms contribute to a generally rich pond or lake area. Thus slope characteristics and the amount of shoal become valuable parameters. A lake with a gradually sloping basin or with broad shoals could normally be expected to be more productive biologically than a deep lake with steep sides.

The mean slope is an expression of the proximity of bathymetric contours to one another. This parameter may be expressed quantitatively as per cent slope of the basin from the formula:

$$\overline{S} = 1/n(\frac{1}{2}L_0 + L_1 + L_2 + L_3 \ldots + L_{m-1} + \frac{1}{2}L_n)\frac{D_m}{A}$$

where $\overline{S}$ is the mean slope, L is the length of each contour, n is the number of contours on the bathymetric map, D_m is the maximum depth, and A is the area of the lake surface.

The shoal area describes the extent of shallow water. The depth to which the shallow area extends may be arbitrary and the area described as per cent of the total bottom area.

Elevation of Lakes with Respect to Sea Level

Of the great number of extant lakes, most occupy basins located above sea level. A relatively few lakes, however, are situated in depressions, the deepest portions of which are below sea level. These depressions are known as *cryptodepressions*, and are worth noting. The depth of that portion of the basin below sea level is termed the depth of the cryptodepression. For example, we have seen that Lake Baikal has a maximum depth of 1741 m. Of this total depth, some 73 per cent (1279 m) occupies a cryptodepression. Lake Chelan, in Washington, is contained in a cryptodepression nearly 129 m in depth. The basins of four of the Great Lakes (Superior, Michigan, Huron, and Ontario) extend below sea level.

The surfaces of some lakes are below sea level. The surface of the Dead Sea, for example, lies 392 m, and that of the Caspian Sea 25 m, below sea level.

Many lakes are found high in mountainous regions resulting, of course, in basins and surfaces considerably above sea level. Lake Titicaca, in South America, is situated at an elevation of 3843 m. Several lakes in Inyo National Forest of California have surface elevations slightly over 3500 m, and a few Colorado lakes lie above 3750 m. The importance of elevation to lake processes, both physical and biological, are many. Lakes at higher elevations are exposed to longer periods of cold and ice cover than those of lower elevation. The biota of higher lakes is normally different from that of lakes lying at lower elevations.

LAKE SHORES AND SHORE ZONES

Lakes and lake basins are by no means static and permanent. Modifications of shore outline and depth are inherent in lakes as the result of dynamics operating within the lake and processes occurring outside the lake.

Newly filled lakes are seldom in equilibrium. Waves and currents of the water (internal factors) work against the shores. The shores offer resistance to the lake forces to varying degrees depending upon the nature of the rock or soil and the slope of the basin. A basin formed tectonically in an area of igneous geologic formations often has steep sides, and these sides will be little affected by the work of waves. Conversely, gradually sloping rims of basins in regions of less resistant rock are early subjected to erosion by hydromechanical processes. The rate and extent of erosion are dependent upon the size of the lake, the magnitude of waves, the depth of the water near shore as it influences currents and the breaking of waves, and, as mentioned above, the composition of the shore material.

The shores and lake bottoms near shore of many older lakes in the Rocky Mountains are composed typically of coarse gravel and rocks, mud and fine materials occurring in the deeper regions of the lakes. Unconsolidated or soluble materials such as sands and limestone are rather

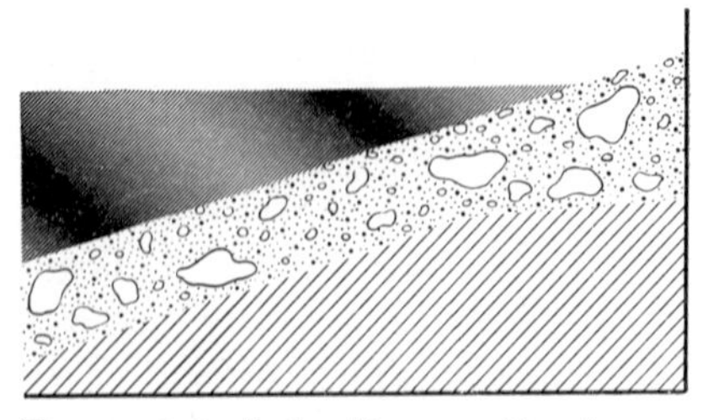

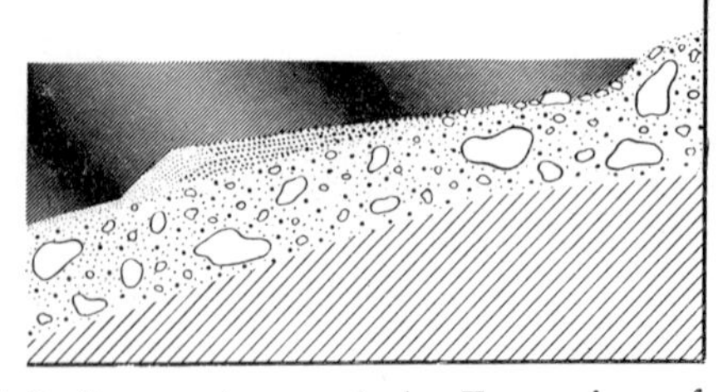

FIGURE 2·9. Lake Bottom Erosion and Sedimentation, and the Formation of a Littoral Shelf. (From Macan and Worthington, 1951. "Life in Lakes and Rivers," *New Naturalist Series*, Wm. Collins Sons & Co. Ltd., London.)

quickly eroded to develop gently sloping lake shores of sand and near-shore bottoms of finer particles. The erosional processes result in the formation of a terrace on the periphery of the lake. The exposed portion of the terrace is termed the *beach*. In the cutting of the terrace by water action some of the lighter, more easily transportable, materials will be deposited along shore as a lakeward extension of the beach; this submerged portion is called the *littoral shelf* (Figure 2·9). As a result of the differences in size and mass of the shore materials, there will be a gradation in texture from the water's edge to the high-water mark. Fine sands occur near the water while progressively larger sands and gravels are more distant.

External factors which contribute to the modification of lake shorelines are, in the main, those associated with various agents of erosion and transport of sediments from the land areas. Wind as a producer of certain currents and waves in lakes might be included in this category. Longshore currents created by winds may carry wave-stirred sediments to some point in the shallow zone and there drop the sediments. As this transport and dumping continues, a *spit* is built from shore out into the lake, usually at a point of some indentation or embayment of the shoreline. If the spit develops across the embayment to separate the bay and lake, the sedimentary structure is termed a *bar*. Figure 2·10 illustrates spits and bars.

FIGURE 2·10. Bay-mouth Bar and Lagoon in St. Mary's Lake, Glacier National Park. The bar is the result of growth of a sand spit originating at the shore. (Photograph courtesy U.S. Geological Survey.)

Streams often fall into the category of external factors which shape the morphology of a lake through alteration of the shoreline and bottom. Depending upon the velocity of the stream and the material through which it is flowing (see below), even a small stream may contribute sediments to a lake. Heavier particles will usually settle out near the mouth of the stream and form a delta. Lighter sediments may be carried farther into the lake where they sink to the bottom. The growth of a delta is dependent upon the stream contributing sediments in greater quantity than lake currents and waves can distribute. Obviously a delta will not be

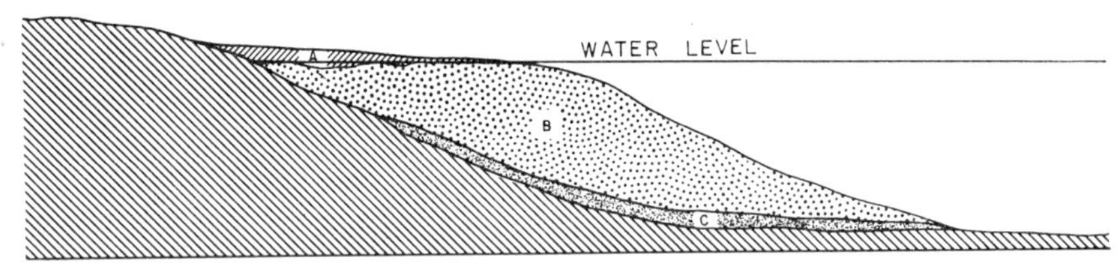

FIGURE 2·11. An Idealized Section of a Delta Showing the Arrangement of Sediments. (A) topset beds, (B) foreset beds, (C) bottomset beds.

formed if the reverse holds. Ideally, a delta is triangular in shape with the apex projecting against the river current, and with the deposited materials stratified in a characteristic pattern. As the coarser sediments are deposited, they form a series of reposing beds known as the *foreset beds* (Figure 2·11). The lighter sediments are carried lakeward and settle out on the bottom of the lake to form the *bottomset beds.* Continued transport and deposition of materials results in the lakeward growth of the delta, and

FIGURE 2·12. The Delta of the Missisquoi River of Vermont, in Lake Champlain. (Photograph courtesy of Soil Conservation Service, U.S.D.A.)

consequently the stream must extend its mouth to the base of the delta (Figure 2·12). The extension of the river channel deposits material over the foreset beds, this material being termed the *topset beds.* Various factors, including wind intensity and direction, wave action, and lake currents, act to modify the form of deltas. Thus few perfect deltas are found.

LAKE SEDIMENTS AND THE NATURE OF THE BOTTOM

Many factors contribute to the composition of a lake bottom and for this reason great variation exists. Even within a restricted area two lakes may differ significantly in bottom type (and bottom associated features also). In Florida, Orange Lake is connected with Lochloosa through narrow Cross Creek which is about 3 km long. The former is surrounded by extensive marsh, and the lake bottom is composed of organic detritus several feet thick in places; Lochloosa has sand bottom over much of its area and lacks the broad marsh areas.

Among the factors which determine the nature of lake bottoms are:

(1) *Age of the lake:* Young lakes are more apt to have rocky or sandy bottoms with little deposition of sediments and organic materials. As the lake matures, more sediments are accumulated.

(2) *Size of the lake:* Increased surface area permits greater wave action which erodes shores, the eroded materials being deposited on the bottom. Similarly, subsurface currents operate to greater extent in large, deep lakes than in small ones. These currents serve to distribute sediments.

(3) *Latitude and climate of the locality:* These factors are often complexly related to a number of other features such as the chemical nature and size of the basin, and together influence the nature of the bottom in a variety of ways. Seasonal temperatures and day length are, of course, correlated with latitude. Thus, chemically rich lakes in southern regions of long growing seasons often contain considerable organic material in the bottom. However, we also find similar bottom conditions in northern climes, for here many species of organisms are adapted to rapid growth during shorter growing seasons. Chemically poor or geologically "young" lakes in both regions often possess bottoms of meager organic content, being rocky or sandy instead. The rainfall pattern of the locality, in terms of intensity, annual amount, and seasonal variation, may influence the nature of lake bottoms by causing fluctuation in lake level and introducing sediments.

(4) *Soils and underlying rock formation:* Lakes occupying basins in sandy soils or in geologic formations which are easily fractured may deposit considerable sediments over the bottom. This condition is contrasted with lakes in basins of highly resistant rocks wherein little erosion would occur.

Generally, lake sediments may be made up of variously sized rock fragments and soils ranging from clays and silts to sands, gravels, and boulders; chemical precipitations and compounds including marl, tufa, ferric hydroxide, ferric carbonate, and silicon dioxide; and organic deposits such as peat.

The coarser sediments are widespread in lakes. As a rule the larger sized grains are found in the shallow zones, while clay and silt may occur over any part of the bottom at any depth. The color of the clay and silt may grade from white through shades of blue and green to black. Sands may range from gray to black.

Chemically, lake marl is essentially calcium carbonate. It is often precipitated by certain bacteria and algae and is mostly characteristic of small lakes and ponds, and some streams. Marl formation occurs in high-carbonate waters chiefly as a result of photosynthesis. The typical color of marl is white to grayish, although blue and black are not unusual. Tufa is a porous lime carbonate formed primarily by some of the algae. Ferric hydroxide, particularly as limonite, is known from bottom deposits in lakes in northern Europe, Canada, and other regions.

In some organically poor and in some relatively productive (but usually clear) lakes, the sediments are composed of a gray or reddish-gray, highly viscous material known as *gyttja* (Swedish: pronounced yüttya). This fine-textured mud is essentially a mixture of the remains of plants and animals, chemical precipitations, and mineral substances. In terms of acidity, gyttja is circumneutral. The humus matter of the sediment usually contains less than 50 per cent organic carbon. Decomposition of gyttja under aerobic conditions contributes much elemental nutrient material to organic production within the lakes. Fossilization of gyttja leads to the formation of anthracite.

Under anaerobic conditions in regions of peat formation, lake sediments may be brown or blackish brown due to the addition of brown humus colloids to gyttja-like material. This sediment, called *dy*, is deposited in lakes with brown water, and is typically acid. The organic carbon content of the humus in dy is usually greater than 50 per cent.

During anaerobic conditions, particularly in summer, organically rich sediments may decompose to form *sapropel.* This is a blue-black substance containing hydrogen sulfide and methane. It has been suggested that fossilization of sapropel results in the formation of oil.

Peat in various stages of compaction and decomposition occurs rather commonly in lakes. Of plant origin, peat is normally thickest near the shore in the regions of extensive plant growth. Lignite and medium-ranked, bituminous coal are derived through peat decomposition.

Whatever the factors and materials concerned in the composition of lake bottoms, there usually exists a mixture of substances. The bottom of Lake Providence, Louisiana, is composed almost entirely of mud, the shallow shore zone being the only region of sand. The lake shore is heavily wooded, and thus considerable organic detritus is present along the border. Cultus Lake, British Columbia, is described as having a bottom predominantly of gray ooze with a high content of carbonaceous plant remains. Only a few small areas of sand exist. The steep shores are littered with boulders and large rock fragments.

VARVES AND LAKE HISTORY

The rate and amount of sedimentation in lakes is not uniform. It varies with seasonal inflow of sediment-bearing waters and with fluctuations in internal currents. In cold climates melting snow in the spring flushes large amounts of coarse sediments into a lake. These sediments are distributed by lake currents over the bottom. In winter when the waters are frozen over there is little stream discharge into the lake, and lake currents are reduced. At this time fine sediments settle to the bottom to cover the coarse sediments of summer. Ideally, a pair of layers, one coarse and one fine in texture, constitutes a *varve*. Under these climatic and drainage conditions, varves may be valuable in estimating the age of a lake or lake

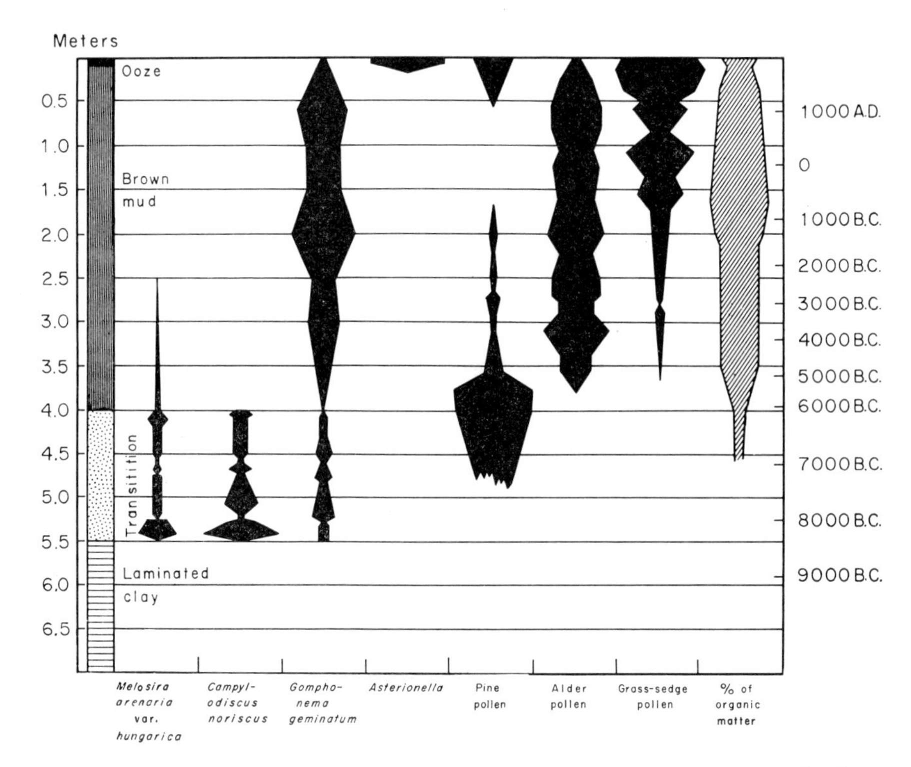

FIGURE 2·13. Some Features in the History of Lake Windermere, England, as Shown in Mud Cores from Deep Water. Relative abundance of the materials is represented by the varying thickness of the histograms. (From Macan and Worthington, 1951. "Life in Lakes and Rivers," *New Naturalist Series*, Wm. Collins Sons & Co. Ltd., London.)

basin. The number of varves in a now extinct lake in the Connecticut Valley, for example, indicates that the lake existed for about 4000 years. More recently, however, varves are being considered as representing seasonal fluctuations in rate and extent of sedimentary deposition. According to this thinking the number of varves deposited from year to year might vary.

Determination of lake basin age is not, however, the only value of lake deposit studies. Remains of plants and animals often occur in the depositional strata of the basin. Since these are arranged chronologically, much can be learned of the past history of the basin and the surroundings from studies of the bottom deposits.

The development and evolution of lake basins together with the succession of biologically different associations of organisms have been studied in a number of lakes. Some of the pertinent details in the history of Lake Windermere in England are shown in Figure 2·13. Although the full story is lengthy and somewhat complicated, we can quickly note some of the more conspicuous features and processes. First, the column at the

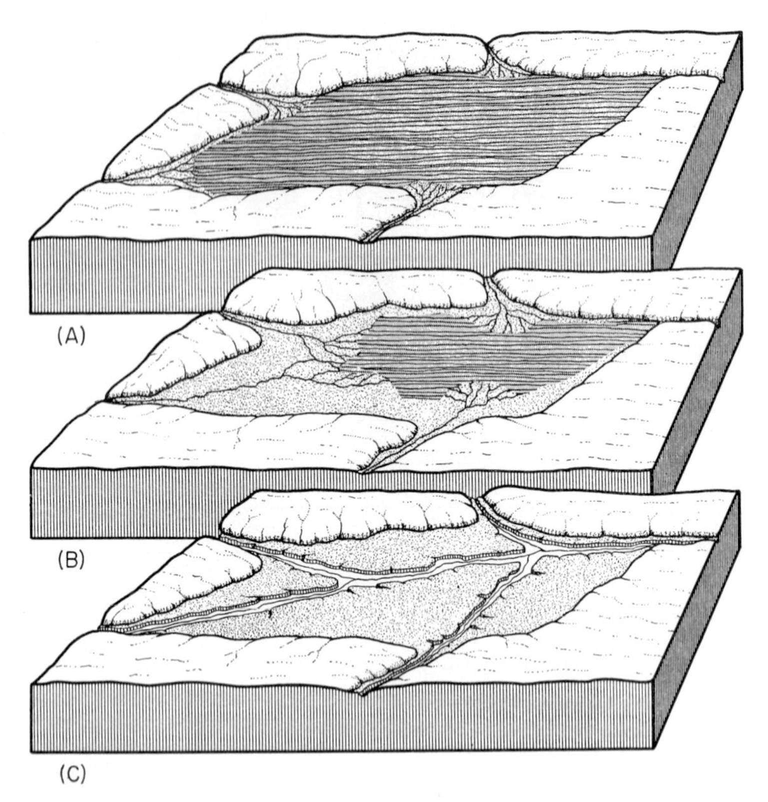

FIGURE 2·14. Stages in the Sedimentation of a Typical Lake. (A), damming of the stream course in the upper right has formed a lake. The lake covers the junctions of the earlier streams and parts of the original stream channels. Streams entering the lake drop their loads and delta building begins at the stream mouths. (B), continued sedimentation gradually fills the lake basin; the deltas join and become essentially continuous. (C), erosion of the outlet by the effluent stream is followed by draining of the basin and subsequent stream-cutting of the deltaic sediments, leaving the deposits as terraces. (After Longwell, Knopf, and Flint, 1939. "Textbook of Geology," 2nd Ed., John Wiley & Sons, Inc., New York, N.Y.)

FIGURE 2·15. Lake Como, an Artificial Impoundment Near Hokah, Minnesota, as it was in 1926.

The same locality in 1935. The clearing of timber from the steep slopes, plowing up and down the hill, and the absence of vegetation to slow down surface runoff has contributed to the rate at which sedimentation has occurred.

By 1950 the former lake bed was being used for gardening, although subject to flooding. (Photographs courtesy Soil Conservation Service, U.S.D.A.)

left of the figure represents a core from the bottom showing the several sediments present. Note the lowermost laminae of clay; these are composed of alternating layers of coarse particles (derived from quickly settling coarse materials washed in during spring) and fine particles (formed in winter by the slowly settling materials received during the previous summer). Thus, two layers represent one year of deposition. Above the clay a transition zone leads to a relatively uniform zone of brown mud and finally to a surface layer of ooze.

The histogram figures in the body of the figure indicate the depth and relative abundance of certain plants in the sediments. The absence of plant and animal remains in the clay certainly means that there was little life in the lake at the time the deposits were formed. Evidence points to the fact that the clays were deposited during glacial times of little life. Observe that certain forms appeared rather abruptly in the lake during the transition period; later changes diminished the numbers of some while *Gomphonema* has persisted and increased. Other forms invaded the lake while the brown mud was being deposited. Note that *Asterionella* becomes most abundant in the ooze. The ooze is attributed to sewage derived from increased human habitation of the lake country beginning about 100 years ago. Variations in pollen figures correlate reasonably well with changing agricultural, forestry, and cultural habits of the early people. The time scale at the right is based on this latter information.

From what has been considered concerning the origin and nature of lake basins and lakes, it should be clear that lakes are not permanent features of the landscape. Once a lake has formed, and indeed even while a newly formed basin is being filled, an array of forces operate to level the shoreline, to fill the basin with sediments, and to erode an outlet. The impermanence of lakes is a basic principle of limnology. Lakes are ephemeral, or as some authorities state it, "lakes are born to die" (Figure 2·14). Although sedimentation alone plays a most important role in the filling and consequent "death" of lakes (Figure 2·15), living plants and animals also contribute to the fate of lakes. Because of the conspicuous part played by organisms in "lake succession," a more detailed discussion of the process is reserved until the inhabitants of lakes and marshes have been surveyed.

Chapter 3

STREAM CHANNELS AND STREAMS

We have seen that in the course of the hydrologic cycle water returns to the surface of the earth in the form of precipitation. Although the amount of rain and snow that falls varies greatly over the world, probably all regions receive some precipitation. Death Valley, in California, receives about 2 in. per year, while parts of eastern North America and western Europe receive from 30 to 60 in. of precipitation per year. Of the water that falls upon the earth, some evaporates, some is absorbed by the soil and sinks in to form subsurface water, and some flows over the ground as *surface runoff* in the form of streams. That precipitation which enters the ground may later issue forth in the form of *ground-water runoff*.

It has been estimated that approximately one third of the precipitation received on the land eventually finds its way into runoff. In local regions the proportion may be considerably higher; it has been shown, for example, that 92 per cent of the winter rainfall in Vermont moves as runoff. Local variation in runoff is dependent upon (1) nature of the soil (porosity and solubility); (2) degree of slope of the surface; (3) development and type of vegetation; (4) local climatic conditions such as temperature, wind, and humidity; and, of course, (5) volume and intensity of experienced precipitation.

Because of inequalities in the nature of surface rock formation and soils, and the fact that land areas are not flat but rather possess some degree of slope, runoff accumulates quickly in small gullies which coalesce to form larger channels for rivulets which finally converge and give rise to brooks and eventually to larger streams. As the water moves, it carries materials which have been picked up along the course of the stream. A *stream* then may be defined as a mass of water with its load moving in a more or less definite pattern and following the course of least resistance toward a lower elevation. The great mass of water in rivers is best emphasized by considering the estimate of some 27,000 km^3 of water carried yearly to the sea by streams. The load of a stream relates to both dissolved and suspended materials; it has been estimated that the Mississippi River annually transports about 136 million tons of dissolved matter and over 340 million tons of suspended material.

The nature of a stream is essentially a reflection of the fluvial processes concerned in the transport of water and materials. As they relate to these processes, certain outstanding qualities differentiate streams and lakes. As suggested previously, the movement (flow) of water in streams is unidirectional and, in larger streams, continuous. Since precipitation varies seasonally in volume and frequency, it follows that wide fluctuations in water volume, rate of flow, size of channel, and rate of land erosion are to be expected in streams, depending upon local climatic conditions and the size of the stream. Because such fluctuations are characteristic of streams, it also follows that bottom and shoreline areas are relatively unstable. The inhabitants of streams exhibit characteristic forms and ways of life as adaptations to their environment. The water of smaller streams is subjected to a greater variety of movements and to more thorough mixing than that of lakes. Streams are more apt to be highly turbid than lakes, at least seasonally. Oxygen content in unpolluted streams is usually higher than in lakes. There is considerable interchange between land and the water of streams. Throughout its linear extent, which may dissect many soil types as well as more than one climatic zone, a stream exhibits a great variety of physical, chemical, and biological conditions. Because of these characteristics, especially that of land-water interchange, a stream is said to constitute an *open ecosystem* in contrast to a lake as a *closed ecosystem.*

The foregoing general aspects are important to our understanding of the dynamics of streams. These features are related to physicochemical and biological fundamentals which must form the basis for stream studies. There are additional fundamentals underlying the appreciation of the stream as a natural community. Each will be considered in more detail. Unfortunately, streams have not received as intensive study as have lakes. Admittedly, much is yet to be learned about the relationships and processes within lakes, but even more remains to be discovered concerning streams.

ORIGIN OF STREAM BEDS AND VALLEYS

Streams may find their original beds in depressions previously developed by an agency other than the stream itself. In areas of volcanism and lava deposition streams occupy folds or surfaces formed by irregular lava flow. Glaciers may scour valleys or trenches later to be occupied by streams. Even within "preformed" valleys, stream forces early begin deepening and widening the valley through erosive processes.

Most streams, however, originate on newly formed land surfaces and excavate their own channels and valleys. As stated above, precipitation falling upon the ground flows into depressions and seeks the least resistant route to lower levels. Gullies are soon excavated and join to form a shallow entrenchment containing the contribution of a number of tributaries. The greater the volume of water and the slope, the faster the erosion (its rate depending upon the type of substrate). With increased erosion the channels become larger and transport more runoff. As erosion continues at the head of the gully, it grows by extending the upper reaches. With headward extension and vertical cutting, a valley is formed. Streams developed upon the initial slope contributing the runoff are termed *consequent streams*. The classification of streams according to hydraulic and erosional processes with respect to the land area includes additional types such as subsequent streams, resequent streams, and others. Reference to a recent textbook in physical geology should give details relating to these types.

We have made free use of the term *erosion* by using it in a general sense. Actually this very significant characteristic of fluvial work is the result of two processes. *Mechanical erosion* is accomplished mainly through abrasion of stream sides and bedrock by mineral sediments carried in suspension or rolled along by the current. *Chemical solution*, or *corrosion*, of channel materials by stream waters is important in regions of soluble substrate, as, for example, in limestone areas.

The rate at which a stream erodes its channel is determined by the nature of the bedrock, composition of the water, climate, and the grade or slope. The greater the slope the greater the capacity to transport abrasive materials through increased velocity. Soft rocks are more susceptible to erosion than hard material is, and the channel is more rapidly deepened in a zone of soft material than in an area of hard bedrock. Hard rock may be cut rather quickly, however, if angular, sharp fragments are moved rapidly by stream flow. It would seem that the time of most rapid erosion is during flood season, for the water level is high, and the increased volume and velocity move more debris and larger particles than during normal stages.

We have seen that in the earliest stages of development of a stream, surface runoff is probably the most important single source of water. But once the valley has been eroded and several streams unite to pool their load and work, maintenance of a stream depends upon additional supplies of water. Because the size and morphology of river valleys are direct results of the work of the waters of the streams, let us consider first the sources of stream waters and later the form of the channels and valleys.

SOURCES OF STREAM WATERS

Nearly everyone has observed the sudden and voluminous rise in stream levels following periods of heavy rainfall or spring thaw of winter snow. We have also seen that the lands in the drainage area of a stream soon dry and the flooded stream returns to "normal" flow within a short time after the rains have subsided or snow has passed. Yet, depending upon its size, the nature of the surrounding soils, and climate in the region, the stream continues to flow. It would appear, then, that the stream waters are being derived from sources other than the immediate surface runoff.

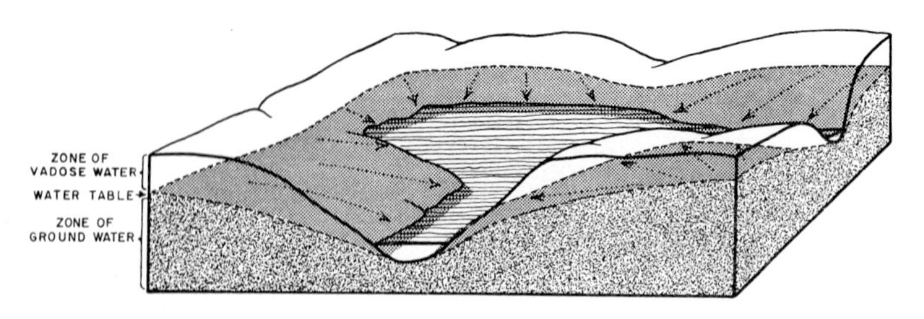

FIGURE 3·1. Diagrammatic Representation of Subsurface Water Zones and Their Relations to Water Flow and Surface Topography. (After Longwell, Knopf, and Flint, 1939. "Textbook of Geology," 2nd Ed., John Wiley & Sons, Inc., New York, N.Y.)

Recall that in the hydrologic cycle varying amounts of precipitation may be absorbed by the soil and held as *subsurface water*. The subsurface distribution of water is dependent upon the local climate, topography, and the porosity and permeability of the underlying soils and rocks. Where permeable rocks occur, water may be taken in and held. The surface of the saturated zone of permeable rocks is called the *water table*. Water in the soil above the water table is *vadose water*, and its volume with respect to the soil is subject to considerable fluctuation. *Ground water* is the water contained in the rocks below the water table and usually is of more uniform volume than vadose water. The water table normally follows the topographic relief of the land, being high below hills and sloping into valleys (Figure 3·1). It is this subsurface water that controls to a great extent the level of lake surfaces, the flow of streams, and the extent of swamps and marshes.

The intercalation of subsurface water into cavities, channels, loose sand, and gravels provides a reservoir for water supply to streams during periods of drought. Whereas surface runoff may be the start of a stream valley and the cutting of the valley may depend upon seasonal rainfall, once the valley has eroded to the water table the stream flow is quite apt to become regular and fairly uniform. For now water can issue as seepage or as springs into the stream channel. Conversely, the stream may also contribute to the supply in the water table. As demonstrated in Figure 3·1, there may be exchange between ground water and the stream bed or lake basin depending upon the nature of the substrate and the slope of the water table. In some instances, however, the stream bed may be perched in impervious rocks above the water table and the interchange between the two waters inhibited.

Although some streams may receive water from springs or from glacial melting, these are not, as a rule, common sources of water in streams. As we have just seen, the major supply is that from ground water, primarily as seepage. Nor should we conclude that ground water enters the stream throughout its entire length. More frequently the inflow is in the region of the source of the stream and its tributaries. The source may be in a marsh, or soggy meadow, or perhaps a periodically dry brook bed. On the basis of continuity of flow, streams may be:

(1) *Permanent streams:* Streams which receive their waters mostly through seepage and springs from subsurface water. In the immediate drainage area the water table usually stands at a higher level than the floor of the stream.

(2) *Intermittent streams:* Streams which receive their waters primarily from surface runoff. Because the runoff is seasonal, stream flow occurs during the wet periods. In regions of considerable rainfall and melting snow, surface runoff may be sufficiently uniform to maintain relatively "permanent" streams.

(3) *Interrupted streams:* Streams which flow alternately on and below the surface. The subsurface flow is usually through coarse sands or gravels (as in portions of the Rio Grande of the southwestern United States), or in limestone; the Santa Fe River of Florida disappears into a limestone sink and follows a subterranean channel for several miles before reappearing on the surface.

FLUVIAL DYNAMICS

Having considered some fundamentals in the formation of stream channels and valleys and the sources of water in the streams, we need now to give attention to some of the dynamics of running water. Consideration of these stream processes necessarily precedes a study of stream valley morphology. For whereas the morphology of lakes is basically attributable to the mode of origin of the basin, stream-bed and valley forms are direct results of stream flow. The processes and parameters with which we shall be concerned belong essentially to the realm of hydrology, and the student of streams should refer to standard texts and references such as the

"Hydrology Handbook" of the American Society of Civil Engineers Committee on Hydrology for further information. Our concern at this time is primarily with those fluvial processes responsible for the shaping of the stream channel, valleys, and ultimately the drainage basin.

Types of Stream Flow

Stream flow may be of three kinds: *laminar*, *turbulent*, and *shooting*. Laminar flow takes place in channels with smooth sides when the rate of flow is low. Water exhibiting laminar flow moves in paths that are straight and parallel to the sides of the channel. Since the channels of streams are usually rough and the rate of flow above a certain critical velocity at which flow becomes turbulent, laminar flow is seldom found in natural currents of any significant magnitude. Laminar flow may occur in shallow streams or in sheets of runoff down smooth slopes.

In most streams the water flows in contact with irregular and rough stream beds. This results in the development of a system of eddies and circular currents giving rise to turbulent flow. Most stream flow is turbulent, and eddying effect is a major force in transporting particles in suspension. More will be said about turbulence in Chapter 6. Shooting, or jet, results when stream flow develops a very high velocity. Turbulence is created and fluctuations in velocity result in spurts of water mass. Such a flow is best observed in narrow channels of considerable gradient, or over waterfalls.

Velocity

Velocity is the distance a mass of water moves per unit of time. The measurement is generally expressed in feet (or meters) per second. Upon this parameter depend movement of dissolved and suspended materials, rate of discharge, erosion (in part), the distribution of animal and plant life, and other aspects of stream economy. Stream velocities may range from near motionless in pools and lower reaches of rivers to relatively high rates of 30 ft (9 m) or more per second. Silver Springs, Florida, a stream of generally uniform flow, was found to have a velocity of 0.7 ft (0.2 m) per second. The velocity of swift-water trout streams in Ontario ranges seasonally from about 1.5 ft (0.45 m) per second in August, to 6.0 ft (1.8 m) per second in April. The velocity of a stream is determined basically by the volume of the water in the stream, the load of suspended sediments, and the gradient. Obviously, a stream flowing on a steep slope has greater velocity than on near-level terrain. Similarly, with a given gradient, an increase in volume will increase the velocity. Friction between water and the sides and bottom of the stream bed tends to reduce velocity. Thus there may be a velocity differential from shore to midstream and from top to bottom. Turbulence develops in the zones of contact between waters of different velocities. It has also been shown that velocity varies inversely as the load.

Discharge

The total volume of stream water passing a point in a given period of time is termed discharge. The measurement of discharge is called gauging and the data describing discharge are usually expressed in cubic

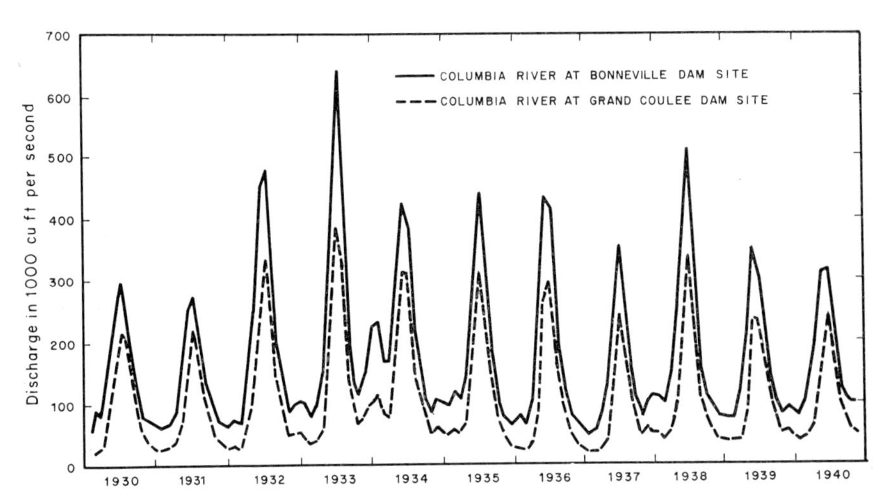

FIGURE 3·2. Seasonal Variation in Discharge of the Columbia River. The annual patterns are affected very little by inflow from tributaries between Grand Coulee Dam up-stream and Bonneville Dam downstream. (After Robeck, G. C., Henderson, C., and Palange, R. C., 1954. "Water Quality Studies on the Columbia River," U. S. Public Health Service.)

feet per second (cfs).* The Niagara River has an annual mean discharge of 219,850 cfs. At New Orleans, the Mississippi discharges at a rate of about 1,740,000 cfs. Figure 3·2 shows the cyclic nature of discharge of the Columbia River.

The rate of discharge is proportional to the rate of flow and volume of water being fed into a stream. Thus the discharge varies with season and contribution by tributaries. Discharge is determined by the shape of the channel, the cross-sectional area of the channel, and the gradient. From these factors a formula for discharge may be stated as follows:

$$Q(\text{cfs}) = WD_mV_m$$

---

* The acre foot and gallon are sometimes used. An acre foot is the amount of water that covers 1 acre to a depth of 1 ft; it is equal to 43,560 cu ft. Metric units are not given here because use of the English system is standard procedure in the United States.

where the discharge (Q) is related to mean channel depth (D_m) in feet, channel width (W) in feet, and mean velocity (V_m), in feet per second.

TRANSPORT AND LOAD

The ability to do work in the form of movement of materials is one of the conspicuous attributes of stream flow. The movement of sus-

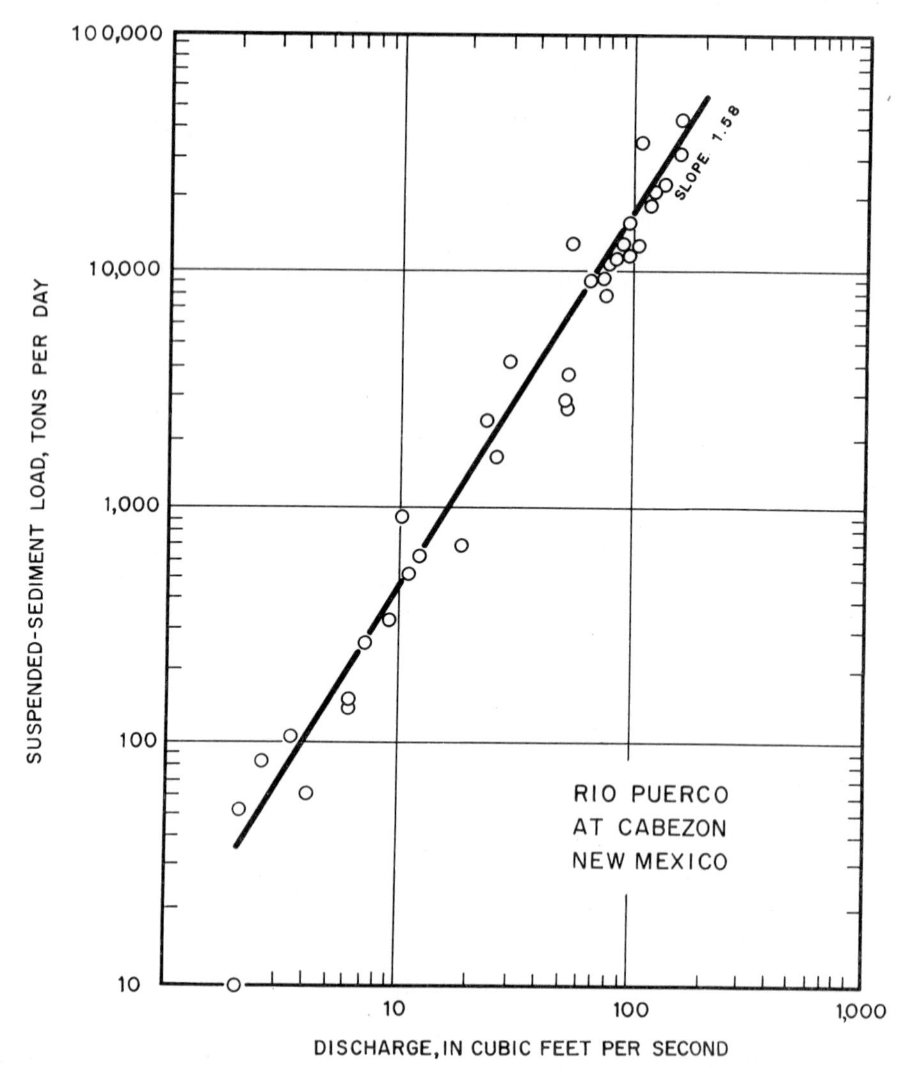

FIGURE 3·3. Stream Load as a Function of Discharge as Measured in the Rio Puerco near Cabezon, New Mexico. (From Leopold, L. B., and Miller, J. P., 1956. "Ephemeral Streams–Hydraulic Factors and Their Relation to the Drainage Net," *U. S. Geol. Survey Profess. Paper*, No. 282-A.)

pended and dissolved substances eroded from the valley or brought in by other means is termed *transport*. The transport of *dissolved* materials is not dependent upon stream velocity. The movement of suspended matter is a function of *competency*, *capacity*, and *load*, the first two aspects being related, in part, to velocity.

Competency pertains to maximum size of sediments transported by a stream and is dependent upon velocity and turbulence. A stream requires a high competency to raise and carry sediments. Also related to competency is the ability of stream currents to roll large particles along the bed. It has been shown that the size of particles transported varies as the square of the velocity. A current of approximately 1 mph can, under proper conditions, move a stone of ¼-in. diameter along a normally smooth bottom.

Capacity of a stream refers to the quantity of material that can be transported under certain optimum circumstances. Capacity is also a function of velocity and turbulence. The capacity of a stream is not constant throughout its course. The result is uneven beds formed from scouring in one portion of the stream and deposition in another.

The volume of sediments actually held by a current at a given time is termed the suspended load. The load seldom equals the capacity. As seen in Figure 3·3, load increases with discharge, although an increased load tends to reduce the capacity of the stream. As the velocity is diminished, the load is deposited on the bottom and along the sides of the stream bed. Deposition, as will be seen later, plays an important role in the form, or morphology, of stream channels and valleys.

Stream Adjustment

In conformance with natural laws of equilibrium and "orderliness," fluvial processes tend toward establishing balance among the several forces operating within the stream proper, and between the stream and its surroundings. In most streams there is, as described above, an apparent increase in discharge from the upper reaches to the mouth of the stream. There is also a decrease in gradient over the course of the stream from source to mouth. This adjustment to opposing phenomena is due primarily to correlations between depth, width, and velocity as one set of factors and discharge as another factor. As shown in Figure 3·4, the correlation is one of straight line; as discharge is increased the dimensions of the stream are increased. Or conversely, the stream bed becomes enlarged to accommodate the greater discharge. The hydrodynamic factors involved in discharge and channel adjustments include erosion, transport, and deposition of the eroded materials along the stream bed.

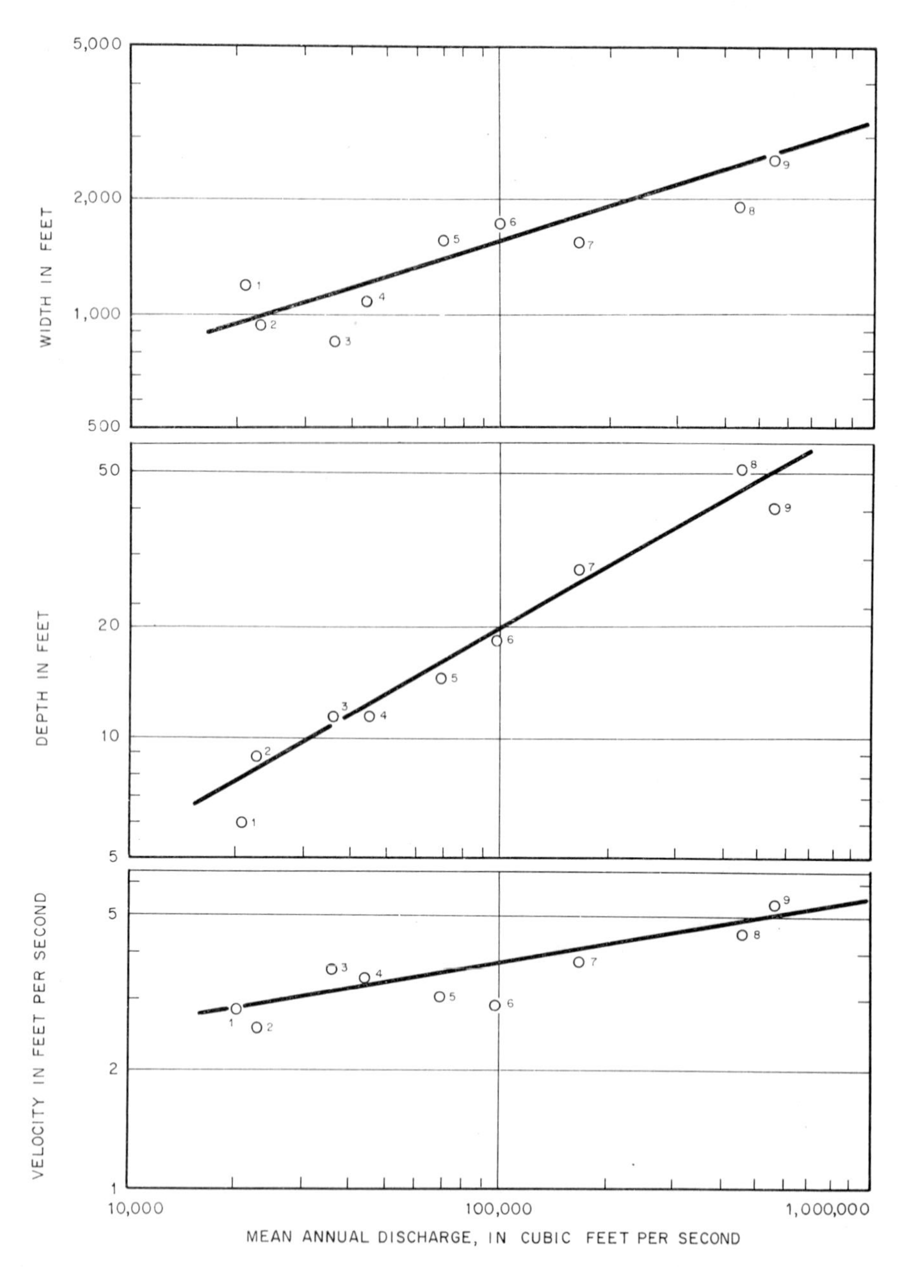

FIGURE 3·4. Relation of Width, Depth, and Velocity to Mean Annual Discharge as It Increases Downstream. Stations 1 through 5 are Missouri River sites from Bismarck, N.D., to Hermann, Mo.; stations 6 through 9 are Mississippi River localities from Alton, Ill., to Vicksburg, Miss. (From Leopold, L. B., and Maddock, T., 1953. "The Hydraulic Geometry of Stream Channels and Some Physiographic Implications," *U. S. Geol. Survey Profess. Paper*, No. 252.)

MORPHOLOGY OF STREAM VALLEYS

In the preceding section we have briefly considered some of the basic dynamics operating in flowing waters. All of these hydraulic processes are of concern, for they relate ultimately to form of stream valleys at a given time as well as to modifications of valley and basin morphology. Bearing in mind the forces responsible for shaping the basins, let us now consider the valley in its linear and transverse aspects, and then the depositional features shaping the valleys.

Stream Length

Streams, especially mountain brooks, creeks, and rivers, show great variations in length. Some mountain streams, important locally to fishermen and to water supply, may be less than a mile in extent. Important rivers may range from relatively short ones such as the 61-km-long Mobile River of Alabama, to the great Nile of Africa, over 6400 km long. The Mississippi-Missouri System of North America is said to be 6233 km in length.

Longitudinal Gradient

As defined, the gradient of a stream is the slope of its longitudinal course. This characteristic is expressed in vertical descent per unit of horizontal distance. Generally the region of steeper slope is near the headwaters, while the more gentle gradient is in the vicinity of the mouth of the stream. In the upper reaches of the Yuba River of California, the gradient averages (in English units) 225 ft per mile for about 12 miles. By contrast, the gradient in the Mississippi River from Cairo, Illinois, to its confluence with the Red River of Arkansas is no greater than 0.5 ft per mile.

In its early evolutionary stages, the gradient of the stream is essentially the degree of slope of the land. This gradient enters into adjustment processes tending toward equilibrium with channel characteristics. Through erosion and deposition the longitudinal profile of the gradient attains the form of a shallow, irregular arc with an upward concavity flattening toward the mouth of the stream. This curve is termed the *long profile*. At points along the stream local equilibria are established through adjustments between channel form, velocity, gradient, and discharge. Erosion of the stream bed is reduced as transport of sediments increases, that is to say, much of the energy formerly directed toward scouring (degrading) the channel is diverted toward deposition, or refilling, of the bed (aggrading). In time, irregularities along the stream slope are more or less leveled. When this stage has been reached the stream is described as *graded*, and the long profile has developed a *profile of equilibrium*.

Stream profiles may be shown graphically by plotting elevation from source to mouth against horizontal distance as in Figure 3·5. Since the gradient of most streams is relatively slight, the vertical distance is considerably exaggerated on the graph. An irregular curve, such as that for

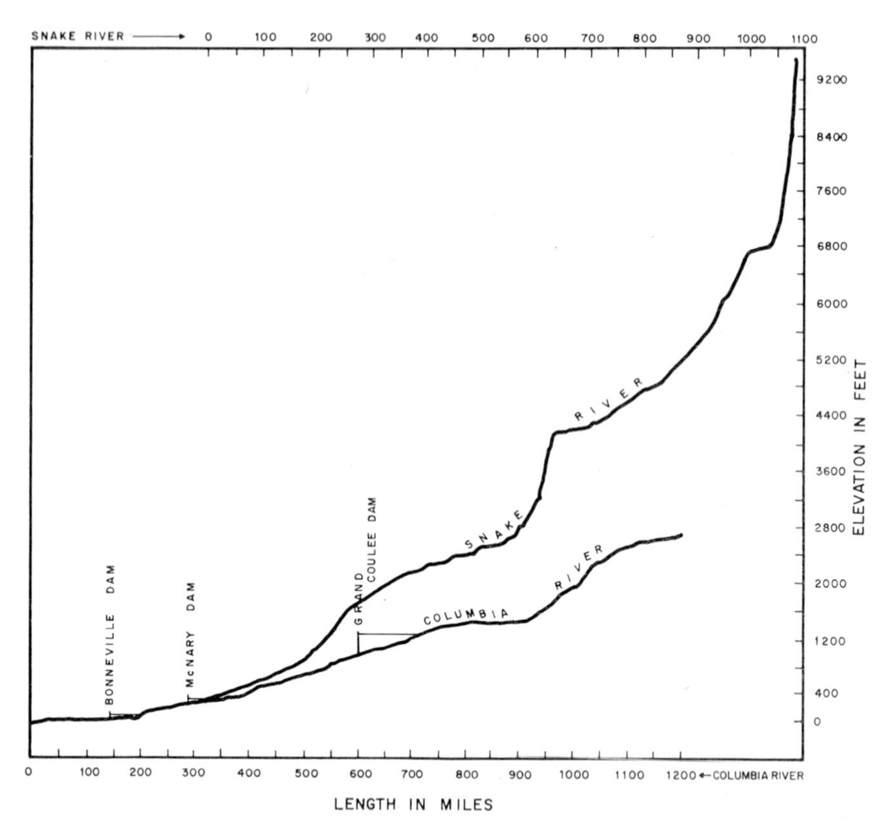

FIGURE 3·5. Condensed Profiles of Snake River and Columbia River. The longitudinal profile (from upper reaches to mouth) of a stream is typically concave to the sky. Variations in erosion rates along the profile account for the irregularities. (After Sylvester, R. O., 1958. "Water Quality Studies in the Columbia River Basin," *U. S. Fish and Wildlife Serv. Spec. Sci. Rept.*, No. 239.)

the Snake River in Figure 3·5, suggests that the stream is still in the process of grading. A smooth curve depicts a stream in equilibrium, one in which local inequalities have been minimized. Ideally, continued erosion and lowering of the land surface result in the curve becoming progressively more flattened. However, various geologic forces such as volcanic activity, uplift, faulting, and change in sea level complicate the ultimate developmental pattern.

There is a limit, although often temporary, to which a stream may erode its channel. This limit is called the *base* level. In the case of streams entering the sea, the sea level is the deterrent to further downcutting, for generally streams do not erode below sea level. Similarly, a stream discharging into a lake is eroded no deeper than the lake level. Since lakes are essentially impermanent, the stream base level controlled by a lake is called a *temporary base level*. The base level of a stream can be manipulated by artificially draining lakes or erecting dams in valleys.

Transverse Profiles

The form of the stream valley in cross-section, or *transverse profile*, may reflect the nature of the area geology, local climatic conditions, and the period of time the stream has been cutting its valley. Ideally, stream erosion should result in a channel having nearly vertical sides. However, due to variations in discharge and water level of the stream proper, and because of surface runoff from the surrounding terrain, the valley in its early stages acquires (in humid regions) a V-shaped transverse profile. Such valleys, characterized by narrowness and steep sides are termed *young*, and are usually found in regions recently uplifted or otherwise newly formed, geologically speaking. Young streams in narrow valleys are typically fast-flowing, the water often encountering falls and rapids in its course. Vertical down-cutting is proceeding at a faster rate than horizontal erosion. Consequently the stream bed may occupy nearly the entire width of the valley floor. Yellowstone Canyon, in Yellowstone National Park, Wyoming, is an example of a young stream and valley.

In time, sediments carried by young streams are deposited as velocity and discharge fluctuate, the valley sides are eroded to more gentle slopes, and irregularities in the channel cause the stream to vary or meander (see below) its channel within the valley. The valley floor becomes widened, and a plain resulting from stream flooding develops. The longitudinal gradient decreases. When these conditions obtain, the valley is said to be *mature*.

A stream valley is *old* when, as a result of continued erosion, the valley sides are reduced to nearly level surfaces and a broad flood plain has been built by deposition resulting from decreased ability of the stream to carry its load. Broad meanders and oxbow lakes occupy the valley floor. The lower Mississippi River is a classic example of a stream and valley in old age.

The Rhone River of Europe and the Colorado River of North America are examples of *rejuvenated streams*. We have seen that streams in old age typically meander over a broad valley floor. Geostrophic processes such as upwarping or bending of the land surface may act to increase the slope of the area, thereby increasing the gradient and erosional power of the stream. Thus the stream is "rejuvenated," and as a result of increased cutting ability comes to occupy deeply cut, winding gorges known as *incised meanders.*

Valley Features Resulting from Stream Dynamics

We have considered some of the major forces of flowing water and how these forces may operate to scour a channel and to modify the form of the valley. Similarly, various hydrodynamic factors, acting singly or together, often result in the formation of valley features which may bear greatly on the characteristics of local segments of the stream as well as the stream generally.

Reference has been made to the cutoff of stream *meanders* as a method of lake formation (Chapter 2) and in connection with mature and old streams. *Meandering* is a mechanism typical of streams that have eroded to base level. As a result of decreased velocity the stream is deflected from one side of its valley to the other, thereby widening the valley in the process. The nature of the substrate also relates to this process. As erosion of the valley banks continues, acute bends develop in the channel. Because of increased turbulence and velocity on the outside of the bend, erosion of the shore imparts a load of sediment which is carried downstream to the next bend. Since the velocity is reduced on the inside of the lower bend, the load is dropped and the channel becomes partially filled, thereby pushing the current farther toward the eroding outer side of the bend (Figure 3·6). The bends become more sinuous, and as they approach and exceed 180°, they are termed meanders. The downstream arc of the bend in the meander erodes at a greater rate than the upstream arc. Thus the meander system migrates downstream, leaving a layer of deposition across the valley. Oxbow lakes and meander scars develop when the downstream migration of one bend proceeds at a faster rate than the one below. The result is the formation of a new channel, or cutoff, between the bends.

Another feature often found in broad valleys develops when scattered masses of debris in the flood plain cause the stream to flow in many small, intermeshed channels. The debris is left when the stream velocity is reduced by seepage or excessive evaporation. A stream exhibiting such a form is termed a *braided stream* (Figure 2·1). Streams occupying glacial outwash plains and alluvial fans often show braiding.

An *alluvial fan* is an expanse of sediments deposited in the form of a broad "fan" by a rapid stream flowing onto a level plain. Although more characteristic of regions of arid climate, for example Death Valley, California, fans may form wherever conducive conditions prevail. Sudden floods in mountainside gullies transport a great load. The load is dropped and the waters usually soak into the fan as the stream flows onto the more level surface. Fans might be thought of as the land form of a delta.

A delta is, of course, another result of fluvial deposition and is formed when the velocity and carrying capacity of a stream is reduced. In this instance the reduction occurs as the stream flows into a standing body of water such as lake or sea. The morphology and structure of a delta were considered earlier as they relate to lake form. Seaward deltas are typical

FIGURE 3·6. Meandering Stream in Big Horn County, Wyoming. Extensive "sand bars" on the inside of each meander result from decreased load-carrying ability of the stream. Meander scars representing former channels are conspicuous to the left of the bend at the lower margin of the figure. Note the dendritic drainage pattern formed by the small streams to the left of the photograph. (Photo courtesy Commodity Stabilization Service, U.S.D.A.)

of the great streams such as the Nile, Rhine, and Ganges. The Mississippi delta covers an area of over 31,000 km^2 on the Gulf Coast of the United States. The depth at New Orleans is said to be nearly 300 m and the advance of the delta is estimated at some 90 m per year. Many streams lack deltas because the streams may not carry sufficient sediments or because currents at their mouths remove the sediments rapidly.

Two additional depositional features of streams in broad valleys are worthy of consideration here because they are often significant factors in determining the nature of the shores of mature and old streams, especially in relation to vegetation and other biological aspects of the streams; these are natural levees and flood-plain deposits. Natural levees are formed by deposition of sediments along the normal channel of a stream as it overflows its channel. These levees take the form of ridges of fine materials which stand higher than the flood plain proper. The areas behind the levees are frequently maintained as swamps or marshes.

FIGURE 3·7. Flood Plain of the Rio Grande North of Espanola, New Mexico. The older flood plain, now under cultivation, is clearly differentiated from the adjacent desert-like areas. The river is at high-water stage and follows a meandering pattern with cutoff streams. (Photograph courtesy Soil Conservation Service, U.S.D.A.)

Along the Mississippi in its delta region natural levees rise to heights of nearly 6 m above the swamps. Flood-plain deposits are those materials left by streams during flood times or as the stream meanders over the flood plain (Figure 3·7). A profile through the flood plain will often give an indication of channel movements, floods, and other events in the history of the valley (Figure 3·8). Often the flood plain may take the form of a very level bed in which the present stream channel is flowing. This feature is a *flood plain terrace* and indeed, depending upon the history of the stream valley, several terraces representing alternating periods of flooding and channel downcutting may be present.

Upstream, towards the headwaters, and in the tributaries of a broad valley we encounter a great variety of streams grading from creeks to brooks. These often contain water only during the wet season. But when flowing, the dynamics of running water are operating as in the larger

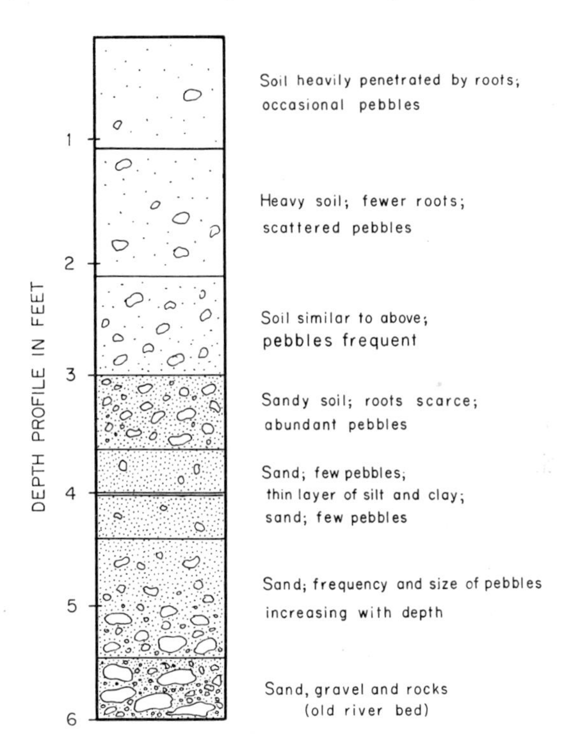

FIGURE 3·8. Diagrammatic Interpretation of Soil Profile of the Ancient Flood Plain of the Raritan River Near New Brunswick, New Jersey. Variation in the nature of the sedimentary strata is attributed to shifting in the course of the river over its bed. (After Wistendahl, W. A., 1958. "The Flood Plain of the Raritan River, New Jersey," *Ecol. Monographs*, Vol. 28, 129-153.)

streams; currents erode exposed surfaces and land areas, sediments are transported and dropped, and the discharge from the smaller streams contributes to the maintenance of the major stream of the system. As we follow a stream toward its source through the headwaters, we are aware of gradations in the structure and morphology of the valley and the stream channel. Along the main stream these gradations may be subtle and we observe that, according to size, velocity, bottom nature, and the like, we follow from brook, to creek, to river. However, in another instance, as a small brook joins a larger stream a more abrupt transition is marked.

STREAM-CHANNEL PARAMETERS

Because of the role of running water in creating and modifying the stream channel, and since the morphology of the channel is a direct consequence of hydraulic factors, certain of the more important stream parameters have been previously considered. It should be evident that certain morphological features described for lakes are also applicable in principle and method to stream studies. On the other hand, the very fact that the water in streams is flowing introduces a number of features not encountered in lakes.

Of the factors common to lakes and streams, let us consider depth, mean depth, length, breadth, area, volume, extent of shoreline, and mean slope with reference to streams.

Depth

The depth of a stream usually refers to the maximum depth within a given segment of the stream. For most streams the parameter is determined by sounding, although navigation charts for major rivers indicate depths. *Mean depth* takes into consideration the transverse bottom slope and is a relationship between cross-sectional area (A) and width (w) of the flowing stream: $\bar{d} = \frac{A}{w}$.

Length

Stream length may refer to the total extent of the stream or to a segment under study. Stream length should describe the "true" length including bends. A straight-line measurement should be so stated.

Breadth

The breadth, or width, of a stream channel may be defined in two ways. It may be (1) the actual cross-channel measurement of water at a given time, or (2) the width at near bankfull stage. The second parameter is evident in regions where well-developed flood plains and V-shaped valleys exist; it is less conspicuous in dry, flat areas.

Area

Stream surface area is simply the measure of exposed water surface. This parameter varies considerably with seasons of drought and flooding. Another expression of areal extent often used in stream studies is that of *cross-section area;* this dimension is obtained roughly by multiplying the mean depth by the mean breadth.

Volume

As in lakes, the volume of a stream is the amount of water held in the basin, or more preferably, the channel. Also as in lakes, stream volume is the sum of the volume in each of the strata bounded by bottom contours. Since the water mass is in motion, the term *discharge* is applied.

Extent of Shoreline

The extent of shoreline is simply a measurement of the length of the shore with respect to the length of the stream or a segment thereof. Features such as main-channel pools, side pools, and embayments increase the shoreline length relative to stream length.

Mean Slope and Area of Shoal

Stream cross sections may be somewhat trapezoidal, semielliptical, or triangular. Depending upon the shape of the channel, the bottom may slope gently or drop off abruptly from shore. The mean slope describes as a per cent the bottom grade with respect to horizontal distance transversely across the stream channel. The area of shoal describes the extent of shallow water as a proportion of the total area of a stream segment. The depth used to designate a shoal is arbitrary.

The major stream-channel parameter not shared with lakes is longitudinal gradient. Because such important hydraulic features as velocity, discharge, load, and erosion relate directly to gradient, it becomes of prime interest in stream studies. This measurement may be obtained from topographic quadrangles where available and of sufficient scale. Field determinations can be taken with stadia rod and transit along the shore, in the channel, or in the stream bed when the channel is dry or the water low.

Thus far, those parameters associated more directly with morphology of stream channels have been discussed. The characteristics of channels and dynamics of running water combine to maintain a stream and drainage basin. Certain relationships among the stream features are established and can be expressed in quantitative terms, as we have seen. In Table 3·1, some of the characteristics of channel-shape and flow are given for selected North American rivers. Note especially, with reference to the Mississippi River data, that the velocity increases downstream. This is in contradiction to the usual impression of higher velocities upstream, and occurs when discharge is uniform along a stream and increased depth compensates for decreased stream gradient.

TABLE 3·1. PARAMETERS OF STREAM-CHANNEL MORPHOLOGY AT STAGE CORRESPONDING TO MEAN ANNUAL DISCHARGE

(Data from Leopold, L. B. and Maddock, T., 1953. "The Hydraulic Geometry of Stream Channels and Some Physiographic Implications," *U. S. Geol. Survey Profess. Paper*, No. 252.)

	Years of Record	Mean Annual Discharge (CFS)	Width (ft)	Area of Cross-Section (sq ft)	Mean Velocity (FPS)	Mean Depth (ft)	Drainage Area (sq miles)
Scioto River at Chillicothe, Ohio	26	3,289	200	1,840	1.8	9.2	3,847
Tennessee River at Knoxville, Tenn.	48	12,820	933	13,450	1.0	14.4	8,934
Yellowstone River at Billings, Mont.	16	6,331	253	1,343	4.8	5.3	11,180
Kansas River at Ogden, Kansas	30	2,514	342	1,300	1.9	3.8	45,240
Kansas River at Wamego, Kansas	28	4,114	525	2,150	1.9	4.1	55,240
Mississippi River at Alton, Ill.	12	96,670	1,750	32,500	3.0	18.6	171,500
Mississippi River at St. Louis, Mo.	14	166,700	1,586	44,440	3.8	28.0	701,000
Mississippi River at Memphis, Tenn.	13	454,900	1,950	99,400	4.6	51.0	932,800
Mississippi River near Vicksburg, Miss.	16	554,600	2,610	105,650	5.3	40.1	1,144,500

STREAM BEDS AND SHORES

The nature and composition of the channel bed and shores of streams are as varied as the terrain through which the streams flow and the internal dynamics of flowing water. From the general survey of fluvial processes just concluded, it would appear that gradient and types of substrata are the major factors in determining the composition of shore and bottom of streams. Rocky to gravelly substrates occur where the stream runs rapidly over a steep gradient, the coarse materials being derived from the mountainous zones (or at least higher regions) where the stream slope would naturally be greater. Sands and muds may be deposited in the upper reaches but usually occur only in wider or deeper regions where the velocity is reduced. Finer sediments are transported to lower regions, and light organic materials do not settle out until the stream reaches near base level. Thus, a gradation in the nature of the bed and shores exists from the fast-water, rough-bed zone to the slow-water, mud-bed region. Associated with this gradation (but related to a number of ecological factors) is a similar gradual change in the distribution of animals and plants, which we shall consider in more detail later.

Lower Stream Course

As a stream nears its base level and approaches its mouth, the bottom typically is composed of loose muds, silt, and organic detritus. Because the stream is operating in a broad flood plain of low relief, the shores are generally poorly defined, the stream being bordered by marshes or swamps. Many backwaters, sloughs, and bayous, such as those of the lower Mississippi, are apt to be present.

Middle Stream Course

The middle stream course typically exhibits a more inclined gradient and greater velocity than are found in the lower course. In this middle zone the bottom is usually characterized by coarser materials, with muds and lighter sediments being found only in pools or sidewaters. The flood plain is, as a rule, less extensive, leading to more discrete shores standing considerably higher than the stream. Wooded stretches rather than marshes and swamps occur along the banks.

Upper Stream Course

Mountain streams and brooks typify the stream in the upper reaches. The current in high velocity rushes and tumbles down through a steep gradient, occasionally slowing in pools. In the fast stretches, the channel and often the shores are strewn with boulders and rubble of various sizes. Where a pool interrupts the course, or the stream flows briefly out onto a meadow, the bottom may contain light deposits of sands and organic detritus.

These impressions of stream bottom and shores are, of course, very generalized. Local climatic conditions in terms of humidity and rainfall, soil types, land elevations and configurations, and age of the area and streams all serve to greatly modify the stream from source to mouth and to contribute to the countless variations exhibited by streams in nature. The topic is introduced at this point to emphasize the importance of the nature of bottom and shore in stream studies as well as in the ecology of a stream or stream segment. In defining a lotic environment and distinguishing it from standing waters, the considerable interchange between a stream and its basin was noted. This relationship becomes increasingly important in terms of contributions of water and nutrients from the valley and shore to the several courses, or regions, of the stream. Contrast the relative productivity of a stream flowing over highly insoluble rock substrates with that of a stream moving through a zone of soluble material rich in chemical nutrients.

DRAINAGE BASINS

The total land surface from which a system of streams receives its waters is termed the drainage basin. We have seen how a system may develop from continued erosion of a land area by water carried first in small trenches and eventually in stream valleys. As a rule, definite patterns of confluence of the various contributaries develop. Regardless of its size, each stream possesses its own drainage basin and the water received usually follows a particular course. The drainage basin of the Mississippi extends over some 3 million square kilometers. Such coverage would demand an impressive number of rivers such as the Tennessee, Ohio, Missouri, and Red with their countless tributaries. The Nile River system serves to drain over 2.6 km^2.

The form of the drainage basin exerts considerable effect upon the regularity of flow in streams within the basin. Two methods of measuring the relationship between stream dynamics and form of the basin have been suggested; these are (1) the *form factor*, and (2) the *coefficient of compactness*. The form factor expresses the ratio of the mean width of the basin to the length from mouth to uppermost part, or

$$\text{Mean width} = \frac{\text{area}}{\text{axial length}}$$

The lower the form factor, the less likely a widespread rainfall over the entire basin. The compactness coefficient relates shape of the basin to a circle of the same area. A high value indicates a less circular form for the basin, therefore reduced chances of receiving heavy rainfall over the entire region.

Within a basin the pattern of streams with respect to orientation and branching usually reflects the composition of the substrate and surrounding terrain. It also indicates the geologic history of the stream and the region.

As far as investigations of streams are concerned, the most important consideration of the drainage basin relates to the present nature of the underlying geologic formations and the surface soils. It is from these sources that the materials carried in solution by the stream are obtained. Thus the physicochemical nature of the stream as well as the richness of plant and animal life can be attributed to a great extent to the drainage basin. Some examples of the accepted designations of basin patterns are given below. There are others in evidence on topographic maps of various regions.

Trellis Pattern

Just west of the old Appalachian Mountains is found a belt of rolling, uniformly crested mountains. These parallel ridges extend northeast-southwest and represent folds in the earth's crust. Streams in this region are remarkably parallel with tributaries joining at right angles to the main stream. Some of the streams join through water gaps eroded across the long ridges. This distinctive pattern is given the name "trellis."

Dendritic Pattern

A *dendritic pattern*, so-called because of its resemblance to tree roots, develops on a relatively smooth, gently sloped surface. Although inequalities in the surface determine the channels, the nature of the surface permits random and rather uniform coverage of the land by stream drainage (Figure 3·6).

Anastomosing Pattern

On flood plains and other flat depositional features, streams deposit their loads according to flood conditions and meanders. These deposits, together with meandering channels, may divide the usual stream channel, forcing the branches into braided networks. The braided stream shown in Figure 2·1 forms an *anastomosing pattern*.

Radial Pattern

A *radial pattern* occurs as streams flow from a symmetrical high region. Drainage on the sides of a volcanic cone such as Mt. Hood, Oregon, typically occurs in a radial pattern.

Stream Piracy

Before closing this section on the origin, dynamics, and morphology of streams, we should give attention to one additional aspect of stream erosion; this is *stream piracy*, or *stream capture*. Stream piracy is a process by which one stream, through more rapid headward erosion, captures the headwaters of a neighboring stream. In the process, the second stream is beheaded.

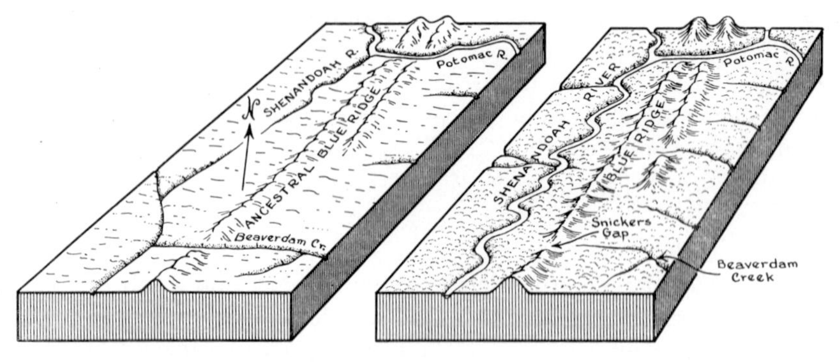

Figure 3·9. Stream Piracy in the Blue Ridge of Virginia. See text for details. (After Longwell, Knopf, and Flint, 1939. "Textbook of Geology," 2nd Ed., John Wiley & Sons, Inc., New York, N.Y.)

Examples of stream piracy are common, and one classic example will suffice. Figure 3·9 illustrates the manner in which an ancestral tributary of the Potomac River captured the upper reaches of Beaverdam Creek, Virginia. As shown in the figure, the Potomac and Beaverdam flowed through separate water gaps in the Blue Ridge Mountains. Because of hard rock in Snickers Gap, the Beaverdam was unable to erode its channel as fast as the Potomac tributary. The result was the beheading of Beaverdam by the tributary to form the present Shenandoah River. In losing its stream, Snickers Gap became a wind gap.

Chapter 4

ESTUARIES

WITH BUT FEW EXCEPTIONS, most brooks and creeks find their way to major rivers, and these in turn wind their way to the sea. At and near the mouth of the lower stream course where sea and river meet, a special and distinctive environment prevails. This ecotone, or "buffer zone," between fresh water of the stream and salt water of the sea, is called an estuary.

An estuary has been defined as a body of water in which river water mixes with and measurably dilutes sea water. It has also been described as the wide mouth of a river or arm of the sea where the tide meets the river currents, or flows and ebbs. It is immediately apparent from the definitions that several factors and processes not encountered in lakes or streams operate in an estuary. These combine to produce communities with highly variable environmental conditions. In the first place, two opposing current systems, the unidirectional stream currents and oscillating tidal currents meet and exert considerable and complicated effects upon sedimentation, water mixing, and other physical features of the estuary. These also greatly influence the biota. Secondly, the mixing of salt and fresh water produces a chemical environment unlike that of the typical sea or river. Thirdly, the diurnal shifting of the more saline waters with tides may necessitate physiological adjustment of the inhabitants of the community. These and other characteristics of estuaries will be considered further under more specific headings.

As in the case of lakes and streams, estuaries exhibit variation, not so much as to origin, but rather in form and extent. Such variation is usually attributable to local modifications subsequent to original formation. Similarly, the amount and distribution of salt water in estuaries are functions of stream inflow, tide and other currents operating within the estuary, and the topography of the area. These attributes also vary. The biota, usually a unique assemblage of organisms, is related to the various physical and chemical features of the estuary. These general considerations should be borne in mind as we study in this chapter the estuary, its origin, morphology, modifying forces, and bottom and shore features.

ORIGIN OF ESTUARIES

The earth's crust is not rigid. It is subject to bending and other deformations. Continental land masses may be raised or lowered, and where the continent is in contact with the sea, conspicuous changes in the shoreline take place. As a land area is depressed, the shores become submerged and the coastline migrates inland. Conversely, as land is up-

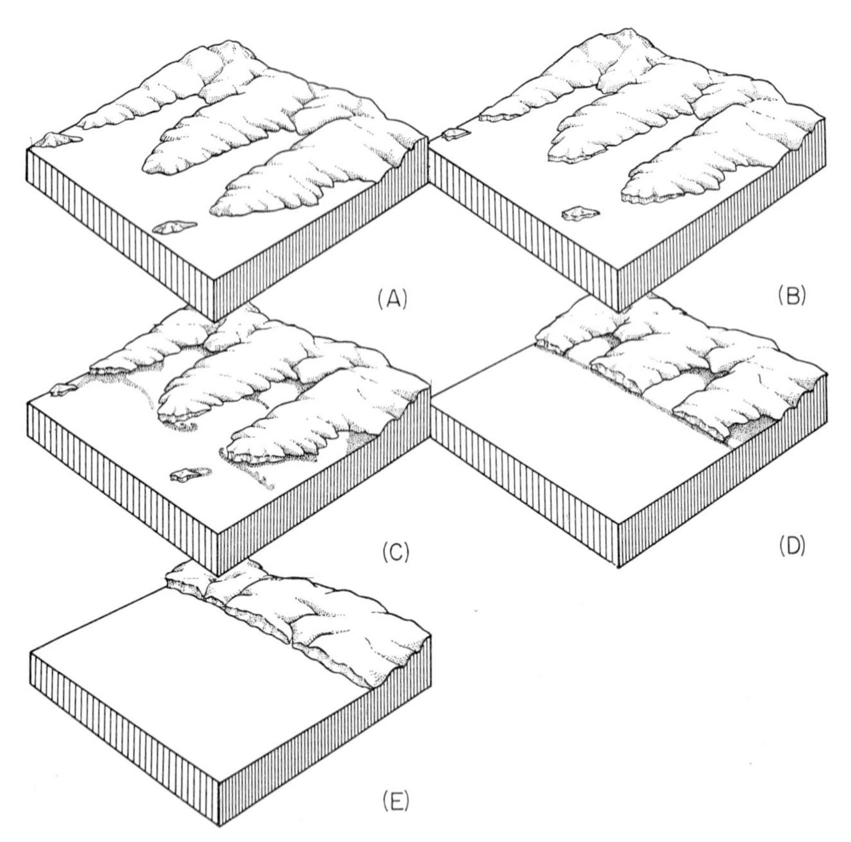

FIGURE 4·1. Developmental Stages in a Shoreline of Submergence. The *initial stage* (A), characterized by headlands projecting seaward, results from drowning of the lower reaches of stream valleys. In *early youth* (B), sea cliffs at the ends of the headlands are typically present. Depositional features, such as bayhead beaches and sandspits, are characteristic of *late youth* (C). In *early maturity* (D), the water between headlands has become somewhat enclosed by the formation of bayhead bars across the entrance resulting in the development of lagoons or "lakes." Continued erosion and filling of the lagoons results in the formation of a more regular shoreline lying back of the original headlands; this stage, essentially one of equilibrium, is termed *full maturity*. (After Johnson, 1919. "Shore Processes and Shoreline Development," John Wiley & Sons, Inc., New York, N.Y.)

lifted the existing shores emerge and the coastline is moved seaward. The basic physiography of coastlines and shores is, therefore, due primarily to *submergence* or *emergence*. Either or both of these may form estuaries. Some shores show the effects of both processes acting at different times in the past. Various local activities, for example, the movement of currents, wave action, tides, stream deposition, glaciation, and wind, operate to modify the original form of shores.

Thus the forms of coastlines may be classified under two headings. Features developed as a result of movements of sea and land in relation to each other are called *initial* forms. Tectonic factors, glaciation, and climate are examples of forces that result in coastlines of initial form. Patterns of shoreline resulting from the action of marine forces on land masses (initial forms) are termed *sequential* forms.

Shoreline submergence results in a very irregular coastline of peninsulas, islands, and uneven bottoms (Figure 4·1). The eastern shore of the United States is primarily one of submergence, although subsequent emergence has altered local areas. The region from the Carolinas to New Jersey gives evidence of having been subjected to periods of both emergence and submergence. It is along submerged shorelines that many well-developed estuaries occur as *drowned stream valleys*. The Chesapeake Bay represents such an estuary.

Most streams entering the sea flow near base level across coastal plains or some other region of low relief. Since the stream lies at or very near sea level, almost any perceptible submergence of the coast results in the encroachment of the sea and the flooding, or drowning, of the stream's mouth. The extent to which the sea invades the stream valley is determined by the gradient and size of the valley, stream discharge, and the range and force of tides of the adjacent sea. Ideally, a typical estuary lies perpendicular to the shoreline.

MORPHOLOGY OF ESTUARINE BASINS

As the mouth of a stream is drowned by the sea, the young estuary assumes a rectangular or triangular shape, depending upon size, shape, and gradient of the stream channel, the volume of discharge of the stream itself, and the nature of the substrate and surrounding topography. The early form of the estuary is rapidly altered by fluvial and marine processes.

Where the drowned stream valley consists mainly of a single channel, the form of the basin is fairly regular and the estuary is said to be *simple*. The estuaries of the Delaware and the Hudson are simple. The flooding of a channel with numerous tributaries results in an *irregular estuary* such as Chesapeake Bay.

Estuaries along the Mediterranean Sea are essentially triangular, broadening from the stream valley toward the mouth of the estuary. Deltas are usually present. In this region the sea is generally calm and tidal range is slight. Along the English coast, estuaries are mostly broadened and tidal mud flats are common. This morphology is associated with moderate

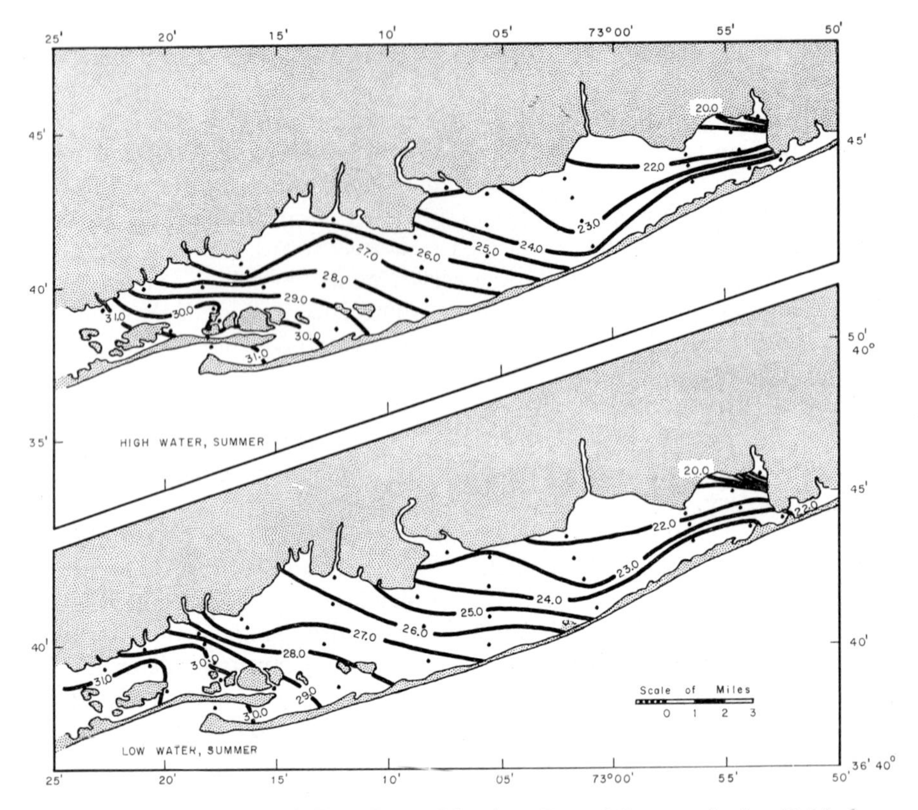

FIGURE 4·2. Great South Bay, Long Island, a Coastal Lagoon Lying Behind a Barrier Beach. Distribution of salinity (in parts per thousand) at mid-depth is shown for high and low water in summer, 1950. Reference will be made to salinity features in Chapter 10. (From report of Woods Hole Oceanographic Institution, Woods Hole, Mass., 1951.)

wave action and considerable tidal range. Stream channels are scoured by flow during the recession of tides. In South Africa, the shores experience great wave action, but slight tidal effects. Here sand bars develop in the mouths of the estuaries, causing the water to spread out broadly behind the bars.

Most of the estuaries along the coasts of temperate and subtropical North America deviate to varying degrees from the "classical." From New England southward many estuaries have been deprived of direct entrance into the sea by the formation of barrier beaches. Through the same process embayments lacking a conspicuous stream inflow have been partially enclosed, resulting in the development of a salinity gradient within the bay. Thus, something of an estuarine nature is imparted to these bodies of water. Great South Bay (Figure 4·2) on the Atlantic Coast of the United States represents such a body.*

Delaware Bay is an example of a relatively "pure" estuary. The Delaware River flows directly into the Atlantic Ocean. In this estuary, the salinity ranges from zero in the river to that of ocean water (approxi-

FIGURE 4·3. A Fjord on the Coast of Alaska. (Photograph courtesy U.S. Geological Survey.)

---

*These coastal bodies of water partly separated from the sea by barrier beaches or bars of marine origin are more properly termed *lagoons*. As a rule, lagoons are elongate and lie parallel to the shoreline. They are usually characteristic of, but not restricted to, shores of emergence. Lagoons are generally more shallow and more saline than typical estuaries.

Great South Bay, on the southern shore of Long Island, is a lagoon developed behind a barrier beach. The beach is primarily a reflection of slight tidal range with considerable wave action. The Bay averages about 4.8 km in width and is approximately 40 km long; mean depth at low tide is about 1.2 m. The deeper areas range to nearly 7 m. Fire Island Inlet, less than 1 km wide, is the only connection between the Bay and sea. The salinity grades generally from 20 to 30 parts per thousand (‰).

mately 35 ‰) in the Bay. Depending upon stream flow, tides, and winds, salt may be detected upstream to near Wilmington, Delaware.

Apalachicola Bay, Florida, is a shallow coastal plains estuary behind a barrier beach. The estuarine waters cover some 362 km^2. Connections with the Gulf of Mexico are through "passes" between barrier beaches. East Bay, an arm of Galveston Bay, Texas, represents a "lagoon-like" estuary. This is described as lagoon-like because, although the salinity of the water grades from fresh to moderately salty, the bay lies parallel to the shore behind a marine depositional feature, probably a spit. The bay is nearly 37 km long and averages about 3.2 km in width. Depths range to nearly 4 m at mean low tide.

Spectacular estuaries occupy glaciated mountainous coastlines. In Norway, southern Chile, and at Puget Sound, for example, elongate, steep headlands alternate with deep U-shaped valleys. Estuaries formed in such areas are called *fjords* (Figure 4·3). Sogne Fjord, of Norway, is 179 km long, averages 6.4 km in width, and has a maximum depth of 1200 m. Puget Sound is 300 m deep in places.

MODIFICATION OF ORIGINAL ESTUARY

Once the shoreline is submerged, stream and sea processes of erosion, transport, and deposition begin to modify the topography of the immediate region and eventually of the entire coast. The dropping of sediments by the stream as it meets the encroached sea initiates the building of a delta in the upper reaches of the drowned mouth. Continued building by sedimentation gives rise to broad, level mud and silt deposits eventually becoming *tidal flats*. Meanwhile, currents and tidal action of the sea erode the peninsulas, or headlands, depositing the materials on the bottom of the seaward region of the estuary. These deposits may develop as bars and spits in various positions relative to the mouth of the estuary. In time, the estuary may become filled by stream and tidal deposits.

The rate and extent of sedimentation and filling of an estuary are primarily dependent upon the original size of the estuary, its age, the present rate of erosion upstream and deposition by the stream at its mouth, and marine forces such as tides and longshore currents. Where stream discharge and tidal currents are relatively slight, coastal currents tend to build depositional features such as spits and baymouth bars across the mouth of the estuary, restricting the entrance. In such an instance, the process of filling is apt to be accelerated. On the other hand, if the scouring action of stream and tidal currents is sufficiently strong, the deposition of entrance barriers is inhibited, fluvial sediments are carried farther from the estuary, and the filling process is slowed.

A great portion of the volume of water, representing the difference between low and high tides, flows into an estuary and out again through the entrance in a short time. If the stream discharge is great and the entrance narrow, a deep channel is cut. If the entrance to the estuary is through easily eroded sediments, the depth is such as to be in equilibrium with the water volume moving through. Once the equilibrium is estab-

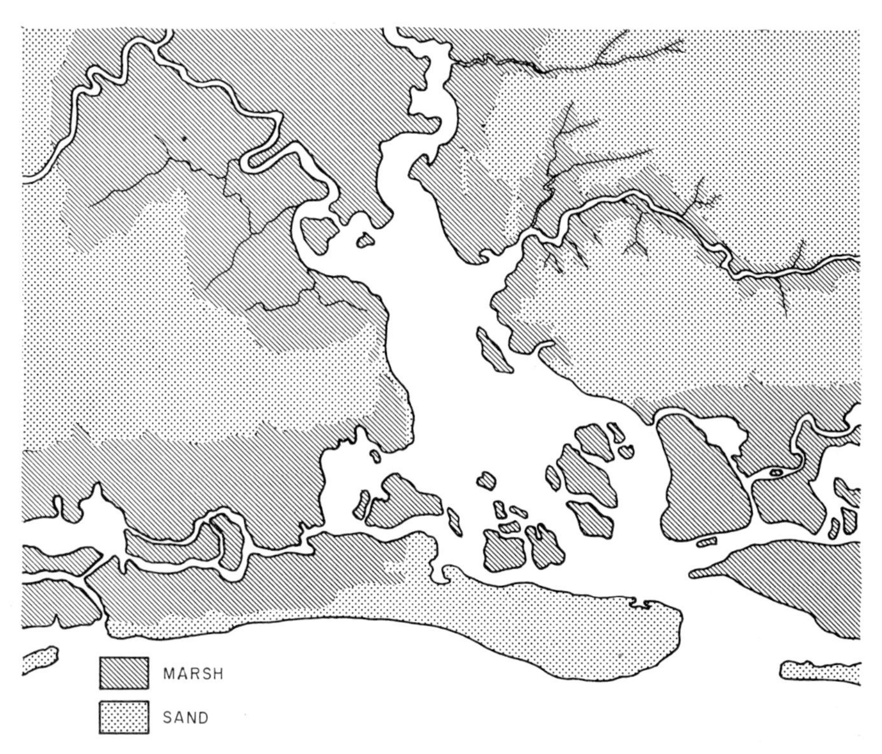

FIGURE 4·4. An Estuary Formed by Meandering Marsh Streams and Partially Blocked by a Barrier Beach. A marsh with tidal streams lies between the beach and the mainland. Coastal geomorphology such as this is characteristic of later stages in the development of a shoreline of emergence. Note that the seaward slope of the barrier beach is sandy, while the landward side grades toward marsh conditions.

lished, there follows only slight scouring or deposition. As filling of the estuary continues, the entrance is made more shallow, gradually decreasing the volume of water passing through in the tidal prism. In some regions of mountainous coastline and great tidal range, restricted entrances result in tidal flow of considerable velocity. Because of the eroding ability of the current, the entrance is made deeper.

The order of events in the filling of an estuary and erosion of the shoreline of a region of high relief such as the New England coast is shown in Figure 4·1. The illustration is quite generalized since the manner and rate of erosion are dependent upon the nature of the materials composing the shore. Erosion of a homogeneous substrate results in steep, straight cliffs; differential cutting of a heterogeneous substrate leads to an irregular shore.

In regions of low coastal relief, erosional effects of the sea are often buffered by the inflowing stream. Deposition of sediments exceeds removal by ocean currents, and barrier beaches and other features are formed. From New Jersey southward, the Atlantic coast of North America is characterized by a sea island system of barrier beaches and other land forms. Some of these lie opposite the mouths of estuaries; others contribute to the development of brackish lagoons lying between islands and the mainland (Figure 4·4).

PARAMETERS OF ESTUARINE BASINS

The fact that estuaries generally are drowned river valleys at the edge of the sea suggests a great variety of morphological features. These features are combinations of certain parameters typical of lake and stream basins made complex by virtue of association with the sea and its characteristics. The origin and location of the estuary determine the original form, but as we have seen, this young form is gradually changed.

The shape of an estuarine basin is of primary importance in determining the nature of the hydrologic forces operating within the estuary. These dynamics in turn often bear upon the animal and plant inhabitants and their relationships with the environment. A triangular estuary with a wide and deep mouth permits incursion of marine waters to considerable distances upstream depending upon amplitude of tide and stream gradient. This results generally in increased mixing of waters from stream and sea, greater over-all circulation, and often the development of strong currents. A narrow-mouthed estuary, on the other hand, is characterized by decreased circulation, more pronounced longitudinal and vertical salinity gradients, and more rapid development of sedimentation features such as spits and bars.

With respect to more specific morphologic features, such as depth, length, breadth, and area, it becomes immediately apparent that instability and fluctuation are inherently conspicuous in each. Daily ebb and flood of tide enlarge and contract each of the characteristics. Shorelines, normally clearly discernible in most lakes and along the upper regions of streams, are often obscured by marsh and depositional features and also fluctuate with tides.

Parameters useful in describing and delimiting static and dynamic attributes of estuaries are essentially those discussed previously for lakes and streams. Depth, mean depth, areal features such as length and breadth, shoreline development, and volume are especially useful from the purely descriptive standpoint. These parameters take on an even more important function, however, in determining and describing dynamic qualities such as rate of flushing and flushing number. These attributes (to be con-

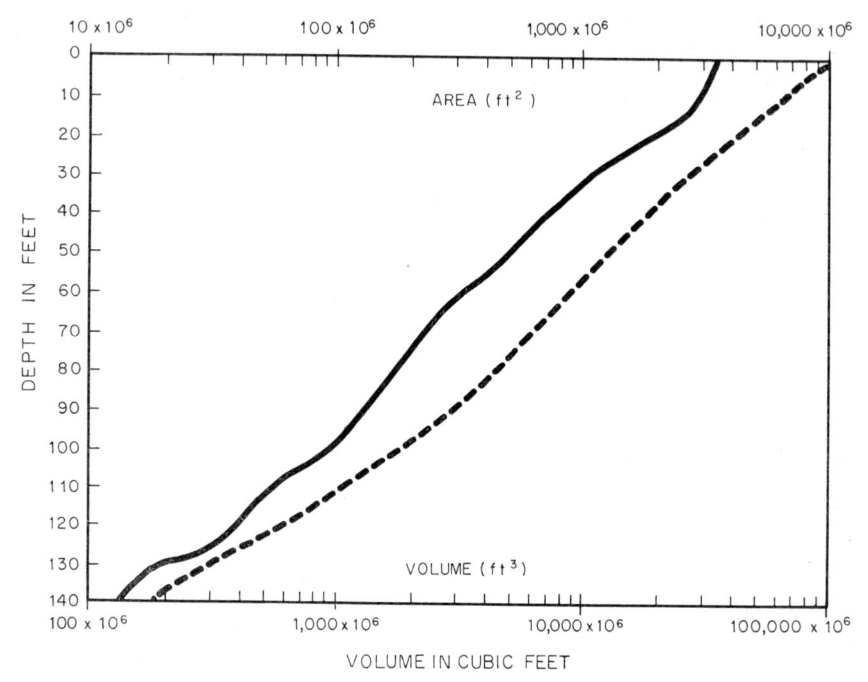

FIGURE 4·5. Area of Narragansett Bay, Rhode Island, at Any Given Depth (solid line) and Volume Below Any Given Depth (dashed line). Depths are corrected to mean low water. (After Hicks, S. D., 1959. "The Physical Oceanography of Narragansett Bay," *Limnol. Oceanog.*, Vol. 4, 316-327.)

sidered later in relation to currents in estuaries) result from the conspicuous and highly important action of tides in estuaries.

The presence and action of such tides demand extended expression of the rather standard parameters listed above. It now becomes necessary to picture the body of water under various conditions of tidal range, for many coastal areas are greatly modified during successive ebb and flood of the tide. A relatively moderate range of conditions and variations is shown in the data pertaining to Biscayne Bay, Florida, a shallow, bar-built estuary, or lagoon.

The northern part of Biscayne Bay has an area of approximately 7.04×10^8 sq ft. At low tide the mean depth is 6.9 ft, thereby resulting in a low-water volume of 48.3×10^8 cu ft. Because of the size of the

Bay and the nature of the entrances through which the waters flow, the Bay does not experience a uniform high or low tide. Allowances for this give an "adjusted" tidal range of 1.85 ft. Thus the volume during a mean high tide is estimated at 61.4×10^8 cu ft. The difference between the high-tide volume and that at low tide suggests that about 13.1×10^8 cu ft of water is introduced into the Bay during a given flood tide. The volume brought in is termed the *tidal prism volume* and, in the case of Biscayne Bay, increases the Bay volume by approximately 27 per cent.

It is interesting to note that the area of Biscayne Bay is not greatly affected by the tide range. This is doubtless due to the fact that a major portion of the "shoreline" is walled with concrete or otherwise retained. In an estuary of low relief and shores the area would fluctuate with tidal prism volume. The relationship between area, volume, and depth in Narragansett Bay is shown in Figure 4·5.

ESTUARINE SHORES AND SUBSTRATES

The shores and substrates of estuaries attest to the vigorous, rapid, and complex sedimentation processes characteristic of most coastal regions of low relief. These sediments are derived through the hydrologic processes of erosion, transport, and deposition described for other aquatic environments. In estuaries, however, these processes are being carried on by both the sea and the stream, thereby complicating the nature of the estuary.

The shores of a coastal-plains estuary are composed, in the main, of mixtures of silt, mud, and sand in varying proportions and degrees of compaction. Near the mouth of the estuary where predominating forces of the sea build spits or other depositional features, the shores and substrate of the estuary are conspicuously sandy. Just inside the entrance, the sand contains considerable quantities of mud. Indeed, marked zonation from the sand shores of the seaward slope to the mud flats or tidal marshes of the inner slope of bay-mouth bars and spits is typical of most estuaries. This is especially noticeable in estuaries lying parallel to the coastline and in lagoons formed behind barrier beaches.

From the estuary mouth with its generally coarse bottom sediments, there usually exists a gradation toward finer materials in the head of the estuary. This sorting is associated with current action. In the head region and other zones of reduced flow, fine muds are deposited, while in the main channel and in the mouth where stream and tidal flow carry greater loads, coarser sediments make up the bottom.

The nature of the substrate is known to exert considerable influence on the plant and animal inhabitants of the estuary floor and shore. Both pure sand and pure mud present problems in the maintenance of living organisms. Some studies have indicated that mixtures of sand and mud support the richer faunas. The muds of the bottom of estuaries tend to hold the more saline waters as the tides ebb. Thus bottom-dwelling plants and animals that require higher salinities are able to exist farther into the estuary than ecologically similar forms in the fluctuating water above the bottom. These and other relationships will be discussed further in later chapters.

Tidal, or mud, flats are commonly built up in estuarine basins. These depositional features, composed of loose, soft mud, or a mixture of mud

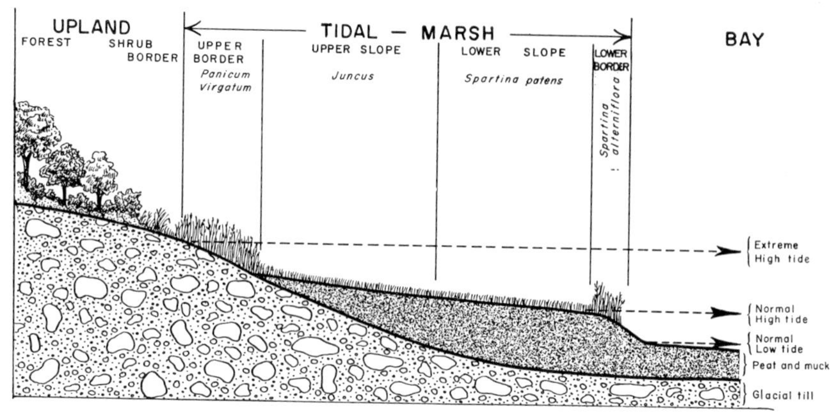

FIGURE 4·6. Diagrammatic Section of New England Tidal Marsh Showing the Major Features of Plant Zonation from Upland to Sea. Vertical scale considerably exaggerated. (After Miller, W. R., and Egler, F. E., 1950. "Vegetation of the Wequetequock-Pawcatuck Tidal Marshes, Connecticut," *Ecol. Monographs,* Vol. 20, 141-172.)

and sand, often develop in the estuary and divide, or braid, the original channel. Depending upon the composition of the substrate and tidal action, vegetation may eventually occupy the flats. Otherwise, the broad, flat, or slightly arched areas remain as barren features of the estuarine basin. Barring pollution, these tidal flats provide a habitat for an abundant fauna that feeds upon materials brought in by the tide or upon organic detritus of the substrate. Similar flats characterize the low-tide shoreline of estuaries.

One of the most significant features of estuaries throughout a major portion of the world is the oyster reef. This assemblage of organisms is usually found near the mouth of the estuary in a zone of moderate wave action, salt content, and turbidity. Because of its location, depth, and areal dimensions, an oyster reef is often a salient factor in modifying estuarine current systems and sedimentation. The form and position of the reefs vary depending upon the nature of the substrate, currents, and salinity.

The reef may occur as an elongate island or peninsula oriented across the main current, or it may develop parallel to the direction of the current. In shallow coastal areas, reefs may grow as islands, often exposed for considerable periods of time during low tides. Biologically, the oyster reef is a unique and interesting community of various mollusks and other organisms and will be considered in a later section.

The high-tide shoreline of estuaries is frequently the lower margin of tidal marshes. Marshes support an abundant flora of grasses, sedges, and other aquatic plants. As a result of the rich plant growth, the marsh substrate is often peat-like in composition. From the seaward margin of the marsh to the higher-land elevations, zonation of plants is a marked feature of the coastal marsh and estuary shore (Figure 4·6). The plant zonation appears to be related to moisture and salt tolerance. Along the estuary-stream shore from the saltwater zone inland, the salt marsh grades into a freshwater marsh or swamp. Thus the subtle transition between fresh water and the brackish estuary is re-emphasized.

PART II

Water

Chapter 5

THE NATURE OF WATER

It is truistic but nonetheless meaningful to say that the most important single feature of lakes and streams is water. Water is a biological necessity of plants and animals. It is required by plants for photosynthesis and by both plants and animals in general metabolism. Water in the lake basin or stream channel is of basic import ecologically, for it is the medium in which the members of the community live. This is to say that the inhabitants must be adapted for finding shelter and food, utilizing dissolved gases, and reproducing in an environment very different from the terrestrial one. As an environmental medium, water enters into and maintains the total economy of the ecosystem. Therefore, before considering the physical and chemical features of inland waters and estuaries it seems proper to give brief attention to some of the properties of water as such. These attributes are basic to the environmental processes and characteristics of waters in lakes, streams, and estuaries, to be considered further in later chapters.

THE WATER MOLECULE

The water molecule is constructed from covalent bonding of two atoms of hydrogen (H) and one atom of oxygen (O). In combination these give the familiar H_2O. Two chemical bonds unite the two H atoms to one O atom in the form H-O-H. The nucleus of each atom is positively charged; negative electrons revolve, or orbit, around the central core. The path of an electron is called an *orbital*, or energy level, and the bonds between a hydrogen and oxygen atom are formed by the overlapping of the orbitals in a fashion such that each atom contributes one electron which, in effect, completes the orbital of both atoms (Figure 5·1). Four of the unshared electrons of the oxygen atom, important in some reactions, are located considerably farther from the nucleus than are the remaining two. Because of the greater positive charge on the oxygen nucleus, the bonding electrons are not distributed equally between the hydrogen and oxygen nuclei, but are relatively closer to the oxygen atom. This results in a partial positive charge on the hydrogen atoms and a partial negative charge on the oxygen atom, or polarity.

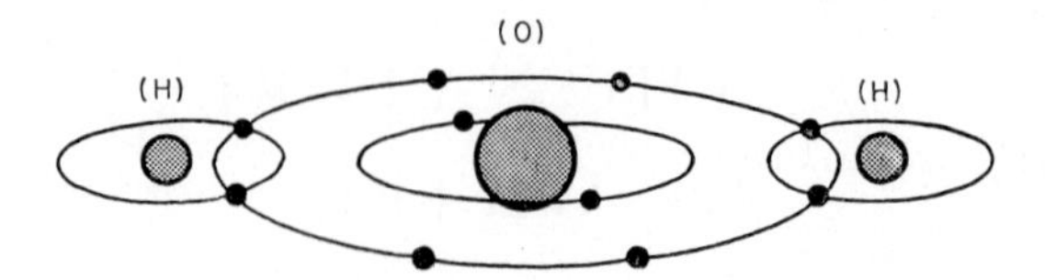

FIGURE 5·1. Diagram of the Water Molecule Showing the Overlapping Orbitals and the Shared Electrons.

Water is a highly organized liquid. The molecules are arranged in a definite configuration. The association of the molecules by hydrogen bonds into chains (Figure 5·2) giving liquid water or ice, and the dissociation of the molecules to form vapor, account, in large part, for the versatility of water. Chemical reactions such as oxidation and hydrolysis

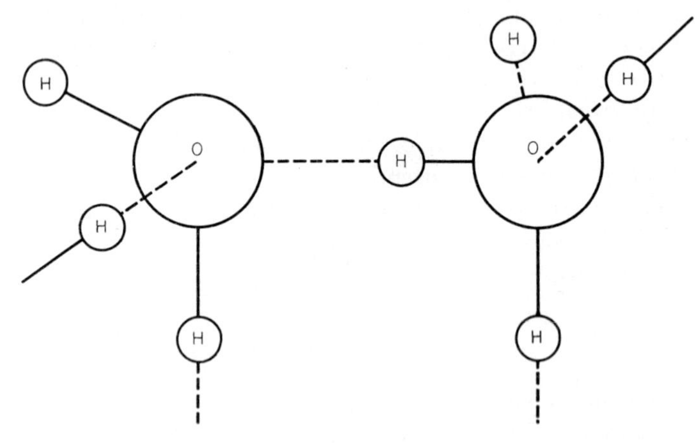

FIGURE 5·2. Two Molecules of Water Joined by Hydrogen Bonding. Chemical bonds uniting two hydrogen atoms with one oxygen atom are shown as solid lines. Hydrogen bonds are represented by dashed lines. This is but a portion of a vast system of variously oriented molecules in liquid water.

occur when chemical bonds between the hydrogen and oxygen are broken. Physical processes, such as the melting of ice and evaporation, take place when the attractive forces between water molecules are broken without affecting the molecule.

Ordinary water is not merely an aggregation of individual water molecules composed of two hydrogens of atomic weight 1, and an oxygen of atomic weight 16. Water also contains isotopic substances such as "heavy water" in which two deuterium (H^2) atoms are combined with an oxygen. There are also lesser amounts of isotopic molecules in which O^{17} and O^{18} replace O^{16}, and very small quantities of tritium (H^3).

PHYSICAL AND CHEMICAL FEATURES OF PURE WATER

Heat of Vaporization

Compared with other liquid compounds of similar, simple molecular composition, water vaporizes very slowly when heated, i.e., water has a very high latent heat of vaporization (evaporation). Vaporization is a process in which thermal energy is supplied to overcome the attractive forces between water molecules (mainly hydrogen bonds in liquid water). A given volume of liquid water contains a larger number of hydrogen bonds per unit volume than most other common solutions.

We have just seen that there are two hydrogen bonds linking each water molecule to its neighbors; therefore the heat required for evaporation is approximately twice the hydrogen bond energy (4.85 kcal per molecular number of H bonds, or 9.7 kcal per gram molecular weight). Depending upon temperature, the amount of heat required for the vaporization of 1 g of water ranges from about 500 to 600 cal (9700 cal per gram molecular weight). In other words, as much heat is used to vaporize 1 g of water as to raise the temperature of 540 g by 1°.

Specific Heat

The specific heat of a substance refers to its capacity to absorb thermal energy in relation to temperature change at constant volume. Water holds a great amount of heat with a relatively small change in temperature. The unit of measure of specific heat is the gram-calorie, or small calorie, which is defined as the amount of heat required to raise the temperature of 1 g of water from 14.5° to 15.5°C. Thus, water is the basis of this parameter and is said to have a specific heat of 1; the specific heat of other compounds is calculated as the ratio of their heat capacity to that of water. Only a few substances, including lithium at high temperatures and ammonia, have a higher specific heat than that of water.

This heat capacity of water has far-reaching implications. For one thing, it acts as a buffer against wide fluctuations in temperatures, thereby ameliorating terrestrial climates near large bodies of water. Secondly, an aquatic organism is subjected to much narrower ranges of temperatures than a land form: whereas land areas may reach 38°C or more, lakes seldom exceed 27°C. Furthermore, the slow rate of seasonal cooling and warming of lakes and streams is attributable to the high specific heat of water. For the same reason, changes in lake temperatures lag behind atmospheric fluctuations. In natural waters, dissolved substances lower the specific heat.

Much heat is necessary to bring water to boiling because a considerable amount of heat is dissipated in energy of vaporization before the boiling point is reached. The "reluctance" of water molecules to separate and emerge as vapor is due, as we have seen, to the hydrogen bonding of the molecules. Methane (CH_4) has nearly the same molecular weight (16) as water (18) but boils at −161°C as compared with 100°C for water. The difference is found in the fact that the outer electrons of the methane molecule are bonded chemically, there being no free pairs of electrons to enter into hydrogen bonding as in water.

Latent Heat of Fusion

The heat taken up in the change of water from a solid state to a liquid phase with no change in temperature is termed the latent heat of fusion. Specifically, 79.7 cal of heat are required to melt 1 g of ice at 0°C. This is about 15 per cent of the heat necessary to separate the hydrogen bonds in the vaporization process. As heat is applied to ice, the molecules are set into motion. Increased motion parts the hydrogen bonds until finally the molecular lattice collapses and brings about melting. There is evidence that some few bonds may be broken before melting begins, but even so, only about 15 per cent of the remainder is separated in the actual melting of ice; the fission of the remaining bonds accounts for the high specific heat and heat of vaporization of water.

Density Qualities

Most liquids, including water, contract and become heavier with cooling: water is less dense, or lighter, at high temperatures, and becomes more dense as temperature is decreased. Fresh water is unique, however, in that it reaches its maximum density not, as with other liquids, just at the freezing point, but rather at 4°C. As it cools below this point the density decreases, that is, the water again becomes lighter (Figure 5·3) as the crystalline structure develops.* Upon freezing, water adds approximately 1/11 to its liquid mass. Thus water pipes burst and ice floats. Water is one of very few substances whose solid state is lighter than the liquid. This quality is important limnologically, for it accounts for the fact

* In this respect, the characteristics of sea water are quite different. The freezing point of sea water is about −2°C, and contraction as well as increase in density associated with decrease in temperature are uniform to that point. Thus sea water is heaviest just before freezing and at that time is also some 6° cooler than fresh water. The sinking of this cold heavy water is a factor in circulation in the marine environment.

that lakes freeze from the surface downward and that only small bodies of water, and lakes in the coldest regions, freeze solidly in winter; thus in most lakes and streams life goes on underneath the ice—albeit at a reduced rate.

Density is not only temperature-related, it is also a function of pressure, and of the concentration of substances dissolved or suspended in the

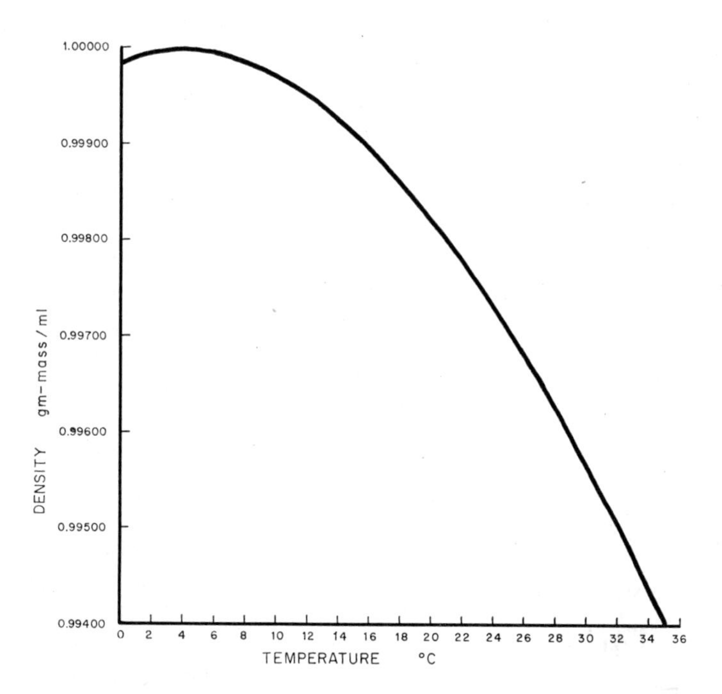

Figure 5·3. The Effect of Temperature Upon Density of Pure Water. Maximum density occurs near 4°C. Above and below this temperature water becomes progressively lighter.

water. High concentrations of dissolved salts account for increased density of ocean waters and the waters of unique types of lakes such as Great Salt Lake. Organic material such as finely divided detritus and organic silt, being heavier than water, will, in sufficient quantities, increase the density. Similarly, clays and fine muds increase the density of waters in lakes and streams.

Viscosity is a measure of the internal, molecular friction of a liquid and is concerned with mobility and flow. In order to flow, the bonding forces holding molecules of a liquid in form must be overcome, and there must be a void in the lattice pattern into which a loosened molecule can move. Compared to many liquids, gasoline for example, water is highly resistant to flow. This resistance is due to the relatively great energy contained in the hydrogen bonds of water molecules. This quality of water limits stream discharge, and also relates to living habits, morphology, and energy expenditure of aquatic animals inhabiting a dense environment which offers considerable hindrance to movement. From what has been shown of the activity of molecules in relation to temperature, we should expect viscosity to diminish with increase in temperature. Actually, viscosity changes about 3.5 per cent with each degree of temperature, although the relationship is not uniform.

Mass, or weight, of water is also related to density. By general standards, water has considerable weight. Although it varies with temperature, the weight of 1 cu ft of water is generally given as 62.4 lb (998.4 kg/m^3). If determinations are made below the surface, atmospheric pressure must also be considered. At a depth of 100 ft (30.5 m) the pressure is approximately 58 lb per sq in. (4 kg/cm^2) or about 4 atms (pressure of water increases approximately 1 atm per 10 m depth).

Adhesion, Cohesion, and Surface Tension

Adhesion is the tendency of a liquid to cling to the surfaces of some materials by means of bonds established between the hydrogen atoms of the water molecule and oxygen atoms of the other substance. Cellulose in wood, for example, contains a great number of oxygen atoms and is, therefore, very wettable.

Cohesion is the property of liquids which offers resistance to being pulled apart or to the formation of new surfaces. It has been calculated that a force of 210,000 lb would be required to break a column of pure, perfectly formed water of 1 sq in. (6.45 cm^2) cross section. At the surface of a water mass the force mainly responsible for such resistance is termed *surface tension* and results from the unsymmetrical activity of water molecules at and below the surface. Whereas the internal molecules are bonded on both sides, those at the surface are attached only to the molecules below, there being none above. A force is thus imparted inwardly giving rise to a surface phenomenon analogous to a taut sheet over the water. The surface tension at the zone of contact between air and water is 73.5 dynes at 15°C. Surface tension varies inversely as the water temperature, and is lowered by the presence of organic substances. The addition of inorganic salts tends to increase surface tension.

Solvent Action

The phenomenon whereby substances may be dissolved and retained in solution is exhibited to a greater extent by water than by any other liquid in nature. Of the known chemical elements, about 50 per cent have been reported in water and it seems probable that at least traces of all elements may occur. This property is, of course, of prime importance to the maintenance of life in the aquatic environment. Basically two types of processes involving hydrogen bonds are concerned in the dissolving action of water.

One type of solvent process may be described as *inert*, for, by means of hydrogen bonding, the dissolved substances are relatively unaffected by the solvent. Many compounds of ammonia, nitrate, and phosphate, as well as sugars, alcohols, and various organic acids, are bonded to the hydrogen atom of the water molecule through oxygen atoms, hydroxyl groups, or through nitrogen bound to hydrogen. It should be pointed out that these compounds are some of the important ones involved in energy transfer and storage in biological systems. Therefore, through hydrogen bonding the substances are delivered unmodified and readily accessible, not only from environment to plant or animal, but within the physiological systems of the organisms as well.

The second type of solvent process involves the separation of the electric charge between the hydrogen and oxygen atoms in the H-O-H molecule. Water is characterized by a very high charge separation; thus various salts such as sodium chloride and the salts of potassium are retained in solution.

In still another activity, water has a remarkable ability to ionize dissolved compounds through the separation of the bonds in the H-O-H molecule. This process is due to the high dielectric constant of water. The molecule shows polarity, and total separation of one H from OH results in two charged ions, H^+ and OH^-. The H^+ becomes the hydrogen ion of acids and the OH^- becomes the hydroxyl radical of bases. A simple, reversible ionization reaction is that involving the solution of carbon dioxide in water: $CO_2 + H_2O \rightleftharpoons H_2CO_3$, or more completely:

$$O{=}C{=}O + HOH \longrightarrow O{=}C\begin{matrix}/OH \\ \backslash OH\end{matrix} + HOH \longrightarrow O{=}C\begin{matrix}/OH \\ \backslash O^-\end{matrix} + H_2O^+$$

TRANSPARENCY

Pure water is quite transparent. Since the extent of penetration and absorption of light-wave components, so important in plant and animal relationships, is a function of transparency, this characteristic of water is of great interest to limnology. Pure water absorbs solar radiation selectively, the major absorption being at the red end of the spectrum. Minimum absorption is in the blue range at about 4700 Ångstrom units (Å). Approximately 90 per cent of the radiation above 7500 Å is absorbed by 1 m of water. Transparency and light in relation to natural waters are considered more fully in Chapter 6.

LIQUID NATURE

The attribute of being liquid constitutes one of the more unique properties of water, for liquids in nature are not common. In addition to water, only one other inorganic substance occurs in liquid form at the earth's crust; this is mercury. It has been suggested that liquid carbon dioxide may be found in quartz, but this is apparently not fully confirmed. Petroleums are liquids, but are of organic origin.

PART III

Natural Waters as Environment

Chapter 6

SOLAR RADIATION AND NATURAL WATERS

AT THIS POINT it might be well to refocus our attention on an idea suggested in the Introduction to our study of inland waters and estuaries. Recall that there appear to be three major themes running through our approach to the ecosystem: (1) fitness of the environment, (2) evolutionary adaptations of the living components of the systems, and (3) traffic in energy. Of the three, two are more or less directly dependent upon solar radiation in order to maintain their importance and to fulfill their roles in the ecosystem; we refer specifically to the nature of the environment and to energy transfer. As we shall presently see, temperature, as a characteristic of natural waters, is of paramount importance in regulating many chemical, physical, and biological reactions; it is, of course, derived mainly from solar radiation. Green plants constitute the major source of energy-containing substances available to animals in natural waters. From this relationship stem the intricate food webs involving energy traffic; the ultimate source of this energy is solar radiation.

During the course of a year the amount of radiant energy received by the earth from the sun approaches the staggering figure of 1.3×10^{21} kcal. Some of the solar radiation is reflected from the earth's surface by living features such as vegetation, and by nonliving components represented by land forms, lakes, and seas. A portion of the spectral array is utilized by green plants in synthesizing energy-containing substances necessary for the sustenance of plant and animal life. In addition, heat enters into the development of meteorological processes such as wind and rain which, in themselves, create new habitats and modify old ones.

The known range of radiations in the electromagnetic spectrum extends from extremely short-wave lengths of gamma rays of the order of 0.001 Å ($1\ \text{Å} = 1 \times 10^{-8}$ cm) to exceedingly long Hertzian rays of 3×10^{14} Å. Much of this total radiation is distributed outside the earth's surface. The direct solar radiations that impinge on the surface of the earth range from about 135,000 to 2860 Å; the most intense radiation, however, extends between 3000 and 13,000 Å. The peak of radiation distribution is in the blue-green range, about 5500 Å. Of those radiations that arrive at the water surfaces on the earth, the infrared components (about 0.1 mm to 7700 Å) contribute largely to heating. Within a narrow segment of the spectrum lie rays of the so-called light, or visible, spectrum. These range from 7700

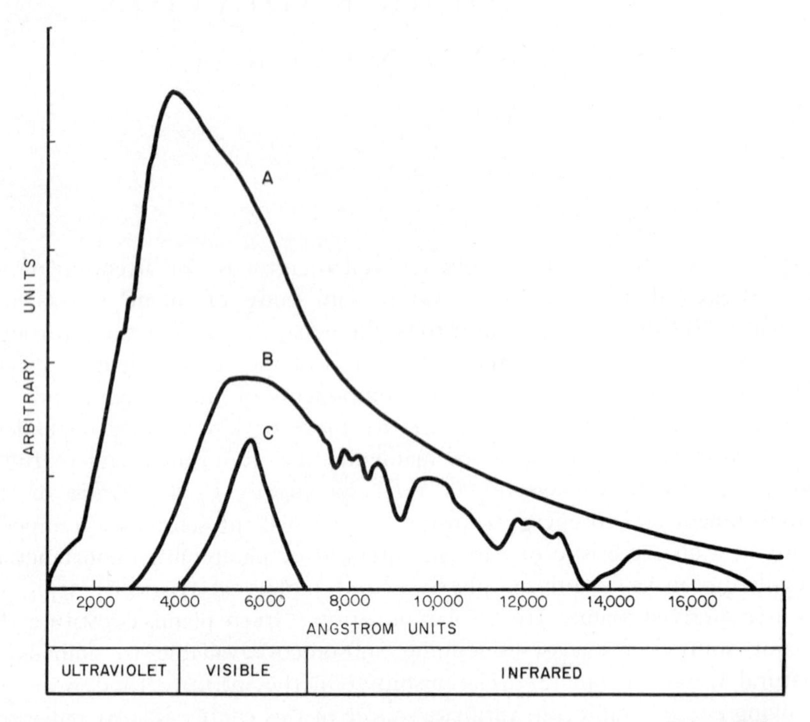

FIGURE 6·1. Distribution of Energy of the Solar Spectrum Above the Earth's Atmosphere (A), and at the Earth's Surface (B). The range of the spectrum to which the human eye is sensitive is shown at (C). (After Hutchinson, G. E., 1957. "A Treatise on Limnology, Vol. I. Geography, Physics, and Chemistry," John Wiley & Sons, Inc., New York, N.Y., after Perl.)

to 4000 Å. Below the visible spectrum are found the components of the ultraviolet segment of the solar spectrum, the range being from 4000 to about 2860 Å. That segment of the electromagnetic spectrum distributed at the earth's surface is shown in Figure 6·1B.

The earth also radiates, the emissions falling in the infrared range of long waves because the earth is a cool body. The earth's radiations are not entirely lost to outer space because of absorption of some of the radiation by moisture in the earth's atmosphere. At the same time, the atmosphere differentially absorbs solar radiation (on the order of 15 per cent).

The net result is the retention of a great portion of the earth's radiation and the transmission of about 85 per cent of solar radiation by atmospheric water vapor.

The phenomenon whereby the earth's atmospheric water vapor transmits most of the radiant energy of sunlight while absorbing the radiant energy of the earth, has been called the "greenhouse effect." The principle of the botanical greenhouse is based on the fact that glass of the sides and roof permits the passage of solar energy, but inhibits the transmission of a great part of the "black body" energy given off from the plants and other structures inside the greenhouse. In other words, the contents of the greenhouse absorb solar energy which passes through the glass; the contents in turn radiate an energy of longer wave length which is not transmitted through the glass. This long-wave heat is therefore retained inside the greenhouse. Actually, greenhouse thermodynamics is more complicated than here indicated, but this simplified statement should point up the analogous roles of greenhouse glass and the layer of atmospheric water vapor in retaining heat.

Of interest to us in terms of light available for photosynthesis and for heat in maintaining the environment is the mean value of solar radiation falling upon the earth's surface. It has been calculated that an average of approximately 1.5 g-cal/cm^2/min of radiation is incident upon the earth at sea level. This value is to be compared with the so-called *solar constant* of 1.92 g-cal/cm^2/min reckoned as the radiation received on the outer surfaces of the earth's atmosphere. Actually, the light falling upon any area of the surface of the earth may be composed of direct sunlight and scattered light from the sky, or indirect solar radiation. Both of these are subject to considerable variation. The magnitude of direct solar radiation striking a given point depends upon season, geographical location with respect to latitude, time of day, the angular height of the sun and elevation of the point under observation, and the transmission quality of the atmosphere. The importance of latitude in the amount of direct radiation is shown in Figure 6·2. The quantity of indirect solar radiation received at a given point is quite variable, but generally constitutes about 20 per cent of the total radiation.

All the light received at the surface of a body of water does not enter the water. A portion of the radiation is reflected, that amount being a function of the angle at which the light strikes the surface of the water and the condition of the surface. From an undisturbed surface, the greater the angle of incidence from the perpendicular, the greater the reflection. When the light rays strike the surface at a very low angle, as much as 35 per cent of the light may be reflected. At high angles the reflectivity is on the order of 5 to 10 per cent. We are concerned limnologically with the light that penetrates the water surface.

In Chapter 5 we noted that one of the conspicuous features of pure water is its high transparency. Pure water does not exist in nature, however; yet the penetration of light into water, and the fate of the spectral components within water, are important in ecosystem relationships, and therefore of primary importance to students of aquatic communities. The deviation of transparency of natural waters from pure, laboratory water ranges widely, both quantitatively and qualitatively, being determined primarily by dissolved substances, suspended materials, organisms,

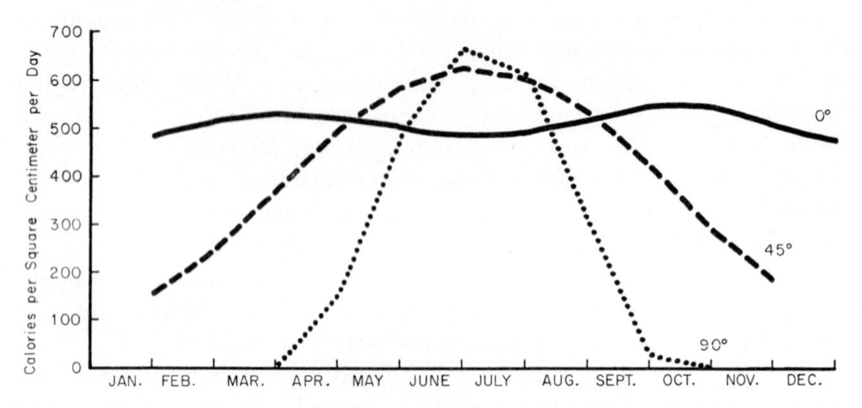

FIGURE 6·2. Total Energy Received from Direct Solar Radiation on Fifteenth Day of Each Month at Sea Level at the Equator and North Latitudes 45° and 90°. (Data from Hutchinson, G. E., 1957. "A Treatise on Limnology. Vol. I. Geography, Physics, and Chemistry," John Wiley & Sons, Inc., New York, N.Y.)

latitude, season, and the angle and intensity of the entering light. In order to more fully appreciate optical phenomena in, and optical qualities of, natural water, let us first consider some principles derived from laboratory studies of pure water.

The total light entering and passing through a given column of pure water undergoes a reduction in total intensity and change in spectral composition with depth. These processes of reduction and change result from scattering and differential absorption of light by water. The reduction in intensity, or *quenching*, of the light entering the surface is expressed in the following relationships based upon Lambert's Law:

$$I_z = I_o e^{-\eta_z} \quad \text{or} \quad \frac{I_z}{I_o} = e^{-\eta_z}$$

in which I_o is the original intensity of the entering light, I_z the measurement of intensity incident at depth z, and e is the basis of natural logarithms (taken as 2.7 in this instance); η becomes a constant for a particular wave length, known as the *extinction coefficient*, or the percentage of the original light held back at depth z. The percentage of light passing through z is termed the *transmission coefficient*. These equations are based on the use of pure water and monochromatic light. Since neither of the conditions occurs in nature, factors which further reduce transmission through the distance $I_o - I_z$ must be applied for natural waters. Such factors would include, in addition to water, suspended materials and dissolved substances.

It has been demonstrated that in pure water at a depth of 70 m the original intensity of blue light is reduced some 70 per cent, and the yellow component shows only 6 per cent transmission beyond 70 m. The red component is not transmitted beyond 4 m, and orange is quenched in about 17 m. In general, approximately 53 per cent of the total incident light is transformed into heat and undergoes extinction in the first meter of water. The longer wave lengths (red and orange) and the shorter rays (ultraviolet and violet) are reduced more quickly than the middle-range wave lengths of blue, green, and yellow.

LIGHT IN LAKES

In natural waters, the blue segment is transmitted farther than all others. At depths beyond about 100 m blue light is the sole illumination. In highly colored or stained waters, orange and red are transmitted more deeply than other components, but all are rather quickly reduced. In many moderately transparent waters, the greatest transmission is in the yellow.

The most accurate and precise measurements of solar radiation in natural bodies of water are best made with photoelectric apparatus, employing various color filters. A photocell contained in a waterproof compartment is connected with a galvanometer and the cell lowered to desired depths. Direct readings of under-water radiation falling upon the cell are made from the galvanometer at the surface.

An American, Professor E. A. Birge, pioneered in the use of electrical equipment for measuring light in lakes. This instrument, developed by Birge and physicists at the University of Wisconsin, was first used in 1912. Called by Birge a *pyrlimnometer*, it measured the total radiation which impinged upon the surface of a thermopile, the thermal current produced being read from a galvanometer. In some ways, particularly with respect to uniform sensitivity and the expression of energy units, the thermopile remains unsurpassed. But it is costly, and thus the relatively inexpensive photocell has come to be used widely in limnological work. Welch (1948, 1952) explains in great detail this and other methods of light measurement.

About 1905, Birge was joined in Wisconsin by Chancey Juday and together their researches and reports gave impetus to limnology. Some of their data pertaining to four Wisconsin lakes are illustrated in Figure 6·3.* The effects of several of the previously discussed factors which relate to transparency of natural waters and comparison with pure water are seen in studying this figure. Total solar radiation is shown as incoming sunlight by the circle with a shaded area at the bottom right. The invisible part of the spectrum, mostly infrared, is represented by shading and includes about half of the entire area. In this, and the other circles, the portion representing the visible spectrum is divided into sectors proportional to the quantity of each of the six colors measured by Birge and Juday. Beginning at the heavy, horizontal line in the "three o'clock" position, observe that the segments are labeled red, orange, yellow-green, blue, and violet in order of decreasing wave length. The enclosure of the red, orange, and yellow segments in a heavy line, and the arrow at "nine o'clock" will be referred to later.

The circles representing pure water and the four lake examples at given depths are drawn to scale proportionate to the total light present at the depths indicated and to the circle representing incoming sunlight. The unabsorbed light remaining at each depth is given as a percentage outside the circle. Variations in each of the six components of the spectrum are indicated by the area of the circle segments.

Immediately apparent is the rate at which transmission decreases with depth, and the variations among the lakes. Similarly, great differences in color changes are seen. It is interesting to note that the over-all rate of light extinction and the proportion of the color segments for Crystal Lake are not too different from pure water. In both Crystal Lake and pure water, the yellow-orange-red segments decrease, and the green and blue increase with depth. Accompanying this is a swing of the arrow in a counterclockwise direction. The data indicate that red waves are absorbed rapidly in the first 3 or 4 m and that with increasing depth the light becomes greenish-blue. In the series for pure water the proportions represent the absorption of direct sunlight; the lake determinations include mixed light from sun and sky. Since mixed light contains proportionately more blue, this light would be transmitted more effectively than sunlight alone in pure water.

---

* Dr. Birge died on June 9, 1950, just slightly over one year short of his hundredth birthday. For nearly 75 years he served the University of Wisconsin "as a teacher, scientist, scholar, dean, and president." A most interesting and enlightening account of Birge, in each of these capacities, is contained in a fairly recent book by Dean Emeritus of Wisconsin, G. C. Sellery ("E. A. Birge," 1956. University of Wisconsin Press, Madison, Wis.). A section of the book, "An Explorer of Lakes," was written by Professor C. H. Mortimer of Great Britain, himself an outstanding limnologist. This section gives an excellent account of certain limnological principles and Birge's contributions to them. Figure 6·3 and the accompanying discussion of it were taken from Mortimer's essay on Birge.

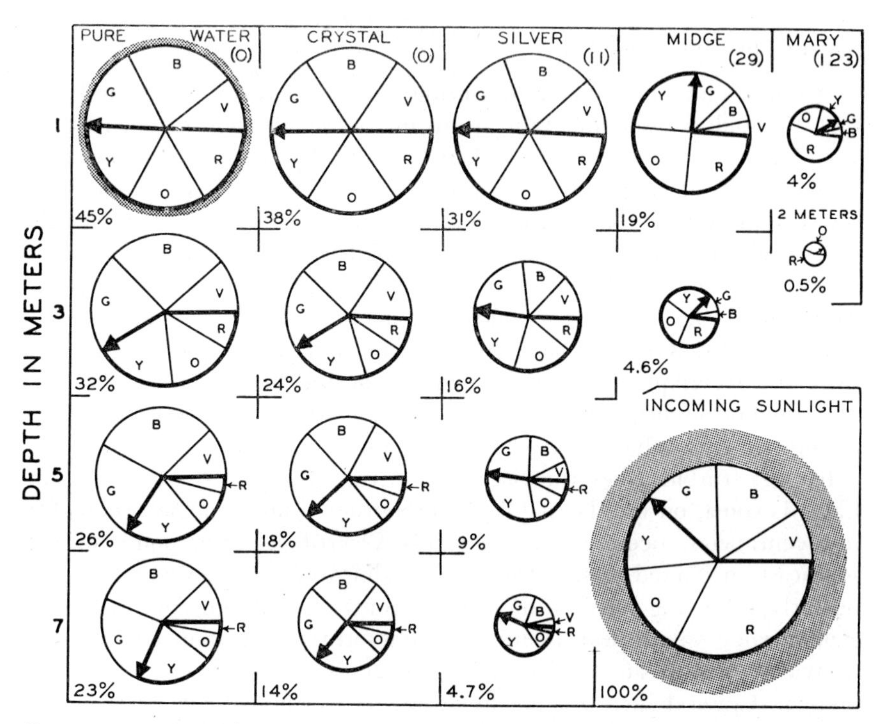

FIGURE 6·3. Graphic Representation of Solar Radiation at a Lake Surface (inset, lower right), and Transmission in Pure Water (left-hand column) at Four Wisconsin Lakes. Data for incoming sunlight and pure water are approximate. Circle areas and percentages represent total light at indicated depths. The relative proportions of each of the spectral components are shown as circle sectors. Figures in parentheses refer to color as measured on platinum-cobalt scale of U. S. Geological Survey. The shaded area surrounding the graph of incoming sunlight and pure water at 1 m of depth indicate the invisible spectrum; note that it is completely absorbed in the first meter of lake waters. (From C. H. Mortimer *in* "E. A. Birge," by G. C. Sellery, 1956, Univ. of Wisconsin Press, Madison, Wis.)

As the diagrams for pure water and Crystal Lake would indicate, these waters have no stain. The figures for the other lakes, Silver, Midge, and Mary, indicate increasing amounts of stain and color and a proportionate decrease in transparency and light transmission. There is also an increase in absorption and reduction of the green, blue, and violet components. In contrast to pure water and Crystal Lake, the arrow swings clockwise in these stained lakes, especially in Midge and Mary. The last two are bog lakes and colored dark brown by dissolved and suspended organic materials from the surrounding bogs. In these lakes, orange and brown are the conspicuous spectral components, but nearly all of the incoming light is absorbed in the upper 2 m.

Secchi Disk Transparency

A much older method of measuring transparency of water, and one that still holds considerable value, is the use of the *Secchi disk*. The method was devised by A. Secchi, an Italian, in 1865. In practice, a white disk, 20 cm in diameter, is lowered into the water by means of a line with measured intervals. The arithmetical mean of the distance at which the disk disappears from view in descent and that at which it reappears in ascent is given as the *Secchi disk transparency*. In employing this device, care should be exerted to standardize techniques and conditions (see Welch, 1948). Reflected light from the bottom will introduce an error since the technique involves comparison of the brightness of the disk with the bottom brightness. It has been calculated that the disk disappears at approximately the region of transmission of 5 per cent sunlight.

In very turbid lakes the Secchi disk transparency is quite low. In Lake Texoma, on the Texas-Oklahoma border, values on the order of a few centimeters occur in places. Similarly, artificial farm ponds of the southeastern United States often give readings of less than a meter, particularly following fertilization when blooms of plankton appear. Secchi transparency values near 18 m are characteristic of certain lakes in the Convict Creek Basin of California. Generally, values above 30 m are unusual. Crater Lake, Oregon, however, has a transparency near 40 m.

COLOR AND TURBIDITY OF LAKES

Color and turbidity are here treated together because of their somewhat similar roles in giving various hues and other optical qualities to lakes; one distinction will be made, however. Both of these qualities determine light transmission in natural waters and consequently "regulate" biological processes within the bodies of water. To varying degrees, both give some qualitative indication of the productivity of the waters when simply viewed from above. The general chemical nature of lakes may often be deduced from the color and turbidity of the water.

Before proceeding further, we should define our terms in order to avoid confusion.

Color

The color that we perceive is made up of unabsorbed light rays remaining from the original entering light, but now passing out of the lake. Completely pure water should absorb all light components and appear nearly black. This is not seen in natural bodies of water, however, for lakes containing little suspended materials usually appear blue. The blue hue is probably the result of scattering of light by water molecules in motion, the effect being proportional to the fourth power of the wave length of the light components. Since the scattering of light of long wave lengths is less than that of short, the blue predominates.

Such hues as we normally see, however, may range from green-blue through blue-green, green, yellow, yellow-brown, to brown. The vivid, poetic, and not necessarily imaginative impressions such as the "sky-blue waters" of Cadman's song, Thoreau's "pure sea-green Walden water," and the "silver" of Seneca Lake as seen by James Gates Percival, all suggest the rich variety of color interpretations possible, depending upon one's point of view.

Limnologically, *true color* (sometimes called *specific color*) of natural waters is derived from substances in solution or from materials in colloidal state. *Apparent color,* on the other hand, is usually the result of interplay of light on suspended particulate materials together with such factors as bottom or sky reflection. In order to realize true color, a sample of the water should be filtered or centrifuged, thereby freeing the water of sources of apparent color.

In Figure 6·3 the numbers in parentheses near the names of the lakes refer to the color of the water in units of the *platinum-cobalt scale* of the United States Geological Survey. The basic technique involves comparison of lake waters with a series of dilutions of a solution of potassium chloroplatinate (K_2PtCl_4) and crystalline cobaltous chloride ($CoCl_2 \cdot H_2O$). The units are called platinum-cobalt units, based upon 1 mg Pt/l as a standard, and range from zero in clear waters (Crystal Lake) to over 300 units in very dark waters of bog communities (Mary Lake: 132). The U. S. Geological Survey set of glass disks corresponding to the platinum-cobalt series is now rather widely used in place of the liquid dilutions (see Welch, 1948, for methods).

The major classes of dissolved and particulate matter found in typical lake waters may be stated as follows:

DISSOLVED SUBSTANCES

- Proteins and related compounds
- Fats and related compounds
- Carbohydrates and related compounds
- Break-down derivatives of all three of the above

PARTICULATE ORGANIC MATTER (*Seston*):

- Living organisms (*Plankton*): mostly microscopic forms
  - Phytoplankton: plant types (Chapter 11)
  - Zooplankton: animal types (Chapter 12)
- Nonliving particles (*Tripton*): dead organisms, detritus, colloidal substances

The average year-round percentage composition (by weight) of these categories in a "typical" lake is shown in Figure 6·4. For any particular lake the bar lengths would doubtless vary greatly; swamp or bog lakes would show a higher percentage of dissolved organic matter; alkaline lakes would be richer in dissolved salts. The important point here, however, is that all elements of these classes, individually or in concert, contribute to the color and turbidity of natural waters.

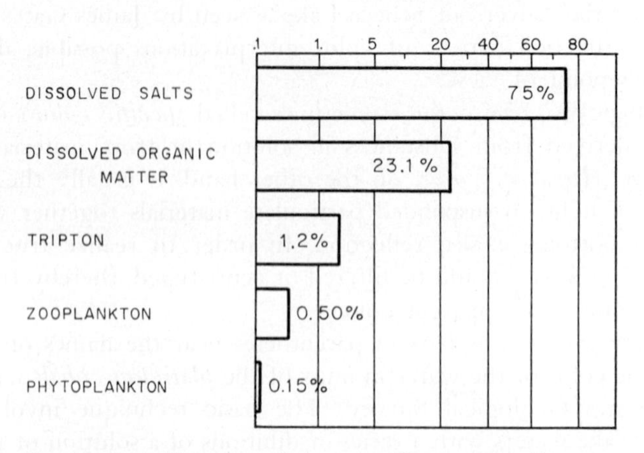

FIGURE 6·4. Year-round Average Percentage Composition of Sestonic Components and Dissolved Organic and Inorganic Material in a "Typical" Colorado Lake. Logarithmic Plot. (Data from Pennak, R. W., 1955. "Comparative Limnology of Eight Colorado Mountain Lakes," *Univ. of Colorado Ser. in Biol.*, No. 2.)

For the most part, lake colors are determined by the predominating components of the *seston*, the mass of various living and nonliving substances in the water. Plankton algae are most frequently responsible for certain colors; an abundance of blue-green algae imparts a dark greenish hue; diatoms give a yellowish or yellow-brown color. Zooplankton, particularly certain of the microcrustaceans, may tint the water red. Humus often causes water to be green, or yellow-brown, the darkest brown coming from extractives of peat.

Suspended colloidal inorganic substances may account for certain tints. Calcium carbonate in lakes of limestone regions results in a greenish color. Volcanic lakes may be yellow-green from sulfur, or red from ferric hydroxide.

Generally, a rich, highly productive lake may appear yellow or gray-blue or brown due to quantities of organic matter. Less productive lakes tend toward blue or green.

We have just seen that the color of lake waters differs widely, depending upon the nature and quantity of dissolved and suspended materials, the quality of the light, and other factors. It is worth noting also that color within a given lake may not be uniform from surface to bottom. Studies in Weber Lake, Wisconsin, revealed that in August the epilimnion was transparent and contained but little organic material; the hypolimnion, on the other hand, was highly colored. At that time, the surface water had a color rating of 12 units, the lower limits of the metalimnion (about 8 m) 81, and at the bottom (nearly 12 m) the color had increased to 93 Pt-Co units. A similar phenomenon has been observed in a New Jersey stone quarry lake. It is suggested that the increase in color in the depths of such waters is due to organic substances derived from bottom sediments and also to increased phytoplankton populations supported by such substances.

In a similar vein, the color of lakes may change periodically. Seasonal increases in surface runoff contribute great quantities of inorganic and organic substances which, as we have seen, impart various colors. Summer or early autumn production of phytoplankton "blooms" causes a lake to become a "soupy green," which disappears later in the season. Exposure to light causes the bleaching of certain colors in natural waters. This process also results in variations in the color of a given lake, such variations appearing seasonally with cyclic fluctuations in light intensity and angle of incidence. The bleaching effect should be felt vertically within the lake, depending upon transparency.

Turbidity

We have seen that color in natural waters is attributable in great part to suspended materials of varying particle size and composition. *Turbidity*, on the other hand, is the term used to describe the degree of opaqueness produced in water by suspended particulate matter. While the nature of the materials contributing to the turbidity is mainly responsible for the color quality, the concentration of the substances, if sufficiently high, determines the transparency of the water by limiting the light transmission within it.

The kinds of materials creating turbid conditions in a given body of water are as varied as the biotic and abiotic composition of the surrounding terrain, inflowing streams, and the lake itself. Substances such as various grades of humus, silt, organic detritus, colloidal matter, and plants and animals produced outside and brought into the lake, are termed *allochthonous*. Turbidity-creating matter produced within the lake is said to be *autochthonous*. Both contribute to the total quantity and quality of lake turbidity.

Although several methods for measuring turbidity have been devised, two are perhaps most widely used today. One technique involves the U. S. Geological Survey turbidity rod, which consists of a marked staff calibrated against known concentrations of a standard material (usually fuller's earth) with a length of platinum wire at one end and an eyepiece at the other. The end bearing the platinum wire is lowered into the water until the wire vanishes from view. The calibration mark at the level of the water surface gives the turbidity in parts per million (ppm). Devices such as the Jackson Turbidimeter and the Hellige Turbidimeter make use of light penetration through a sample of water. In the former apparatus, the disappearance of the flame of a standard candle when viewed vertically through a column of water is a measure (ppm) of turbidity. The Hellige instrument compares a vertical beam of light through a column with the Tyndall effect produced by lateral lighting from the same source, usually an electric bulb. An adjustable aperture regulates the light reaching the eye of the observer and the aperture size is calibrated for turbidity and determined in parts per million. For details of these instruments refer to Welch (1948).

Turbidity is not a uniform parameter even within a specific lake. Seasonal increases in stream discharge, for example, may introduce considerable amounts of silt and other sediments and materials, thereby altering the lake color and turbidity. With decrease in stream discharge, much of the allochthonous matter begins to settle in the lake basin. The rate of settling is not the same for all classes of materials, grading from sand (1 ft in 3 sec in pure, calm water) to colloidal particles (1 ft in 63 yr).

In natural waters the settling process is complicated by certain attributes of water itself and by the dynamics of water movement, some of which have already been considered. In accordance with Stokes' equation, the rate of settling (velocity of fall of a spherical body through a liquid) is dependent upon gravity acceleration, the radius of the body, the viscosity of the liquid, and the specific gravity of the body and the liquid. Recall that viscosity is related to temperature and that temperature is not necessarily uniform throughout a lake, this being especially important during summer stratification. Turbulence and other water movements contribute to the rate of settling in lakes, particularly during spring and fall overturns.

Some classes of matter which exist in lakes do not settle under normal conditions. These include true colloidal systems or very fine particles of favorable specific gravity, and animal and plant forms capable of modifying their specific gravity and employing other mechanisms (flotation devices, locomotion) to maintain position. The particles in colloidal systems of silt are kept in suspension by the combined effects of turbulence and negative charge on the colloidal particles. It has been observed that the addition of acids to turbid ponds causes flocculation and precipitation of silt particles. These processes are brought about by neutralization of the negatively charged particles by the positively charged hydrogen ions of the acids. The mechanisms and dynamics of settling and the relationships of particle size to distribution by water movements in lakes are not well known—this in spite of the obvious importance of turbidity in such processes as energy circulation of nutrients, and the effects of sedimentation in lakes.

One most important role of turbidity in the interrelationships of the aquatic ecosystem is the effect, previously mentioned, of suspended materials on transmission of light. This aspect has received considerable attention, especially as it relates to productivity and energy flow within the community. At this point we wish to consider only the effects of turbidity upon light; turbidity as a factor in productivity will be considered later.

The damping effect of suspended particles on the transmission of solar radiation, also termed "light-quenching," can be determined effectively from the use of the formula expressing Lambert-Beer's Law,

$$\frac{I_z}{I_o} = e^{-kdc}$$

in which the concentration (mg/l) of light-quenching material c, and the thickness of the column in which light is quenched d, are related to the ratio of observed light I_z to incident light I_o to determine the partial extinction coefficient k of suspended particles in natural waters. (This k is not to be confused with that representing over-all coefficient in some formulas.) The parameter I_z may be measured with an underwater photometer. The value of c is the dry weight of suspended matter per unit volume of water after centrifuging. The advantage in the use of Lambert-Beer's expression over Lambert's Law lies in the recognition of variable c; in Lambert's law this factor is included in the extinction coefficient k.

The Lambert-Beer expression has been used to interpret data on light quenching in western Lake Erie and other lakes. In the Lake Erie investigation, I_o/I_z was set at 100 (observed light equals 1 per cent of surface light), and d represented the depth at the incidence of 1 per cent of surface light. During the study the highest value for c was found to be 35.9 mg/l with a d value of 1.2 m; thus k became 0.107. The highest observed value for d was 7.6 and for c was 5.4; a k value of 0.112 was then derived. A significant datum was obtained in spring when it was found that the average depth associated with 1 per cent of the surface light was 3.5 m; this degree of turbidity represented a load of suspended matter near 10 mg/l. The rapid absorption of light and the quenching effect resulting from turbidity factors are vividly demonstrated in these figures.

It might appear that color and turbidity act together to exert relative effects on light penetration. Such may not be the case in all instances, however. Studies in Atwood Lake, a reservoir in Ohio, have indicated that color and turbidity are very probably independent variables exhibiting no interaction with respect to transparency. It was further found that color may be the major factor affecting light penetration, except of course during periods of introduction of large amounts of silt during heavy rainfall.

COLOR AND TURBIDITY OF STREAMS

Color

Basically, color in stream waters derives from the same physical laws of light transmission, differential absorption by substances in the water, reflection, and back-radiation as operate in other natural waters. However, as a result of several factors streams generally fail to exhibit the great variety of color that may be seen in lakes. The upper reaches of most streams are characterized by clear waters, at least during the nonflood season. In these regions the streams lack a true plankton which, as we have seen, may be responsible for certain apparent colors in lakes. Relatively "pure" shallow streams appear clear due, in part, to the fact that light is absorbed quite rapidly in the first meter or so.

In many areas, particularly in the southeastern United States, small, shallow streams which drain "flatwoods" and swamps are often colored a light to dark amber (about 160 on the U. S. Geological Survey Scale). This color is probably attributable to dissolved plant substances such as tannin; the streams are typically acid. Similarly, certain small creeks in southern Indiana become "inky black" in the autumn due to extractives from large accumulations of leaves in the water. Various extrinsic factors may also account for apparent color in streams. Rich growths of diatoms on rocks on the stream bed may lend to the water a brownish hue; algae may impart a greenish tint, and sulfur bacteria a yellow color. Certainly pollution, from a variety of sources, can contribute to both true and apparent color in streams and should not be disregarded in field investigations.

Turbidity

In the lower stream courses (and in spring in upper stream courses) turbidity becomes a dominant and characteristic feature of most running waters. Depending upon the chemical nature of the material in suspension and the particle size, colors may range from near white through red and brown. In streams where turbidity is not excessive, a plankton may develop and lend to the stream a greenish color.

In the monumental report of his investigations of the plankton of the

Illinois River from 1894 to 1899, C. A. Kofoid characterized lake and stream turbidity thus:

> In this matter of silt and turbidity the river as a unit of environment stands in sharp contrast to the lake. Deposition of solids and clear water are normal to the environment of the lake, while solids in suspension and marked turbidity are the rule with river waters. Owing to their varied occurrence these elements, silt and turbidity, also add to the instability of fluviatile, as contrasted with lacustrine, conditions.

As we have seen in Chapter 3, most streams flowing near base level carry considerable loads of silt and other fine particles. Such high turbidity states result in a decrease in phytoplankton due to quenching of light penetration. Throughout most of the lower Mississippi, light at depths of 200 to 400 mm is only one millionth of that entering at the surface. In many of the larger rivers of the interior of the United States, the turbidity frequently exceeds 3000 ppm. Such concentrations of particulate matter often serve to absorb heat, thereby raising the temperature of the water.

COLOR AND TURBIDITY OF ESTUARIES

Color

The causes and chemistry of colors in marine waters are not thoroughly known. It is likely, however, that dissolved substances impart certain hues or enter into color shifts. For example, it is known that water-soluble yellow pigments are common in coastal areas and may contribute to the various shades of green in off-shore waters. The blue of the sea results, as in inland waters, from the scattering of light by water molecules. Suspended detritus and living organisms give colors ranging from brown through red and green. The Red Sea derives its name from a brownish color due to great numbers of particular algae.

As we move into coastal estuaries we find that the brilliant colors of the open seas are generally not apparent. In some estuaries of moderate tide and current action, blue and green hues are noted during the non-flooding periods. Most estuaries, however, are characterized by dark colors resulting from typically high turbidity. On occasion various planktonic forms become so numerous that they give a reddish or greenish tint to the water. Dissolved materials such as tannic acid delivered by the inflowing stream of the estuary or from local decomposition of organic substances cause estuaries to be light brown.

Turbidity

The major component of turbidity in estuaries is, of course, silt. The volume of silt transported into estuaries by streams fluctuates seasonally, with the maximum discharge taking place during the wet season. Some materials may be brought in from the sea, but these are usually minimal. In addition to allochthonous silt and other materials, much matter which contributes to turbidity originates from erosion within the estuary itself; the magnitude of autochthonous substances appears to be dependent upon the shape of the basin and prevailing currents.

Throughout the year the amount of materials in suspension in an estuary decreases from the upper reaches to the mouth of the basin, that is, transparency increases downstream. This is due to diminution in velocity and carrying capacity of the inflowing stream current, and to the electrolytic effect of sea-water salts. The latter process involves the coagulation of negatively charged particles of colloidal silt by positive ions of certain metals present in sea water.

Even though there is typically a decrease in turbidity seaward along the axis of an estuary, its waters are decidedly more turbid than the sea. Evidence of this is seen in large areas of highly discolored water lying opposite the mouths of estuaries, contrasting vividly with the clearer sea water. Such masses of estuarine waters often extend many miles seaward or, depending upon currents, for great distances along the coast. The major effects of high turbidity in estuaries are (1) the quenching of light penetration, thereby inhibiting photosynthesis and the production of plants, and (2) the building of deep zones of mud, silt, other sediments, and detritus. In many estuaries, notably during periods of considerable stream inflow, light is reduced to 1 per cent of surface radiation at less than 3 m. In certain regions of the open sea the yellow-green components are not diminished to 1 per cent until a depth of nearly 100 m is reached. Transparency decreases shoreward such that in coastal waters the 1 per cent illumination level is generally between 15 and 30 m.

Chapter 7

THERMAL RELATIONS IN FRESH AND ESTUARINE WATERS

FROM THE BROAD and basically ecological point of view, the thermal properties of water and the attending relationships are doubtless the most important factors in maintaining the fitness of water as an environment. We have already considered (in Chapter 5) some of the physical properties which contribute to the total state of the hydroclimate. Even so, it would be worth recalling the fact that the specific heat of water is among the highest of all substances. This very great absorption capacity accounts for many conducive (and nonconducive) features of the aquatic environment. Remember also that water has the unique attribute of reaching its maximum density at 3.98°C (see Chapter 5) rather than at freezing. These two characteristics will be important in considerations of thermal dynamics in fresh and estuarine waters.

At the beginning of the preceding section we learned of the range of solar radiation that impinges upon the earth's surface. We directed our attention mainly toward those components responsible for illumination so important to photosynthetic activity of plants. At the same time we were aware of the presence of light of wave lengths longer than those in the visible range, namely the infrared rays which range from about 7000 Å upward. We know from Lambert's Law that light absorption by water increases exponentially with the light path. In a column of water 1 m in length 91 per cent of light with a wave length of 8200 Å is absorbed; only 9 per cent is transmitted through the column. In 2 m some 99 per cent of this light is absorbed. Over 50 per cent of the solar radiation is absorbed within a depth of 2 m. It is this light with which we shall now be concerned, for the heating of natural waters results from this absorbed radiation. In other words, we should consider *temperature* as an intensity factor of heat energy. Another limnologically important factor of heat energy is the capacity factor; we have considered this in Chapter 5 in our discussion of specific heat.

TEMPERATURE AND HEAT IN LAKES

One of the most outstanding and biologically significant phenomena of lakes is found in the relationship between water and temperature as expressed by seasonal variations. In many lakes, these variations take the form of pronounced changes in the over-all thermal structure and dynamics. During winter, the temperature of the water in moderately deep to deep lakes is relatively uniform from surface to bottom; or if ice forms, this colder layer floats on the underlying waters. In spring, circulation and mixing of water results, typically, in a uniform temperature from

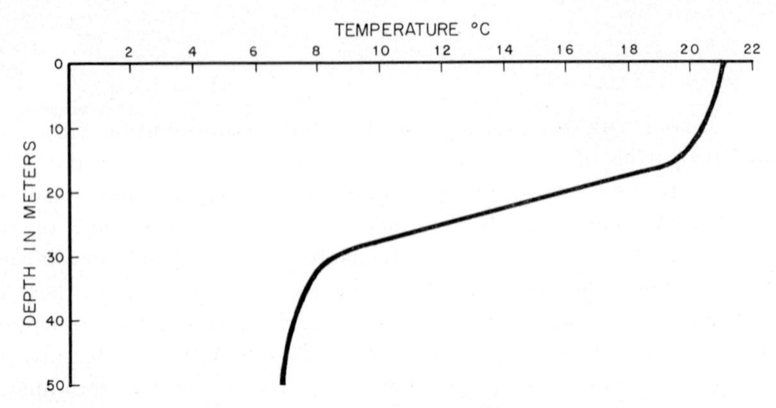

FIGURE 7·1. Summer Temperature Conditions in a Lake of the Temperate Regions.

surface to bottom. During summer, the vertical distribution of temperature may come to resemble that shown in Figure 7·1. In this case, the lake is essentially stratified. From the surface to about 15 m, temperature changes little with depth. Between 15 and 30 m, the temperature drops rapidly. The region of the lake below approximately 30 m is rather uniformly cool. In the fall, the summer condition is broken up by circulation and mixing similar to that of spring, resulting once more in uniform temperature of the water. How do these conditions come about? The answers are found in light absorption, heat dynamics, density phenomena, and wind action. These are best appreciated by considering in more detail the *annual temperature cycle* of a lake.

SEASONS IN LAKES

In winter, as shown for March at the extreme left of Figure 7·2, ice at near 0°C covers the surface of the lake. Recall that ice floats because it has a density less than that of water. Below the ice the temperature to the bottom is relatively uniform and, as suggested by the temperature of

1.5°C at about 15 m, gives evidence of being rather thoroughly mixed—the mixing having occurred in the fall before the ice formed. The density is nearly uniform below the ice. Because of the low angle of the winter sun and the shading of the water by the snow and ice cover, photosynthesis is inhibited. This, together with respiration of organisms and lack

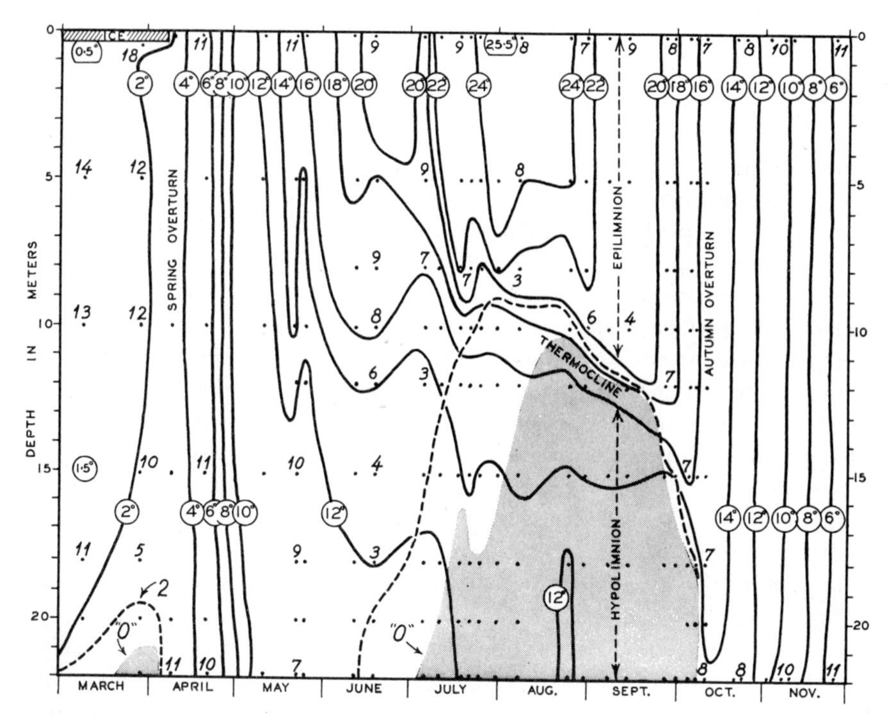

FIGURE 7·2. A Seasonal Cycle of Temperature and Oxygen Conditions in Lake Mendota During 1906. Dissolved oxygen in parts per million are shown in italics. The shaded areas depict the seasonal distribution of near-anaerobic conditions (less than 0.2 ppm of oxygen). The dashed line represents oxygen concentration of two parts per million. (From C. H. Mortimer *in* "E. A. Birge," by G. C. Sellery, 1956, Univ. of Wisconsin Press, Madison, Wis.)

of oxygen replenishment from the atmosphere, results in low oxygen content of the water. Indeed, by late March oxygen is nearly depleted toward the bottom.

In the spring, a higher sun and increased day length brings about the melting of the ice. As the surface waters warm up to 4°C and become more dense, a slight, and temporary, stratification exists which sets up convection currents. These currents, aided by wind, serve to mix lake water throughout until it is uniformly at 4°C and at its maximum density. Note from the vertical isotherms in Figure 7·2, that in Lake Mendota the mixing continued into early May even while the water was warming to 10°C. It is apparent that the amount of dissolved oxygen had increased throughout the lake. This vernal process of mixing has been termed the *spring overturn*, or *spring circulation period.*

Now we have seen that heat is absorbed very rapidly in the surface waters and, therefore, cannot be responsible for heating the deeper waters. Thus some other agent must contribute to the circulation of warm waters into the deeper regions. *Heat transfer in lakes is accomplished mainly by winds.* It is only by this force that mixing can continue beyond 4°C. Winds blowing across the surface of water set up a current resulting from frictional differences between moving air and water. Upon reaching the shore this current moves downward and across the bottom, thereby setting up the spring circulation.

As summer approaches, the weather warms, the longer days mean longer periods of insolation, and the brisk spring winds subside. Under these conditions the surface waters warm rapidly, expand, and become lighter than the lower waters. Although the wind may continue to blow, its contribution to mixing of lake waters diminishes, for now the thermal density gradient opposes the energy of the wind. With the progression of the summer season the resistance to mixing between two layers of different density (resulting from increased temperatures) becomes greater than the force of winds—the significant density differences having been built up during periods of summer calm. A condition called *thermal stratification* is now evident and the temperature curve resembles that in Figure 7·1.

Reference to Figure 7·2 reveals that Lake Mendota became thermally stratified during June and July. In late July, the uppermost region of warm, homothermal water extended downward some 8 m. This region is termed the *epilimnion.* Below the epilimnion there developed a zone of rapid drop in temperature with depth. This zone, only 2 to 3 m thick (Figure 7·2), is the *thermocline.* In modern limnological parlance the thermocline is defined as "the plane of maximum rate of decrease in temperature." The *zone* of rapid drop in temperature, including a gradient on either side of the thermocline, has recently been termed the *metalimnion.* Observe in Figure 7·2 that the temperature in this zone decreases as much as six degrees. The zone below the metalimnion is designated the *hypolimnion.* In many lakes the hypolimnion becomes devoid of dissolved oxygen and high in carbon dioxide during a portion of the summer.

The condition of thermal stratification is very stable. The metalimnion constitutes an effective barrier between the epilimnion and the hypolimnion. Currents stimulated by wind, as well as convection currents derived from cooling at the surface, are limited in depth by the metalimnion. This means that heat and nutrients present in the region of maximum light are prevented from mixing throughout the body of the lake. Similarly, hypolimnetic water substances and many nonmotile organisms and those of limited mobility are restricted to the region. For photosynthetic organisms this state of affairs is particularly critical. Depending upon the depth of the metalimnion, of course, photosynthesis is essentially inhibited in the hypolimnion. This means that oxygen available for animal respiration is reduced, carbon dioxide is increased, and that microscopic algae (phytoplankton) as food for herbivorous animals is scarce. Indeed, it has been found in many lakes, Fayetteville Green Lake, New York, for example, that minute animals (zooplankton) eat great quantities of purple bacteria which are abundant in the deep, oxygenless waters.

The "impedance" to over-all lake current flow presented by the metalimnion has been called the *thermal resistance to mixing*. What, we might ask, is the character of this resistance, and how is such a barrier maintained? The answer to the first question is found in the density-temperature relationships mentioned briefly a few paragraphs back. In Figure 7·1 we see that between depths of about 15 m and 30 m the temperature declines from near 20°C to approximately 8°C, or about 12°C in 15 m. Recall that the density of water is dependent upon temperature, and that the difference in density per degree change is fairly great. For example, as water warms from 4° to 5°C the density changes 8×10^{-5} gm mass/ml. Therefore within the rapid and wide temperature change in the metalimnion a considerable range of density also exists. Consider also the density difference between the temperature at the epilimnion-metalimnion interface and that at the hypolimnion-metalimnion interface in Figure 7·1. This difference represents a strong value of 0.00164 gm mass/ml. Where temperature-depth data are available a graph of the *relative thermal resistance* can be drawn. This parameter represents the ratio of the density difference between water at the upper and lower faces of each stratum to the difference between water at 5° and 4°C.

We might conclude then, that resistance to mixing presented by the metalimnian region is primarily the result of temperature decrease with depth and the related density differences within the zone. Now, what perpetuates the metalimnion and its resistance to mixing? We have already learned that the epilimnion is a region of relatively free circulation and considerable turbulence, and further, that the metalimnion acts to limit the downward extent of water motion. As wind-blown surface currents move against the shore of the lake they are deflected downward and encounter the metalimnion. The metalimnion barrier in turn causes the current mass to spread out horizontally at the upper zone of the metalimnion. Each increment of warming of surface waters in the summer adds a proportionate measure of relative thermal resistance between the epilimnetic and metalimnetic waters. Thus the barrier effect is somewhat self-perpetuating in the sense that density differences between the two regions are increased as the continually warming waters circulate toward the cooler metalimnion.

Throughout the great range of lake morphology, climate, and topography, the various attributes of the thermocline are subject to considerable variation. The thickness of the metalimnion may fluctuate with season, becoming thinner as summer progresses. This is shown rather clearly in Figure 7·2, but in other lakes the effect may be more pronounced, the

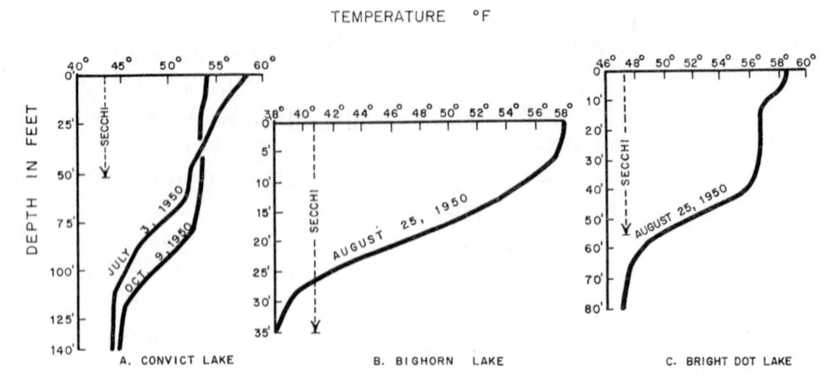

FIGURE 7·3. Temperature Profiles for Three Lakes of Various Depths in the Convict Creek Basin of the Sierra Nevada of California. Secchi disk transparency is also shown. (After Reimers, N., Maciolek, J. A., and Pister, E. P., 1955. "Limnological Study of the Lakes in Convict Creek Basin, Mono, California," *U. S. Fish and Wildlife Serv. Fishery Bull.*, No. 103.)

thickness ranging from about 12 m in early summer to around 5 m in late summer. Recently a very thin metalimnion of some 2 to 5 mm in Barber's Pond, Rhode Island, was reported. Across this layer, visible to underwater observers, the temperature decreased almost 8°C. The depth of a metalimnion is determined by a number of influences such as length and time of seasons, and by basin morphology. Within a particular lake the metalimnion usually becomes deeper in late summer (Figure 7·3A). Local variations in metalimnion level in response to wind action and to the temperature of incoming waters may be expected. One additional point should be considered here; smooth temperature curves are not always obtained. Typically a late-developing stratification results in the formation of a thick epilimnion and more than one metalimnion. In Figure 7·3C note that two such regions are present. More will be said about the metalimnion subsequently with respect to some general aspects of stratification.

With the approach of autumn, the angle of incident light decreases, day length shortens, and cooling begins. This is to say that the lake loses heat faster than it is absorbed. As the cooling extends toward the deeper regions of the lake, the density differences between isothermal strata become less. Thus the thermal resistance to mixing is weakened. Wind-generated currents carrying cooler, oxygenated waters reach ever more deeply into the lake. The metalimnion sinks rapidly. As shown in Figure 7·2, overturn during the period of *autumnal circulation* began in Lake Mendota in early October and continued into December. Observe in the figure the uniform, and relatively high, oxygen content from surface to bottom accompanied by a temperature difference of less than 1°. The period required for autumnal circulation is dependent upon local climate and the depth and morphology of the lake. The deeper the lake, the greater the time required for uniform cooling. Two points are important here: (1) cooling (as with heating) takes place only at the surface, and (2) cooling to 4°C is accomplished primarily by convection currents.

An *inverse temperature stratification*, not depicted in our figure, may exist during winter as the water cools below 4°C. Remembering that water reaches its most dense state at 4°C, we can easily appreciate that as surface waters cool beyond that point another "density difference" occurs in which the lighter, cooler waters float on more dense, "warmer" waters. Although the winter temperature conditions of stratification are reversed with respect to summer, the density relationships are similar—the less dense mass occupies the upper part of the lake. The inverse stratification is most pronounced following the formation of the ice cover. Immediately below the ice (0°C) the temperature of the water rises sharply to near 4°C, or whatever the temperature of the water body. Inverse stratification is usually of rather short duration and, indeed, may not occur every year in a given lake. The period of *winter stagnation* now exists.

The sources and processes involved in the heating of lake waters beneath ice have received considerable attention. It appears, however, that heat from direct solar radiation through the ice, and heat derived from bottom muds are the major sources of energy for warming the lake. The heat-distributing mechanisms are apparently density currents of two types: (1) the temperature-dependent currents which are derived from heating of water through the ice in the shallow zones and which flow into the deeper parts of the lake, and (2) chemical density currents derived from dissolution of bottom substances in water; more dense by virtue of the dissolved materials, these currents may also transport heat.

Thermal Classification of Lakes

As a result of extensive studies of temperature phenomena, heat, and stratification, since the time of Forel, a great store of terminology and schemes of classification of lakes based on thermal characteristics has accumulated. In an attempt to keep terms and classifications at a minimum we have adopted a scheme proposed by Hutchinson (1957). By no means universally applicable, by virtue of the variable nature of lakes, the system is useful and points to the more modern concepts and approaches to thermal properties and dynamics in lakes. Earlier terms and classifications, certainly not useless, are to be found in older texts and other works on ecology and limnology.

The following types are proposed for lakes occupying basins of sufficient depth to allow for stratification mixing ("mixis"), and the formation of a hypolimnion. Consideration has been given to altitude, geographical location with respect to latitude, and to depth of the basin.

AMICTIC

Lakes insulated and protected by ice cover from outside influences of weather and other factors; poorly known lakes of the Antarctic and high altitudes.

COLD MONOMICTIC

Lakes of the polar regions in which the waters at any depth never exceed a temperature of 4°C; ice-covered and exhibiting an inverse temperature stratification in winter; one mixing ("monomixis") at temperatures not greater than 4°C in summer.

DIMICTIC

Lakes in which two circulations take place each year in spring and autumn; thermal stratification is inverse in winter and direct in summer; typical of lakes in the temperate zone and in higher altitudes in subtropical regions.

WARM MONOMICTIC

Lakes of the warmer latitudes in which the temperature of the water never falls below 4°C at any depth; one circulation each year in winter, directly stratified during the summer. Lake Providence, Louisiana, stratifies from May through September, and circulates continuously from October to April.

OLIGOMICTIC

Warm lakes in which the water temperature is considerably higher than 4°C; circulation occurs rarely at irregular periods; found typically at low elevations in tropical zones.

POLYMICTIC

Lakes in which mixing is continuous but at low temperatures, usually just over 4°C; characteristic of high mountains in the equatorial regions.

Stratification does not develop because of heat loss to a relatively uniform environmental temperature.

Holomixis and Meromixis

With respect to circulation, or mixing, generally, most lakes are said to be *holomictic* ("wholly mixing"); that is, if circulation takes place it is complete and extends the entire depth of the lake. In these lakes the temperature of the hypolimnion usually decreases uniformly to the bottom. However, some lakes are known in which the summer temperature curve resembles that of a holomictic lake except that temperature in-

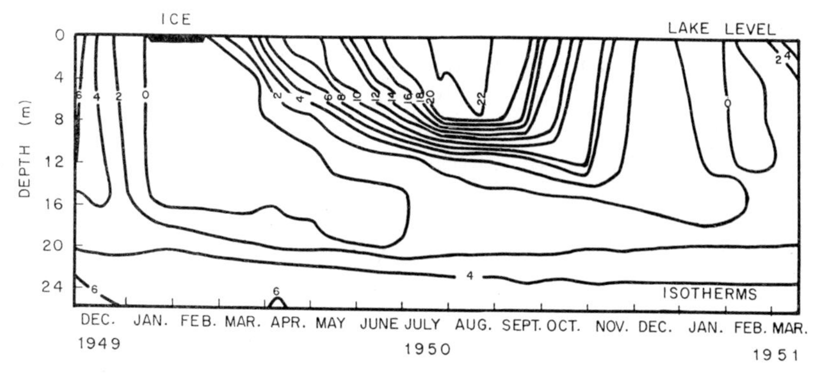

FIGURE 7·4. Seasonal Variations of Temperature in Meromictic Soap Lake, Washington. The period of complete ice cover is indicated by the heavy bar. (From Anderson, G. C., 1958. "Some Limnological Features of a Shallow Saline Meromictic Lake," *Limnol. Oceanog.*, Vol. 3, 259-269.)

creases slightly through a considerable distance. How is it that warm water is lying underneath cold water? What of the density relationships? This bottom water is not less dense than the upper water. It has been found that the bottom stratum usually contains large quantities of dissolved salts, and these increase the density sufficiently to prevent mixing with the upper layers. We have, then, a lake in which circulation is not complete; such lakes are called *meromictic* ("partly circulating"). The bottom, noncirculating layer is called the *monimolimnion*, and the density gradient becomes a *chemocline*. The waters above the chemocline in which thermal stratification such as in holomictic lakes can occur is termed the *mixolimnion*. A lake in western Austria (Längsee) possesses a monimolimnion that has not mixed with the upper waters for over 2000 years. The monimolimnion contains a heavy concentration of salts derived from the sediments by biochemical processes. Heat in the bottom layer is apparently gained from bacterial activity and from insolation.

Since there is no circulation by currents, heat is lost mainly through conduction. The annual temperature cycle of a meromictic lake is shown in Figure 7·4.

Heat Budgets and Lake Stability

Having considered the annual temperature cycle in a typical lake of the temperate zone, as well as some unusual cases of stratification, let us now give attention to some of the broader aspects of heat and heating. As stated earlier, temperature is the expression of the intensity phase of heat energy, while heat, as such, is an energy factor.

For purposes of comparison of lake qualities and expression of the nature of the lake waters as environment with respect to organisms, the calculation of heat income and heat budgets becomes a useful tool. *Summer heat income* is the quantity of heat delivered to a lake necessary to warm the waters from the homothermal spring condition of 4°C to the summer maximum. Although direct solar radiation contributes essentially all of this heat, the distribution of it is accomplished by wind-driven mixing of waters of differing densities. The work of the wind in summer heat income is done against gravity because of density differences arising from heating of surface layers by direct insolation. The *annual heat budget* takes into account the total quantity of heat taken into the lake to warm the waters from the lowest temperature of winter to the maximum summer temperature. Either parameter can be determined in absolute terms such as calories, or gram-calories. Usually, however, we wish to compare heat characteristics of several lakes, in which case it is well to introduce area. The result is then expressed in calories per square centimeter of water surface. The heat content per unit surface area of a lake at any given time can be determined by totaling the product of temperature times volume for declared depth intervals, and then dividing by the surface area of the lake. The gain in heat content per unit area beyond the homothermous 4°C state is the summer heat income, and is more a function of mean depth than of temperature differences. The annual heat budget may be derived in the same way, except that the lowest winter temperature is substituted for 4°C. Table 7·1 gives morphological and thermal data for five lakes in the Convict Basin of California, and shows the relationship between depth and annual heat budget. Comparisons of annual heat budgets of lakes must consider depth differences. Shallow lakes are not able to store all the heat transmitted into them. During the period of maximum temperature, the lakes become vertically homothermous and heat is transferred and lost to bottom sediments.

Within temperate regions the annual heat budgets of typical lakes range from about 30,000 to 40,000 cal/cm^2. Much of this heat (33 to 75 per cent) enters the lake after the spring circulation period, and work energy is involved in the circulation of the heat to lower lake levels. Lake Michigan, with an annual heat budget of 52,400 cal/cm^2, and Lake Baikal, whose annual budget amounts to 65,500 cal/cm^2 are the highest known. The summer heat income for Lake Michigan is 40,800 cal/cm^2, and Lake Baikal receives 42,300 cal/cm^2 between spring isothermal 4°C and summer maximum. To bring about the thermal conditions shown in Figure 7·2 for Lake Mendota an annual budget of about 23,500 cal/cm^2 and a summer income of some 18,240 cal/cm^2 are required.

TABLE 7·1. MAJOR PHYSICAL FEATURES OF CERTAIN LAKES IN CONVICT CREEK BASIN, CALIFORNIA, IN RELATION TO TEMPERATURE AND HEAT DYNAMICS

(Data from Reimers, N., Maciolek, J. A., and Pister, E. P., 1955. "Limnological Study of the Lakes in Convict Creek Basin, Mono County, California," *U. S. Dept. of Interior, Fish and Wildlife Serv. Fishery Bull.*, No. 103.)

					Temperature, °C		Heat Intake, gm cal/cm^2	
Lake	Elevation	Surface Area (ha)	Mean Depth (m)	Transparency (m)	Bottom to Surface	Lake Mean	Summer Heat Income	Annual Heat Budget
Convict	2275	68.6	26.4	15.3	6.6-14.7	10.9	18,173	22,984
Mildred	2970	4.3	8.1	12.0	10.0-12.5	11.1	5,822	7,298
Witsanapah	3228	1.8	8.4	9.0	5.5-10.0	8.5	3,827	5,358
Edith	3030	7.3	17.7	13.8	4.1-14.1	8.7	8,601	11,826
Dorothy	3102	43.8	40.8	20.1	4.4-14.7	8.8	17,967	25,402

In view of the importance assigned to the work of wind in distributing heat in a lake, we should question how much work a wind of a given velocity can do and how much work is required to mix a given quantity of heat. As Hutchinson has pointed out, little work has been done on the problem, and our knowledge is therefore meager. However, some data from various sources as compiled by Hutchinson will serve to convey a general idea. With respect to the first question, Langmuir has stated that winds of velocity 300 to 700 cm/sec developed forces of 0.65 to 6.3 dynes/cm^2 on lake surfaces. The summer heat income of Green Lake, Wisconsin, is 27,316 cal/cm^2 and the heating period is 122 days. Calculations show that the mean rate of work directed toward heating the waters is 0.02 dynes/cm^2. These data indicate that the work of the wind in summer heat distribution is very small in comparison to the force of wind on the lake surface.

The loss of heat during the period of cooling from the summer maximum temperature to the winter minimum must be of the order of the annual heat budget. This loss from the lake is mainly through radiation to the surroundings. Cooling by evaporation also expends heat. The waters of effluent streams remove considerable quantities of heat from lakes, especially from impoundments where warm, surface waters are removed while cooler waters remain behind. Heat exchange processes in a lake and impoundment are shown in Figure 7·5.

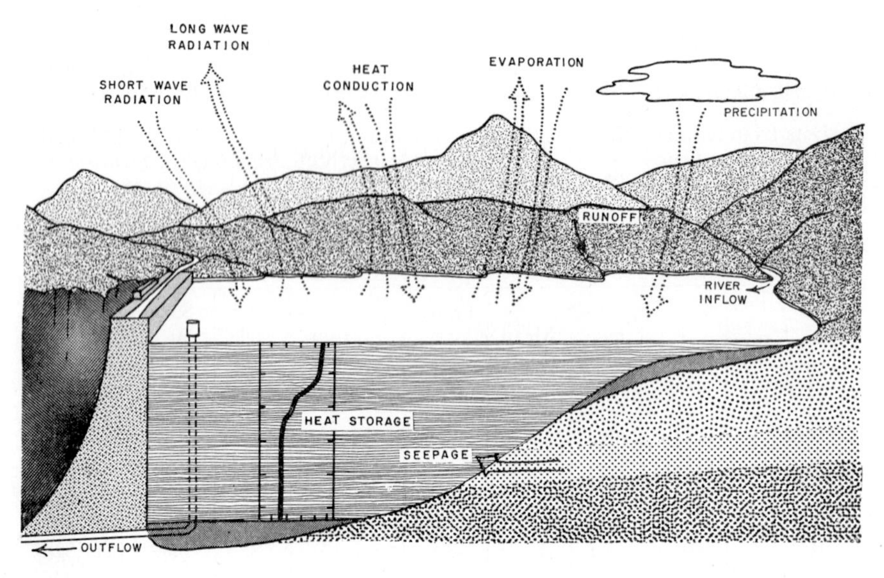

FIGURE 7·5. Major Processes of Heat Transfer in a Lake. A temperature profile is shown as an inset. (After Saur, J. F. T. and Anderson, E. R., 1956. "The Heat Budget of a Body of Water of Varying Volume," *Limnol. Oceanog.*, Vol. 1, 247-251.)

From the foregoing considerations we sense that a certain quantity of energy is expended in developing summer stratification in holomictic lakes. We have also made reference to the stability of the stratification and to the resistance offered by the lake to mixing. This, in essence, is one definition proposed for the term *stability*. More properly, *stability of stratification* is that energy of resistance which the lake offers to oppose upset of density stratification. As warm, less dense water comes to overlay cool, more dense water during stratification the center of gravity is shifted downward. Stability then becomes a measure of the amount of work required to raise the center of gravity or to displace it to its original position. Stability decreases in autumn as the thermocline sinks below the center of gravity and the waters above and below this center attain similar densities. The greatest stability is probably reached just prior to maximum heat content in summer.

While we have been considering temperature and heat relationships in relatively large bodies of standing waters, we should not lose sight of the fact that pools and small ponds also exhibit responses to seasonal and daily temperature influences. Such responses often limit activity of some organisms and should certainly be studied in limnological investigations of small situations. "Microthermoclines" and odd temperature stratifications exist in certain small bodies under given conditions. A reversed stratification has been shown to exist at times in small Burt Pond, Michigan. The data for this pond are given in Table 7·2 and indicate, among

TABLE 7·2. SUMMER POND-WATER TEMPERATURES IN RELATION TO ALGAL MAT ON BURT POND, MICHIGAN. TIME: 1330

(Data from Young, F. N. and Zimmerman, J. R., 1956. "Variations in Temperature in Small Aquatic Situations," *Ecology*, Vol. 37, 609-611.)

	°C
At surface of Mat	21 to 22
2.5 cm Below Mat Surface	23 to 24
5.0 cm Beneath Mat	25
7 to 8 cm Beneath Mat	21 to 23
15 cm Beneath Mat	20

other things, how rapidly the temperature of a small body of water responds to atmospheric conditions. It is suggested that the reversed stratification results from heating of the surface by direct radiation in the morning, followed by cloudiness and winds which cool the surface in the afternoon.

THERMAL PROPERTIES OF STREAMS

The basic thermal characteristics and processes in the water in streams are not different from those of lake waters. Penetration of light, absorption, specific heat, and other factors attributable to the nature of the water molecule operate in determining the fitness of the stream as an environment just as in lakes. As an environment, however, a stream presents a considerably different set of temperature conditions. These conditions derive from variations in velocity, volume, depth, substrate, cover, water source, and a number of additional features operating seasonally, daily, and even longitudinally along the stream course at a given time. Recall that we have termed a stream community an "open system" by virtue of the considerable land-water interrelationships; some of these influence temperatures.

The major factor in the warming of stream waters is direct solar radiation. A series of temperature curves recorded over several successive days of cloudless skies followed by an overcast day will show rather regular fluctuations from late afternoon maxima to early morning minima during the cloudless period, but the cloudy-day curve will be flattened. Additional evidence for the importance of direct insolation is found in the frequent occasions when the temperature of the water exceeds that of the air. This usually occurs on very clear days of intense sunlight. The limit of nocturnal cooling is, of course, regulated in small, shallow streams by atmospheric temperatures. In larger, deeper streams it depends upon the rate of heat loss before warming begins again. Additional heat may be gained from bottom sediments in streams, but little data are available to show the extent of such heating.

On the other hand, the temperature of stream water is a measure of the actions and interactions of a wide variety of factors. Consider a stream rising in a rocky, wooded highland and flowing down over the coastal plain to the sea. In the upper reaches the waters are cooled by the substrate, by the shading provided by vegetation, and possibly by the entrance of spring-fed tributaries. Turbidity may be quite low. As the stream approaches the lowlands it becomes wider and deeper, and more water is exposed to direct sunlight. This, and increased silt content which absorbs considerable heat, result in the development of a segment quite different with respect to temperature from the upper reaches.

Depending upon size and origin of streams, their diurnal and seasonal temperatures follow atmospheric temperatures more closely than do those of lakes. Diurnal temperature variations at a given point in a stream may be related to two major sets of factors: (1) conditions at that point, and (2) conditions upstream from the point. Under (1) we should consider velocity and discharge, season and hour, and the daily range of fluctuations of air temperatures at the point. Factors pertaining to (2) include the nature of upstream environment, substrate (and impoundments, if any), atmospheric conditions and temperatures upstream, and distance and time of flow from critical upstream situations.

With respect to size and water temperature fluctuations, we can conclude that the smaller the stream, the greater the temperature variations and the more rapid the response to environmental fluctuations. Furthermore, these fluctuations exist throughout the year, but are minimized beneath ice cover. The range of daily variation of water temperature is maximal when there is the greatest differential between mean diurnal air temperature and mean water temperature. With increased volume and turbidity the range of fluctuation diminishes. Because of the wide range of temperatures exhibited by small, cold streams, a single temperature reading may not at all approximate the average daily figure.

The source of the stream's waters and the nature of the drainage pattern in some instances determine the thermal properties of the stream. Many spring-fed streams and those originating from surface discharge of subsurface aquifers are essentially thermostatic. Lander Springbrook, New Mexico (see Chapter 14), is a small rheocrene ("flowing") spring. Seventy meters below the source the annual variation in water temperature is 4°C; air temperatures fluctuate from about 14.3° to 33°C. The annual variation of water temperature of Silver Springs, Florida, a large spring run (daily discharge about 600 million gallons mainly from aquifer openings) varies no more than 1° throughout the year. Diurnal fluctuation in Silver Springs is also of the order of 1°. These spring-fed streams do not receive significant quantities of surface water which might otherwise bear on their thermostatic qualities. Surface-fed streams typically show wide seasonal fluctuations corresponding with atmospheric conditions. Figure 10·3F (see Chapter 10) illustrates the nature of seasonal variations and correlation with air temperatures in a stream typical of much of the nonmountainous regions of the United States.

Because of turbulence and the shallow nature of most streams, thermal stratification is not generally an attribute of streams. When stream waters stratify, the process usually takes place in pools along the stream course. In some instances the stratification is a result of inflow of cool spring water at the lower levels of the pool. In Jack's Defeat Creek, Indiana, a pool was found to be stratified, the temperature decreasing 7.5°C from surface to bottom at about 62.5 cm. Other qualities, including free carbon dioxide, dissolved oxygen, and alkalinity were also stratified. Inflow of spring water possibly accounts for the local stratification in this case. If warmed upstream waters are gently introduced into a cool pool of water, the inflowing, less dense waters are apt to flow over the more dense pool waters and on downstream without considerable mixing.

One word on the effects of human endeavors, such as agriculture and industry, on stream temperature and ecology needs to be interjected here. We know that water taken from a stream and used in irrigation is warmed in the process and, upon being dumped back into the stream, increases the downstream temperature for some distance. Similarly, water used in cooling various industrial installations warms the stream. In addition to the simple warming of the stream which, as such, may exceed the temperature tolerance of some of the organisms, the increased heating also decreases the oxygen retention capacity of the water, thereby affecting certain organisms and stream metabolism. Reservoirs, whether for water supply or hydroelectric power, exert considerable influence on the quality of stream water below the impoundment. As spring temperatures of streams approach the favorable point for spawning of warm-water fishes, sudden discharge from the lower levels of an upstream impoundment cools the stream and inhibits reproduction of the fishes. Conversely, the influx of warm, surface waters from an impoundment into a cool trout stream has deleterious effects on the fishes and the stream community. Much precise work needs to be done on these aspects of stream relationships. The results of such research could go far in enlightening the agencies concerned with dam-building projects.

HEAT AND TEMPERATURE IN ESTUARINE WATERS

Before considering some of the more general thermal aspects of the estuary as a body of water, we should give attention to the effect of increased concentration of salts upon certain basic properties of water. Recall that pure water has a specific heat of 1. As the concentration of dissolved salts increases, the specific heat decreases. For sea water at 17.5°C and salinity of 35‰ (parts per thousand) the specific heat is 0.932. This suggests that less heat is required to warm a given volume of salt water than to warm the same amount of fresh water. Whereas for pure water density is a function of temperature alone, being maximum at 4°C, the presence of salts depresses the temperature of maximum density in sea water. At a salinity of 24.70‰ maximum density occurs at the freezing point, −1.33°C. At salinities higher than 24.70‰ the temperature of maximum density is below the freezing point. Unlike pure water, sea water at high salinities undergoes regular increase in density as it cools to freezing. Generally, the freezing point of sea water varies inversely as the salinity. At a salinity of 10‰, commonly experienced in estuaries, the freezing point is near —0.5°C; at 35‰ sea water freezes at about —2.0°C. The complexities in estuarine thermal dynamics are easily appreciated when we remember that within a few miles the salinity in an estuary may grade from near 0‰ to 25‰ or higher.

The heat content of estuarine waters is derived mainly from solar radiation. These waters are directly heated *in situ* as they occupy the estuary basin. Heat is also received indirectly from inflowing stream water and from tidal flow from the sea. The temperature of the estuary is, therefore, primarily a function of the temperatures of entering streams and the sea together with tidal stages. If the estuary empties into a relatively deep sea, the seaward temperature is apt to be more stable than in the upper zone where the stream temperature may fluctuate widely. On the other hand, estuaries of spring-fed streams flowing into shallow sea zones may exhibit more stable features in the headward regions.

In view of the shallow nature of most estuaries we should expect considerable diurnal and seasonal fluctuations in surface temperatures. In the East Bay, an arm of the Galveston Bay system of Texas, during July, surface water warms from an early morning temperature of about 27.0°C, to 33.0°C in late afternoon; atmospheric temperatures during the same period range from near 28.0° to 34.0°C. Seasonal temperature fluctuations depend upon latitude and a number of local factors such as water source, basin morphometry, winds, and tides. In Apalachicola Bay, Florida, the annual temperature range is of the order of 25°. Temperatures in the Sheepscot estuary of Maine range from about freezing to near 25°C in the upper reaches of the estuary, while in the lower region the range is about 15°. These temperatures are determined primarily by atmospheric and climatological conditions. A number of instances have been noted, however, in which sudden changes in the pattern of oceanic currents result in the influx of cold masses of water into an estuary, thereby drastically reducing water temperature.

Another factor that affects temperatures in estuaries is the flooding of marshes and mud flats during high tides. In warm, humid regions the surface of exposed marshes may, during the summer, become quite warm. As the rising tide covers the marshes with estuarine waters, heat is imparted to the waters. In other regions the exposed muds may become cooled through evaporation and thus lower the temperature of incoming water. Studies in the Elkhorn Slough estuary of California have shown that mud gains heat much less rapidly than the over-lying water, even in a depth of only 15 cm.

Temperature distribution in estuaries is largely a function of depth together with the relative effects of stream inflow and tidal exchange. In a shallow, mixing estuary the waters tend toward vertical homothermy. Longitudinal temperature patterns vary seasonally. Depending upon the morphology (surface-volume ratio) of the stream basin and the volume and rate of discharge, the upper waters of the estuary may be cooler in winter and warmer in summer than the lower estuary waters. From the mouth of the Patuxent River of Chesapeake Bay to a point about 74 km upstream, the average surface temperature from March to July grades from about 23° to 21°C; bottom-water temperatures similarly decrease on the order of approximately 3°. From September to January, both surface and bottom temperatures increase slightly over 1° from the bay upstream.

Where conditions of climate and sufficient depth are met, estuaries may exhibit a vertical temperature gradient. In summer the surface waters are usually warmer than the underlying layers. With the cooling weather of autumn, the upper waters cool more rapidly than the deeper masses, resulting in overturn and mixing. In midwinter the temperature of the surface waters may hover near that of maximum density, thereby giving rise to convection currents which continue to circulate the waters. Increasing insolation and winds in spring warm the surface, and a temperature gradient follows. Density stratification derived from cold or warm fresh water flowing over the more dense salt water also contributes to vertical temperature differences in estuaries. Such a stratification may be recognized by salinity measurements from the surface downward.

Chapter 8

NATURAL WATERS IN MOTION

Poetic visions of placidity and quiet notwithstanding, natural waters in their basins or channels are rarely, if ever, perfectly calm. In the smallest pool and the largest lake or stream, movement of some form and degree takes place. Motion in such waters derives from temperature and density differences, gravity, and wind acting together or singly and to varying extent. Directly or indirectly every facet in the aquatic community, be it physical, chemical, or biotic, is in some way affected by movement within the body of water.

In this chapter we wish to single out some of the more conspicuous forms of motion in lakes, streams, and estuaries and to consider them as basic features of natural waters and as physical factors determining the fitness of waters as an environment for habitation by living organisms. This latter consideration will be dealt with more directly in Chapter 14.

MOVEMENT OF WATER IN LAKES

As a general rule, the water of a lake basin is partially or wholly in motion. It is this movement, derived from either internal or external forces, or both, that is responsible for the circulation of heat, dissolved substances, and some organisms in lakes. Turbulent flow is characteristic of the movements of lake water; this results from the action of molecular systems imparting irregular direction and velocity to water mass. Turbulent movement is then incorporated into the more conspicuous forms of motion in lakes. The larger movements of water are called *current systems*, and are frequently defined as two types: nonperiodic, or arhythmic; and periodic, or rhythmic.

Nonperiodic Current Systems

Nonperiodic systems are those often termed simply "currents," implying a unidirectional flow of water. This movement may be caused by differential heat distribution within the lake, the passage of stream waters through the lake, and by winds. Broadly speaking, nonperiodic systems are produced and maintained by external forces.

Wind is doubtless the major external force acting to set lake waters in motion. Sustained winds blowing across the surface serve to pile up water at the down-wind end of the basin. The result is lowering of the water mass in another part of the lake. As the winds subside, the piled-up water begins to flow along a gradient, or down the slope, so to speak. Eventually the over-all lake level reaches equilibrium and the current system comes to rest. We have seen that the action of wind is also important in the development of summer stratification of lakes.

A phenomenon familiar to all who have traveled the ocean or large lakes is that of *wind streaks*, or those linear accumulations of foam, floating debris, or oil. Irving Langmuir, better known for his work in electricity and related fields, demonstrated that surface flow resulting from wind drift is not an even and uniform movement of water containing variable velocities, but rather possesses a definite pattern. Water flow in wind drift is in the form of helices lying parallel to one another oriented in the direction of the wind with the direction of the helices alternating clockwise and counterclockwise, and the effect of streaks is derived from the accumulation of materials in zones of convergencies between clockwise and counterclockwise helices. The width of the zone of convergence varies seasonally, being wider in the late autumn than in spring. It appears that the downward vectors of the helices serve to transfer momentum from the surface into the deeper regions of the lake.

In large lakes the effects from stream inflow, wind, and earth's rotational forces often combine to produce a pattern of large current swirls. The surface currents of Lake Constance, in Europe, have been studied rather intensively, especially with relation to the path of the Rhine River through the lake. The general circulation pattern is shown in Figure 8·1. As a result of geostrophic forces and the shape of the basin, the Rhine waters flow across the upper end of the lake and follow the northern shore. The river current, together with wind patterns, apparently produce the large swirl between Langenargen and Rorschach.

The influence of the Rhine on the currents in Lake Constance serve to point up another kind of nonperiodic current in lakes; such may be termed *density currents*. These currents result from the passage of river water of a given density through a lake of differing density. As suggested previously, the route of the stream water is determined mainly by geostrophic forces, the shape of the lake basin, and local meteorological agents. The depth at which the river water flows is a function of density differences. If the density of the river water is greater than any of the lake waters, the stream flows over the bottom of the lake. Similarly, if the lake is generally more dense than the stream, the latter flows at the lake surface. During the summer months the water of the influent is usually cooler than the lake surface, and therefore more dense. Thus the river water flows downward until it meets a region of greater density, whereupon the river flow becomes horizontal. The "waterfall" at the mouth of the Rhone in Lake Geneva is familiar. In thermally stratified lakes, the river flow is usually above the hypolimnion, thus the lower lake levels receive little of the value of enrichment by the inflowing stream.

In Norris Lake, a Tennessee Valley Authority impoundment, there are indications that density currents move through the lake as the water is

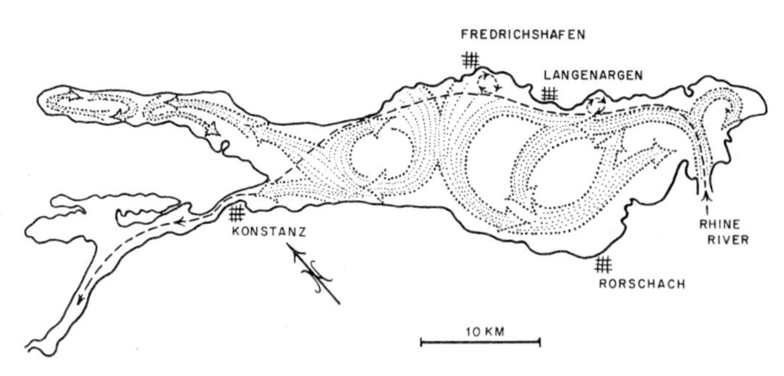

FIGURE 8·1. Schematic Pattern of Surface Currents in Lake Constance, Switzerland-Germany. The route of Rhine River water through the lake as determined by conductivity measurements is indicated by a dashed line. (Modified from Hutchinson, 1957. "A Treatise on Limnology," John Wiley & Sons, Inc., New York, N.Y., after various authors.)

drawn down for hydroelectric power. These currents may be poor in dissolved oxygen content due to organic decomposition, causing movement of fishes to avoid the oxygen minima (see Chapter 14). Density currents may be important in ponds and small lakes, and probably relate to distribution of materials beneath ice cover in winter. In Lake Mead, formed by Hoover Dam on the Colorado River of southwestern United States, a density current resulting from increased silt load forms during the summer. This *turbidity current* may flow into the lake below the level of the river density current.

One more aspect of nonperiodic movement of water should be considered, that of *turbulence* and *eddy effects.* Although the mechanics of these phenomena are best known from laboratory studies, the applicability of the principles to limnology is vast and deserving of more intensive study in the future. To anyone who has watched a stream plunging through riffles or along irregular shores, the intense turbulence and the resultant eddy systems are obvious. Within lakes, however, such turbulence is reduced and certainly less conspicuous, albeit present as the result of currents such as those previously described and to be discussed later. The great importance of eddy effects is in heat transfer within the water mass (*eddy conductivity*), in the diffusion of dissolved materials (*eddy diffusivity*), and momentum transfer (*eddy viscosity*).

In Chapter 3, we distinguished laminar and turbulent flow with respect to streams. It is agreed that laminar flow, wherein sheets of liquid move uniformly without local variations in velocity (but with velocity differences from "sheet" to "sheet"), does not normally occur in nature. The more characteristic flow of natural waters is one of turbulence, in which irregular motion of subsidiary masses is contained within some relatively

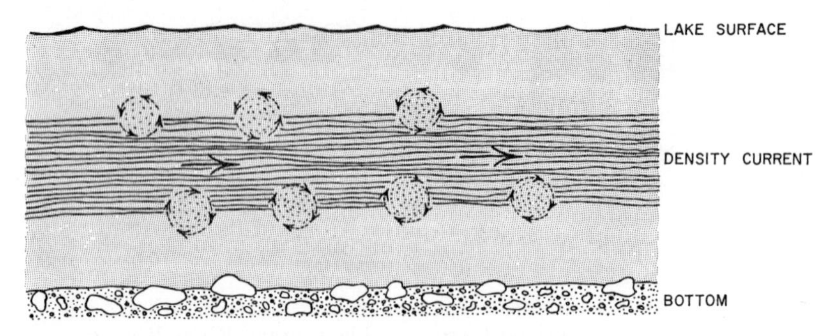

FIGURE 8·2. Highly Schematic Representation of Eddy Systems Set up Along an Interface of Lake Water and Density Current as the Current Flows Below Surface Through a Lake. (See text for discussion.)

simple current. The nature of the turbulence is an expression of various factors including flow velocity, velocity gradients encountered, and the nature of the boundaries of the flowing systems. Figure 8·2 may be interpreted in terms of a density current passing through a lake. In this example, the lake proper serves in somewhat the same sense as a stream bank in offering resistance to moving masses. The velocity and magnitude of the eddy systems operating at the upper and lower zones of the density current, for example, are functions of velocity factors of the lake proper and the current, and boundary characteristics including temperature and chemical composition. It appears, then, that through eddy effects, mass and molecules can be exchanged from one layer to another.

One effect of eddies is that by which masses leaving one layer carry with them momentum from that layer and acquire the momentum of the new layer before returning to the original one. This is a simplified definition of *eddy viscosity*. In practice, a *coefficient of eddy viscosity* may be calculated from:

$$r_s = -A \frac{d\bar{v}}{dn}$$

in which r_s is the Reynolds stress, or shearing stress per unit surface area; $\frac{d\bar{v}}{dn}$ is the shear of the velocities as measured; and A (from "Austausch") now becomes the coefficient of eddy viscosity, that is, the relationship between mass and transverse distance through which a liquid flows per unit time in a path corresponding to that of the main direction of flow. The transfer of motion through eddy viscosity increases with turbulence, thereby giving rise to the fact that eddy viscosity is considerably greater than molecular viscosity, often on the order of 1000 to 1 million times. In natural waters eddy viscosity is a primary force in reducing the settling rate of sediments, and in maintaining the position of microscopic organisms.

In the same sense that eddy viscosity relates the transfer of mass and momentum from one water stratum to another, *eddy conductivity* describes the exchange of heat across surfaces. This parameter takes into account the specific heat of the fluid and the heat gradients encountered between masses. This heat transfer is then proportional to the mass transfer described in the formula for eddy viscosity. Important here is the idea that conduction may be lateral as well as vertical. Eddy conductivity is fundamental in the heating processes of lakes, described in the preceding chapter.

The transfer of dissolved substances, so very important in lake metabolism and maintenance of living organisms, is another function of mass exchange and is termed *eddy diffusivity*. An important concept of eddy effects is that of the transfer of more than one substance, or the *principle of common transport*. This principle holds that a motion that transfers one environmental factor may, simultaneously, carry another property. For example, while dissolved salts are being transported downward, dissolved gases, or heat, may be carried upward in the same eddy system. Some idea of the rate at which substances may be distributed through this process is gained from the fact that summertime eddy diffusivity in the western part of Lake Erie is near 25 cm^2/sec, or probably some 104 times that in water in a quiet bottle.

The basic concern with eddy effects for our purposes relates to the more general ways in which the processes contribute to lake dynamics. For fuller discussions of the details and mathematical derivations of eddy relationships see Hutchinson (1957), or recent works in hydrology.

Periodic Current Systems

To this category belong water movements exhibiting some form of rhythm or periodicity. Two major systems are to be considered; these are: traveling, *surface waves* and standing waves, or *seiches*. As has been shown for nonperiodic currents in the preceding section, external forces, mainly wind and other meteorological factors (atmospheric pressure), are responsible for periodic currents.

SURFACE WAVES

Lake surfaces are never completely smooth; slight irregularities and turbulences are present. Friction occurs between these surfaces, and wind blowing across the lake results in movement of the water. This motion imparts the first phase of wave formation. Subsequently, a crest is built, and increased wind velocity tends to build up the wave; eventually, an eddy is produced by wind on the trailing slope of the wave, giving a certain thrust to its back, thereby accelerating the motion of particles in their orbits within the wave.

There is no essential horizontal movement to waves in open water. A particle of water moves in a vertical circular pattern, or orbit, returning to its original position as the wave rises and falls. The diameter of the orbits and, correspondingly, the velocities, decrease rapidly with depth. It has been calculated that the diameter of orbits at a depth equal to one-half the wave length is on the order of 1/500 that of surface orbits, or quite imperceptible. Not until the wind develops to such a force as to produce a tumbling crest, or whitecap, is there any forward motion of the water. A similar effect is exerted by the bottom in shallow zones where friction directed upon a wave causes the mass to pitch forward, producing *surf*. These breaking waves are called *waves of translation;* open-water swells are termed *waves of oscillation.*

Both velocity of wind and the time that it has been exerting force upon a wave determine the size of waves. In order to build waves of any great amplitude a considerable amount of time is required to transfer the wind energy into the kinetic and potential energy of a wave. Similarly, space is required, for a wave must have area in which to move before the wind in order to gain amplitude. The relationship between wave height and fetch during strong winds is expressed in:

$$h_w = 0.105\sqrt{x}$$

in which x is the fetch, or downwind distance (in centimeters) from shore to location of the wave in question.

Although waves and wave action are of especial interest to physical limnologists, our interest in terms of the over-all ecology of lakes is mainly in the action of waves in circulation of vital materials in lake waters, and the stirring effects of breaking waves in shallow-water zones. From a broader point of view, wave action is most instrumental in sedimentation and erosional processes which serve to modify the morphology of the lake basin and littoral areas (review Chapter 2).

SEICHES

A *seiche* (pronounced sāsh) is a form of periodic current system, described as a standing wave, in which some stratum of the water in a basin oscillates about one or more nodes. This rocking motion can be seen in a bowl of thin soup passed by a shaky waiter, the seiche being set up by movement of the basin. By blowing upon the soup near one side of the bowl one can create a seiche due to pressure changes, although the basin be fixed. Seiches in lakes have been recognized for many years, and since Forel's pioneering work, beginning about 1870, much attention has been given to this intriguing phenomenon. The name, attributed to de Duillier, writing in 1730, is derived from the Latin *siccus* through the French *séche* meaning dry, and referring to the exposure of low shorelines left dry as the lake water recedes during an oscillation.

We have seen previously that sustained blowing of wind across the surface of a lake will cause water to pile up in the down-wind region of the basin. If the wind suddenly subsides a current will naturally flow toward the area of lowered surface. The water mass does not come immediately to equilibrium. As a result of the energy of motion imparted by the gradient current, an oscillation occurs about a stationary node. This oscillation, a *surface seiche*, continues until damped by contact with the basin proper or by meteorological forces.

The origin of surface seiches may be found in a number of natural mechanisms. From the considerable study of lake oscillations, however, it appears that surface depressions resulting from sustained winds, and from local sudden changes in atmospheric pressure, are the primary agents. There are cases in which local rain showers set up a seiche due to the pressure of the impact of the falling rain. Sudden inflow from a contributing stream has been suggested as the origin of a seiche in Loch Earn, Scotland. Earthquakes have been named also as possible causes of seiches.

Since these standing waves operate according to the laws of oscillating systems, the formation of harmonics is a general characteristic. In addition to the simple, uninodal seiche, binodal and trinodal seiches have been reported commonly. In fact there is at least one record in Loch Earn of a seiche exhibiting a nodality of the sixteenth. Normally, the periods and locations of the nodes are determined by morphological features of the basin, i.e., depth, diameter, and form. The amplitude of the seiche depends upon the source and the intensity of energy giving rise to the oscillation and to the form of the basin.

Although seiches occur in all enclosed or partially enclosed bodies of water, in small basins the effect is detectable only by sensitive recording devices. In larger lakes the oscillations may be quite conspicuous. Because of the narrow form of Lake Geneva near the city of Geneva, water-level fluctuations due to seiches often leave the shoreline dry. Here, in 1841, a seiche with an amplitude of 1.87 m was recorded. In the same lake in 1891, Forel observed a seiche which lasted for 7 days, 17 hr. This standing wave underwent 150 oscillations, the amplitude dropping from a maximum of 20 to 7 cm; the period of oscillations was of the order of 73 min. The Great Lakes exhibit oscillations; the periods of longitudinal

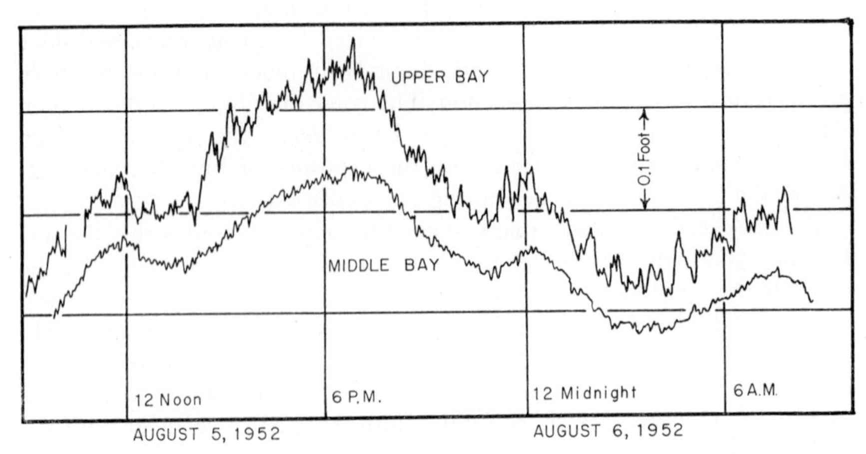

FIGURE 8·3. A Sample of a Seiche Record for Two Stations in South Bay, Manitoulin Island, Ontario, at the North End of Lake Huron. The stations, utilizing water-level recorders, are separated by about seven miles. Note that several oscillations are apparent in the record: one at 9 min, one at about 50 min in the early portion of the record, but diminishing, and one long oscillation at about 6 hr. This seiche persisted only a few days. (From Bryson, R. A. and Stearns, C. R., 1959. "A Mechanism for the Mixing of the Waters of Lake Huron and South Bay, Manitoulin Island," *Limnol. Oceanog.*, Vol. 4, 246-251.)

seiches in Lake Erie are near 790 min, and for Lake Huron 289 min. Transverse seiches are also known and serve to complicate measurement of lake activities.

Amplitude and period of seiches are determined in practice by the use of a float contained in a protective housing. The float is connected to a counterbalanced pointer over an appropriate scale. Termed a *limnometer*, the device may employ a tracing stylus to give a graphic recording or limnograph. A limnograph record is shown in Figure 8·3.

The periods of seiches can be calculated from various, and often complicated, formulas. For a uninodal seiche in a rectangular basin of regular bottom the period t may be calculated from:

$$t = 2L\sqrt{gh}$$

in which L is the length of the basin, h is the water depth, and g is gravity acceleration. Bearing in mind that a seiche is water in motion, it becomes apparent that turbulence will enter into the system as soon as oscillation begins. The effect of turbulence will be to damp out the oscillation as well as to increase the period.

Thus far, our discussion has dealt with the *surface seiche*. Since about the turn of the century, however, it has been known that internal, periodic current systems could be present in lakes under certain conditions. Within a stratified lake of two or more layers of differing densities the strata may oscillate with respect to each other without being apparent at the lake surface. Recent compilations of data relating to *internal seiches* indicate that both the ranges and the periods of such seiches are significantly greater than those of surface seiches. During summer stratification the thermocline may, as a matter of character, oscillate between the lighter layer of the epilimnion and the denser hypolimnion. An oscillation of this nature can be clearly and relatively easily detected through temperature recordings at various depths.

Data derived from laboratory and field studies of displacement of lake layers of equal temperature (isotherms) have led to the general acceptance of two major forms of internal seiches. One type, the thermocline seiche, results from the movement of water masses between the epilimnion and the most dense zone of the metalimnion. Another system, the *hypolimnion seiche*, moves along the boundary between the metalimnion and the hypolimnion. The time of one complete oscillation (period) is usually greater in the case of the hypolimnion seiche than for the thermocline seiche, both, however, being more or less dependent on the length of the lake. The differences in periods are due to unequal densities of the two strata. In Lake Geneva (about 37 km in length), the period of an internal seiche is about four days.

The mechanisms responsible for setting up interval waves are the same as for the surface seiche. Wind-driven waters of the lake surface pile up at the windward end of the basin, causing the deeper waters at the leeward end to compensate by being pushed up. If the lake is stratified, however, the resultant effects of the displacement are different. Because of greater density difference between air and water than between cool and warm water, slight depression of the water surface by air in motion results in a proportionately greater displacement of the lower water stratum toward the surface in the windward end of the lake. As the wind diminishes, the isothermal strata oscillate about a node near the center of the lake. The metalimnion serves as a sort of "cushion layer" on which the epilimnion moves over the hypolimnion. Figure 8·4 shows the oscillations of three isotherms in Cayuga Lake, New York, during summer stratification.

The problem of internal seiches is not a simple one, either in terms of seiche motions and dynamics or definition. All internal oscillations in lakes may not be "true" seiches, that is, motion resulting from free oscillations. Forced oscillations, or those that are sustained by direct wind action, may be present and sometimes mixed with true seiches. In very shallow lakes forced oscillations may damp the effects of seiches, making difficult a distinction between the two.

Internal seiches are of considerable importance in the total economy of the lake, probably more so than surface seiches. In Lake Victoria an internal seiche with a period of about 30 days brings about the displace-

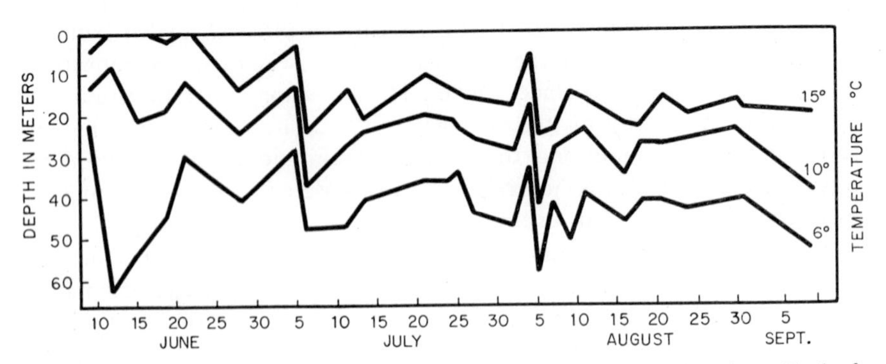

FIGURE 8·4. Oscillations of Three Isotherms at a Point in the Southern End of Cayuga Lake, New York, During Summer, 1951. Observe the development of stratification and the sinking of the isotherms from late June to September. (From Henson, E. B., 1959. "Evidence of Internal Wave Activity in Cayuga Lake, New York," *Limnol. Oceanog.*, Vol. 4, 441-447.)

ment of water into remote parts of the lake basin. The resultant turbulence mixes lake water over the bottom materials and transports dissolved and suspended materials into the shallow zones. In addition to this more or less horizontal transport, internal seiches serve to distribute heat and nutrients vertically within the lake. One particularly interesting problem is created by the action of internal seiches; this is the determination of the position of the thermocline through temperature measurements. Figure 8·5 shows the variations in position of the thermocline when in oscillation. It has become increasingly apparent that the thermocline is not a stable stratum, but rather is subject to much shifting and modification during the season.

True, or lunar tides, have been reported for large lakes, although the amplitude of lake tides would naturally be small. In Lake Superior the range is about 1 in. (3 cm). The difficulty in observing and measuring lake tides arises by seiche complication; a seiche in phase with lunar time (12.5 hr) would magnify the tide range, or if out of phase, would inhibit tidal action.

STREAM CURRENT SYSTEMS

The major processes and properties of flowing waters have been described earlier in Chapter 3 because of the importance of such processes in the formation, maintenance, and alteration of the stream or channel. It is suggested that so much of that chapter as pertains to the dynamics of stream flow be reviewed at this time, for those same processes and attributes contribute to the several generalities to be considered in this section.

We have seen that wind is the major force in creating current systems in lakes. In streams the important factor in producing water movement is gravity. So long as a gradient on the earth's surface exists, water flows in response to gravity, to "seek its lowest level" following the route of least resistance. Because the route is seldom straight, the gradient rarely uniform, and the sides and bottom of the channel not often smooth for any distance, varying velocities and turbulences result. Bearing in mind the importance of these factors and activities in channel structure, let us now consider them with reference to the stream as a living-space.

A conspicuous characteristic of stream flow is transport of dissolved substances, suspended materials, and living plants and animals. This ability

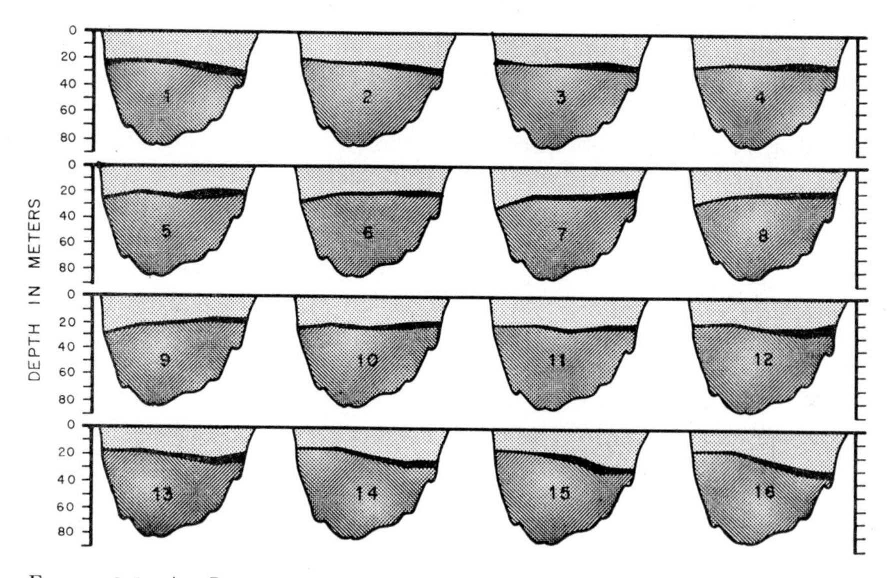

FIGURE 8·5. An Internal Seiche. The Hourly Variations in the Position of the Thermocline (11°C to 9°C) in Loch Earn, Scotland, August 9, 1911. The thermocline is shown in black. (After Mortimer, C. H., 1953. "The Resonant Response of Stratified Lakes to Wind," *Schweiz. Z. Hydrol.*, Vol. 15, 94-151, from data by Wedderburn.)

to move matter also contributes greatly to the unstable nature of bottom types and shore zones, mainly through shifting of bottom materials and through sedimentation. Although these processes and features could result from nonturbulent flow, they are usually of greater magnitude in turbulent waters.

Turbulence in streams is not uniform throughout a cross-sectional area of the stream. Maximum turbulence occurs at some region in the cross section and diminishes from that zone toward the surface or the bottom. Zones of maximum turbulence are usually located on either side of a similarly restricted *thread of maximum velocity*. The number of high-velocity threads and turbulent zones apparently increases with channel width, depth, and nature of the bottom. The result is decrease in stream stability, making for a more rigorous environment for plants and animals. Eventually, however, uniform discharge may contribute toward a balance between the bottom materials and turbulence and the stream bed becomes more tenable for organisms.

Maximum velocity threads and associated turbulent zones do not maintain a constant position in the stream cross section, except perhaps in straight streams occupying uniform channel shape. Most streams curve and bend throughout their courses, and the velocity threads follow courses that curve even more than do the streams. As a stream flows around a curve, the maximum velocity thread is found on the outside of the curve; if the stream bends in the opposite direction with the next curve, the thread moves across the channel between bends. The actions of the maximum velocity thread, coupled with those of the accompanying turbulent zones, are major processes in the erosion of the outside stream bank and deposition of sediments on the inside of the curve.

The presence of one or more maximum velocity threads gives rise to varying horizontal and vertical velocities within the stream proper. Such are in evidence as we view the body of a stream and observe the morphology of the stream bed. We are impressed by the presence of a large sand bar separated by a deep channel from, perhaps, a gravel bed; farther downstream the bottom may be strewn with large rocks or be quite muddy. These features attest to the ever-changing nature of the stream resulting from shifting of the velocity threads and the turbulent zones, and further, to the carrying capacity of the current itself as a function of velocity. A relationship between velocity and the composition of the stream bed is shown in Table 8·1. Obviously, the relationships are not nearly so precise as indicated here. The velocity varies with depth and with obstructions such as boulders. In other words, an organism attached to the bottom or to the downstream side of a large stone may not actually be inhabiting a current velocity indicated by the general nature of the stream bed.

Swirls and eddies of widely ranging proportions constitute another form of current systems in streams. These systems serve to mix stream waters and to effect deposition of organisms, organic debris, and sediments. Many forms of algae are adapted to eddy dispersal of vegetative structures by taking advantage of shallow, well-lighted swirls for rapid reproduction and subsequent distribution as the eddy contributes to the main current.

Density currents due to differences in temperature, chemical composition, or to silt load occur in streams, especially large streams with smooth

TABLE 8·1. THE NATURE OF A RIVER BED AS A FUNCTION OF CURRENT VELOCITY

(Data from Tansley, A. G., 1939. "The British Islands and Their Vegetation," Cambridge University Press, Cambridge.)

Velocity (per sec) (m)	Nature of Bed	Habitat Description
>4 ft (1.21)	rock	torrential
>3 ft (0.91)	heavy shingle	torrential
>2 ft (0.60)	light shingle	non-silted
>1 ft (0.30)	gravel	partly-silted
>8 in. (0.20)	sand	partly-silted
>5 in. (0.12)	silt	silted
<5 in. (0.12)	mud	pond-like

bottoms. Spring-fed tributaries are often cooler than the main stream, and as the colder waters meet with the main stream a stratification results. In the lower reaches of a stream or, more properly, in the estuary, salinity currents move with the tides underneath the fresh water. Near the mouth of the Escambia River, in Pensacola Bay, Florida, salinity measurements made in the autumn indicated a surface salinity of 4.5 parts per thousand while at the bottom (about 4.5 m) the salinity was 24.4 ‰. This sharp salinity stratification was also reflected in the freshwater and marine fish distribution.

Tides exert influence on river flow, often well beyond the brackish-water zone. This effect serves to increase turbulence in the lower stream course, and also to bring about slight periodic flooding of the valley. This latter process increases the exchange between stream and land. The Paumunkey River flows on the Piedmont plateau and coastal plain of Virginia, and salinity effects are felt some 12.8 km from the coast. Tidal influences, however, reach about 64 km upstream.

From the consideration of fluvial processes in Chapter 3, and the brief review and synthesis here, it is immediately apparent that current systems in the various stream regions as well as in the stream as a whole are extremely complex. Actually, these physical aspects of stream dynamics have received little extensive study in the field under natural conditions. A greater amount of study has been carried on in hydraulics laboratories, and considerable theoretical work has been accomplished. For further information, textbooks in hydraulics, and professional publications such as those of the United States Geological Survey, should be consulted.

CURRENTS IN ESTUARIES

The three most important factors operating to produce currents in estuaries are oceanic tides, stream flow, and wind. The interactions of these forces, particularly the somewhat antagonistic processes of oscillating tides (their vertical ranges and lengthwise flow) and unidirectional stream flow (its velocity and volume), serve to make the estuary a restless and complex system of water movements. Additionally, the morphology of the basin of the estuary and the channel of the stream modify and determine the stream and tidal dynamics. It should be borne in mind that these forces are not regular and constant. Stream flow varies seasonally with rainfall (Figure 8·6), while tide height and movement are correlated with lunar effects and wind. Winds may serve to increase either tidal movement or stream flow, depending upon the direction and intensity of the moving air mass.

Within most estuaries there exists a relatively regular and uniform rate of water transport. This is to say that the volume of estuarine water discharged to the sea through the mouth of the estuary essentially compensates for the amount of fresh water introduced by the stream into the upper region of the estuary. In an estuary where the incoming fresh water is mixed with the estuarine water, i.e., where no stratification of the two exists, considerable time may be required for the discharge of a particular mass of the fresh water. This time interval is called the *flushing time*. It is an average figure used to describe the period during which a quantity of fresh water derived from stream or seepage remains in the estuary. A simplified formula for determining flushing time is:

$$t_F = \left(\frac{S_s - S_f}{S_s} V_f\right)\frac{1}{v_D}$$

in which S_s and S_f represent the salinity of sea water at the entrance to the estuary and the salinity of the estuarine water, respectively; V_f is the volume of water in the estuary; v_D is the daily volume movement into and out of the estuary. Flushing time in Great South Bay, Long Island, through Fire Island inlet is of the order of 48 days, or approximately 96 tidal cycles. Obviously a parcel of fresh water entering the estuary near

FIGURE 8·6. In Watersheds in Rocky Regions a Brief Rainfall May Greatly Increase the Discharge of Streams. The upper photograph shows Stony Brook, near Princeton, New Jersey, at low-water stage. The lower photograph, of the same stream, was taken following a short period of rainfall. The leaves in the bush on the left indicate that the water had previously reached an even higher level than that shown. (Photographs by W. J. Woods.)

the mouth will move out faster than one in the upper reaches. The main point is that, on the average, the linear movement of dissolved substances and suspended materials, including living plankton, is a slow process. In many instances the time is sufficient for biological events, important in maintaining the integrity of the estuarine community, to occur.

In a "mixing estuary" the stream flow is held back and often reversed by a strong flood tide. Because of the forces of the two opposing current systems, the reversed flow up the stream is normally slow, ceasing completely as the peak of the flood is reached. Upon the ebb, the outward tidal flow is strengthened by stream flow, the velocity decreasing after the mid-time of the ebb and the movement coming to rest at dead low water. As we have seen for streams, the cross-sectional velocity of an estuary is not constant; a thread of maximum velocity exists. In estuaries, this thread is usually near the surface in the middle of the water course, the precise position depending, however, upon the nature of the shores and bottom.

In many estuaries the typical current pattern is one wherein the lighter, fresh water flows seaward over the upstream movement of denser saline waters (denser by virtue of having a greater concentration of dissolved salts). Under these conditions, a *vertical salinity gradient* exists, and the estuary is said to show stratification. Studies on Chesapeake Bay have indicated that there is a net horizontal seaward flow in the upper layer, and in the lower layer a net horizontal headward flow. The boundary between the two is not level, but slopes toward the right (looking downstream). In the Tamar Estuary of England, it has been found that at some distance upstream both currents may be seen at the surface. During times of considerable fresh-water flow, the more saline waters of the incoming tide move to one side of the basin.

Where stratification and net seaward-headward movements exist, it is possible for certain nutrients and organisms to be transported upstream in the bottom currents, while other materials may be carried to the sea in the upper currents of less saline water. It has been suggested that oyster larvae are moved upstream by lying on the bottom during an ebb tide and then rising into each successive flood tide to be more or less passively transported for the duration of the tide. Figure 8·7 depicts the flow pattern in a typical stratified estuary. Further relationships between currents and salinity distribution will be considered in Chapter 10.

In the preceding paragraphs we have seen some of the effects of oceanic tides on the dynamics of the estuary of a stream. In a great number of instances tidal effects are not restricted to the brackish zone, but occur throughout a considerable length of the stream-estuary-sea continuum. Indeed, in the low-lying course of the Amazon, tides serve to bring about measurable variations in stream depth some 960 km above the salt-water zone. The result is the entrance of tides into the stream before the preceding tide has ebbed. In the Amazon there are said to be eight tides along the stream course at one time.

The morphology of the stream channel and estuary basin regulates, to a great extent, the velocity and magnitude of tidal currents. In a wide, unrestricted estuary, currents are typically slow, without significant turbulence. In a narrow, deep basin, rapid, turbulent currents generally occur. If the mouth of the estuary is restricted by depositional features or land closures, the incoming tide may be held back until it suddenly breaks forth into the basin as a tidal wave, or *bore*. In the Severn Estuary,

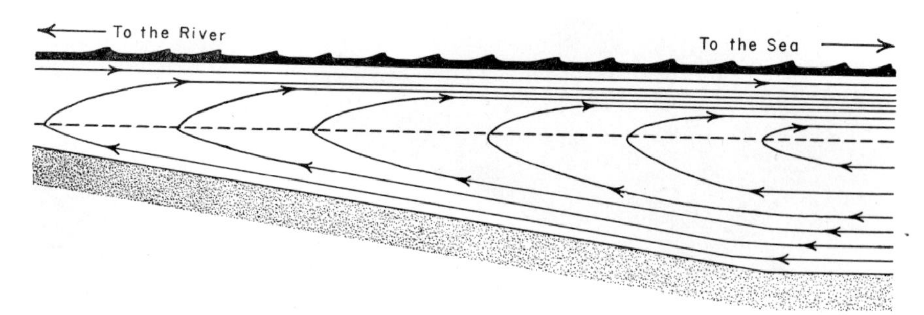

FIGURE 8·7. Schematic Presentation of Current Flow Pattern Along the Central Axis of an Estuary. (After Pritchard, D., 1952. "Salinity Distribution and Circulation in the Chesapeake Bay Estuarine System," *J. Marine Research*, Vol. 11, 106-123.)

England (shown in Figure 8·8), the tidal range may approach 15 m. A bore of 3 m (sometimes reaching 7.5 m) has been reported in the Tsientang River of China. Such currents exert profound effects on the nature of the substrate, turbidity, and biota of the estuary.

Seiches, similar to those described in lakes, may occur in coastal bays and estuaries. The period of these long stationary current systems is determined by the depth and horizontal parameters of the basin. Oscillations are apparently common in many coastal waters, the effects often obscured, however, by tides. A trinodal oscillation has been recognized in San Francisco Bay, and the great water-level fluctuations of the Bay of Fundy result from the synchronization of tide and seiche, the latter having an amplitude of some 15 m. The causes of seiches in bays and estuaries are not completely known. It is probable that forces such as variations in atmospheric pressure and wind can supply sufficient energy for the development of standing waves in coastal bodies. As in lakes, friction operates to damp out the oscillation.

The importance of waves and wave action in the estuary is primarily a function of the morphology of the estuary basin. Wherever the mouth of the estuary is restricted and surface area of the body of water slight, there is little opportunity for the development of waves of any appreciable magnitude. Consequently, the role of wave action in water-mixing and erosion is negligible. On the other hand, broad estuaries having wide mouths are subject to receiving the full effect of oceanic waves in addition to local disturbances. Generally, however, the work of moderate

FIGURE 8·8. The Severn Bore at Gloucester, England. This is a rear view of the wave of translation brought up by high spring tides at the estuary mouth. The bore is usually reduced by the time it reaches Gloucester. (Photograph courtesy of British Information Service.)

waves is that of erosion of the tidal marshes, resulting in the introduction of nutrient materials into the estuary and modification of the shoreline.

Winds are also influential agents in producing certain currents in estuaries. We have already considered the possible role of wind in the development of seiches. Wind contributes greatly to the unusually high waters moved into coastal features during hurricanes and other storms. Depending upon duration and intensity, wind storms may temporarily disrupt the normal circulation patterns in estuaries. Similarly, as the wind is directed, the rate of ebb and flood of tides may be influenced.

Chapter 9

GASES IN LAKE, STREAM, AND ESTUARINE WATERS

Of the many extraordinary properties of water, probably none contribute more to maintaining life in aquatic communities than the capacity of water to hold substances in solution and the ability to enter into chemical reactions. These characteristics result largely from the rather weak nature of the hydrogen bonds linking water molecules. Recall from Chapter 5 that a relatively small amount of energy is required to separate or to unite the molecules. Therefore, natural waters, including rain, always contain some quantities of dissolved gases and inorganic and organic substances. To repeat an oft-stated truism, "pure water does not occur in nature." On this fact rests the physical-chemical-biological structure of natural bodies of water that maintains ecosystem continuity as well as evolutionary mechanisms at the organismal level.

The naturally occurring substances that contribute to the "impurity" of waters are derived from various sources, are present in varying states and quantities, undergo certain transformations, and contribute differently to the over-all metabolism of the water community. Certain gases in proper proportion are essential to respiration and photosynthesis in aquatic situations; other gases are lethal to life. Certain dissolved mineral salts serve as nutrients for free-floating plants; other salts may limit life through osmotic effects. Dissolved organic materials are also present in natural waters, but their reactions are not well known. The nature, quantity, and transformation of substances dissolved in natural bodies of water are generally indicative of the origin of the basin or channel, the climatic regime of the area, and the composition of the substrate in the drainage system.

In this and the following chapter we propose to consider certain properties and reactions of some of the more conspicuous substances contributing to the chemical nature of natural waters as an environment. Although our approach is that of naming the materials and considering them in turn, it is most important to keep in mind the fact that the lake or stream is a dynamic system of chemical interdependencies and interrelationships in association with physical and biological features.

DISSOLVED GASES IN LAKES

Oxygen

Of all the chemical substances in natural waters, oxygen is one of the most significant. It is significant both as a regulator of metabolic processes of community and organism, and as an indicator of lake conditions. Hutchinson (1957) has succinctly and aptly stated the case for oxygen in saying: "A skillful limnologist can probably learn more about the nature of a lake from a series of oxygen determinations than from any other kind of chemical data. If these oxygen determinations are accompanied by observations on Secchi disk transparency, lake color, and some morphometric data, a very great deal is known about the lake."

The oxygen available for metabolic relationships in natural waters is the oxygen held in simple solution. This is not, as some beginning students are given to think, the O in H_2O. The volume of oxygen dissolved in water at any given time is dependent upon (1) the temperature of the water, (2) the partial pressure of the gas in the atmosphere in contact with the water, and (3) the concentration of dissolved salts (salinity) in the water.

The solubility of oxygen in water is increased by lowering the temperature. For example, the solubility increases about 40 per cent as fresh water cools from 25°C to freezing. This can be illustrated by a simple exercise with the nomogram in Figure 9·1. Place a ruler or other straightedge across the figure to connect a chosen temperature on the uppermost scale with a given per cent saturation on the middle, inclined scale, say 10°C and 100 per cent saturation. Now read the lowermost scale of oxygen concentration in cc/l at the point where the ruler intersects the scale. Interpreted in one fashion we see that 7.7 cc/l of oxygen constitute 100 per cent saturation at 10°C. Now connect 5° C and 100 per cent saturation and note that under these conditions a greater concentration of oxygen is dissolved at 100 per cent saturation, 8.7 cc/l to be exact. In other words, solubility increases with decrease in temperature. The original purpose of the nomogram is to determine per cent saturation of a given oxygen concentration at various temperatures and altitudes. Altitude must be considered in order to take pressure into account.

At a given temperature the concentration of a saturated solution of a slightly soluble gas that does not unite chemically with the solvent is very nearly directly proportional to the partial pressure of that gas (Henry's Law). The solubility of oxygen also relates to Dalton's "law of partial pressures" which states that the total pressure of a mixture of gases is equal to the sum of the pressures exerted by each of the component gases. In other words, the solubility of each gas is independent of other gases in the mixture. Under similar conditions of pressure and temperature the solubility of oxygen in water is over twice that of nitrogen and about one third that of carbon dioxide.

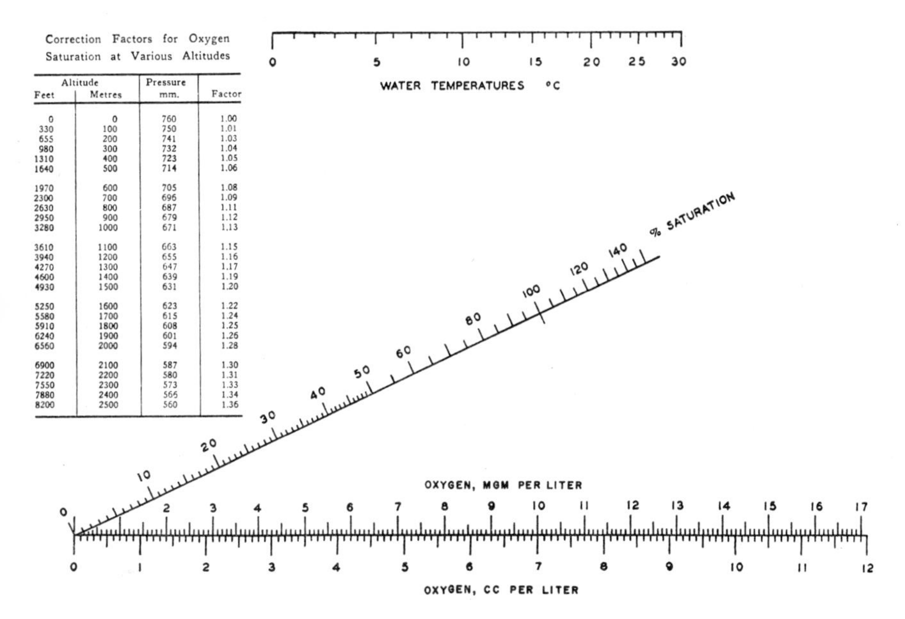

Correction Factors for Oxygen Saturation at Various Altitudes

Altitude Feet	Altitude Metres	Pressure mm.	Factor
0	0	760	1.00
330	100	750	1.01
655	200	741	1.03
980	300	732	1.04
1310	400	723	1.05
1640	500	714	1.06
1970	600	705	1.08
2300	700	696	1.09
2630	800	687	1.11
2950	900	679	1.12
3280	1000	671	1.13
3610	1100	663	1.15
3940	1200	655	1.16
4270	1300	647	1.17
4600	1400	639	1.19
4930	1500	631	1.20
5250	1600	623	1.22
5580	1700	615	1.24
5910	1800	608	1.25
6240	1900	601	1.26
6560	2000	594	1.28
6900	2100	587	1.30
7220	2200	580	1.31
7550	2300	573	1.33
7880	2400	566	1.34
8200	2500	560	1.36

FIGURE 9·1. A Nomogram for Determining Oxygen Saturation Values at Various Temperatures and Altitudes. In practice, a straightedge is used to connect observed temperature and dissolved oxygen concentration. The point of intercept on the inclined scale gives the per cent of saturation. Correction for altitude is made by applying the factors given in the table in the upper left. (From Rawson, D. S., 1944. "The Calculation of Oxygen Saturation Values and Their Correction for Altitude," *Limnol. Soc. Am. Spec. Publ.*, No. 15.)

The effect of the third factor, the concentration of dissolved salts, upon the solubility of oxygen is that of decreasing the oxygen concentration as salinity increases. At 0°C fresh water at saturation contains slightly over 2 cc/l more oxygen than does average sea water (35 ‰ salinity); at 15°C the difference is about 1.5 cc/l.

Having given attention to the major factors responsible in regulating the solution of oxygen in waters, let us now consider the sources of this gas. Obviously the atmosphere in contact with the lake surface is an unlimited source of oxygen. The volume per cent of oxygen in the atmosphere is calculated to be 20.99, or approximately 210 cc of oxygen per liter of air. This is some 25 times the concentration of oxygen in the same volume of fresh water.

The rate at which atmospheric oxygen passes across the air-water interface and becomes dissolved in the water is dependent upon a number of factors. For one thing, increased wave action or other disturbances at the lake surface results in greater passage of the gas into solution. Secondly, the greater the difference in partial pressure between air and water, the greater the rate of solution. Thirdly, the less the moisture content of the gas, the more rapid the solution of that gas. Bear in mind, however, that in all of these processes there may be a "two-way" movement, that gas can be lost from the lake to the atmosphere. The *direction* of movement, as well as the *rate*, is determined by the foregoing factors.

Oxygen in natural waters may also be derived from photosynthetic activity. In shallow, nonstratified ponds lacking significant wave action, oxygen may be derived mainly as a by-product of carbohydrate synthesis by rooted plants and by phytoplankton, viz.:

$$6CO_2 + 12H_2O \rightharpoonup C_6H_{12}O_6 + 6O_2 + 6H_2O$$

or mathematically reduced:

$$6CO_2 + 6H_2O \rightharpoonup C_6H_{12}O_6 + 6O_2$$

Photosynthesis is also a source of dissolved oxygen in large deep lakes, but, in contrast to ponds, this process is restricted to a certain region of the lake. This zone is delimited by the vertical range of transmission of light effective in photosynthesis. This lighted region is called the *euphotic zone*. It extends horizontally from shore to shore and vertically from the surface to a level beyond which photosynthesis-effective light fails to penetrate. As we have seen, turbidity, color, and the absorptive effect of water itself, serve to quench light; thus they essentially determine the euphotic zone. In the shallow shore region (the *littoral*), submerged rooted plants along with phytoplankton contribute oxygen to the lake. In the euphotic zone of the open lake (the *limnetic*), phytoplankton contributes the autochthonous oxygen. In most temperate lakes during summer thermal stratification the euphotic zone corresponds closely with the epilimnion. This accounts mainly for the low-oxygen conditions in the hypolimnion as briefly considered previously. Within the near-surface region of maximum photosynthesis and oxygen gain from the atmosphere, the water often becomes supersaturated with oxygen at the height of diurnal production cycles. Maximum oxygen production usually occurs in the afternoon on clear days, the minimum immediately after dawn. Cyclic daily fluctuations in oxygen concentration, called the *oxygen pulse*, have been observed when conditions are proper. Figure 9·2 illustrates hourly changes in dissolved oxygen concentration along with fluctuations of other physical and chemical features of a lake.

The lower limit of the euphotic zone is marked by a level at which organic respiration and decomposition consume oxygen at a rate equal to that at which oxygen is produced over a 24-hr period. This is the *compensation level;* it will enter significantly into our discussion of productivity in Chapter 14. Below the compensation level in our typical lake lies the *aphotic zone*, a zone in which there is insufficient light to maintain

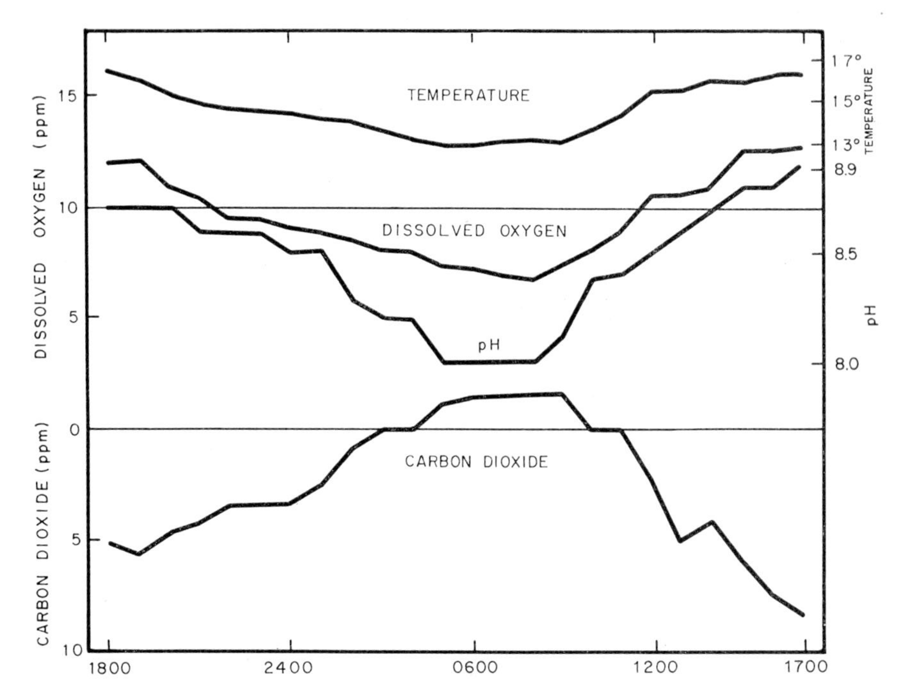

FIGURE 9·2. Diel Fluctuations in Temperature, Dissolved Oxygen Concentration, pH, and Carbon Dioxide Content of Surface Waters of Buckeye Lake, Ohio, in Summer. Negative carbon dioxide values represent the amount of the gas required to make water neutral to phenolphthalein. (After Tressler, W. L., Tiffany, L. H., and Spencer, W. P., 1940. "Limnological Studies of Buckeye Lake, Ohio," *Ohio J. Sci.*, Vol. 40, 261-290.)

oxygen production at compensation. This is the region of low oxygen content, often depleted. It frequently corresponds with the hypolimnion.

A vertical oxygen distribution pattern such as we have been considering is characteristic of moderately productive lakes of small size. These lakes, termed *eutrophic* ("rich food"), typically exhibit a hypolimnetic loss of oxygen, the curve of the oxygen distribution dropping sharply through the metalimnion and described as *clinograde.* Figure 9·3A represents a clinograde oxygen distribution in a eutrophic lake. It seems generally agreed that temperature and oxidation of organic materials are the major factors in creating the oxygen deficiency in the hypolimnion. The rate and magnitude of this oxidative breakdown are dependent upon the volume of organic substance supplied to the process. This substance

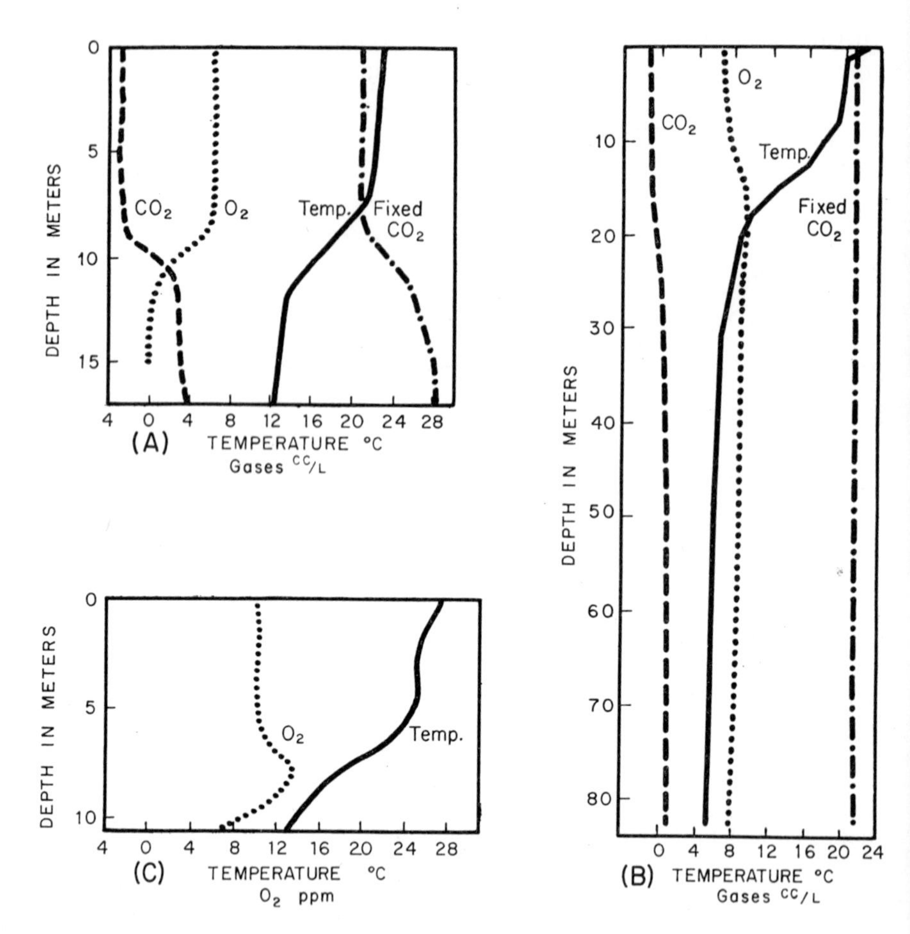

FIGURE 9·3. Various Forms of Oxygen Curves in Lakes. (A) clinograde curve of vertical oxygen distribution in Conesus Lake, New York, August,, 1910; (B) orthograde curve for Skaneateles Lake, New York, August, 1910. (From Birge, E. A. and Juday, C., 1914. "A Limnological Study of the Finger Lakes of New York," *Bull. U. S. Bur. Fish.*, Vol. 32, 525-609.) (C) orthograde curve with decrease in oxygen near the bottom in a New Jersey quarry lake.

for the most part is produced in the upper lighted zone, which, in relation the organic synthesis, is called the *trophogenic region*. From here the materials settle into the zone of decomposition, or *tropholytic zone*. It is now apparent that correlations exist among epilimnetic circulation, light transmission in the euphotic zone, and the production of organic substance in the corresponding trophogenic region.

At somewhat the other extreme of oxygen distribution and biological processes we find relatively deep, unproductive, or *oligotrophic* ("scant food") lakes. Such lakes are usually rather clear for considerable depths and turbidity is low. Since light transmission is high, the euphotic zone is deep and photosynthesis may occur, although reduced, at some distance below the upper strata. As a result of low productivity per unit volume there is little oxidation in the hypolimnion. The oxygen curve

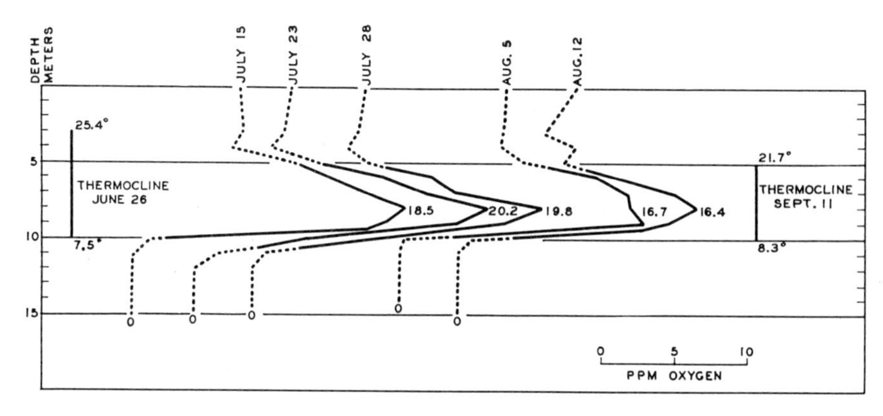

FIGURE 9·4. Metalimnetic (Thermoclinal) Oxygen Maximum in Myers Lake, Indiana, in Summer, 1952. (After Frey, D. J., 1955. "Distributional Ecology of the Cisco (*Coregonus artedii*) in Indiana," *Invest. Indiana Lakes and Streams*, Vol. 4, 177-208.)

may show a decided increase through the metalimnion and a high oxygen content in the hypolimnion. An orthograde oxygen curve in an oligotrophic lake is shown in Figure 9·3B. Between the two extremes in oxygen distribution, production, and uptake described here exists a vast range of intermediate and intergrading patterns, many of them exhibiting various complications in oxygen states and metabolism.

In certain clear lakes during summer stratification very high oxygen concentrations may occur in the metalimnion (Figure 9·4). This great concentration of the gas is apparently derived from levels in and below the epilimnion in which the transparency of water permits a high rate of photosynthesis. The *metalimnetic oxygen maximum* may be followed by a corresponding marked reduction of oxygen in the metalimnion. The mechanism, or mechanisms, responsible for this *oxygen minimum* cannot be described with certainty. It has been suggested that rapid oxidation of slowly settling materials from the epilimnion might account for the low metalimnetic oxygen concentration. Such a process would necessitate that the position of the metalimnion be below the trophogenic zone or at least that photosynthesis proceed at a reduced rate.

The annual cycle of oxygen distribution in a dimictic lake generally follows the pattern shown in Figure 7·2 (Chapter 7). Note that underneath the ice in March the concentration of dissolved oxygen is 14 ppm. This saturation was acquired during autumn circulation accompanied by falling temperatures, the latter, of course, increasing the solubility of the gas. In winter some of the oxygen is consumed by organismal respiration and by organic decomposition. In deep lakes oxygen consumption is relatively small, but in shallow bodies of water the oxygen may become depleted, thereby resulting in mortality of animals. Vernal circulation redistributes the oxygen as shown for late April and early May. With the development of summer stagnation the hypolimnion becomes quite poor, often completely lacking in oxygen.

In meromictic lakes the oxygen distribution reflects the essential two-region structure of the water mass. Depending upon the circulation pattern above the chemocline, the oxygen content may take the form of a clinograde curve. The waters of the monimolimnion are usually anaerobic. Lakes of the tropical and subtropical regions exhibit widely diverse oxygen relationships. Polymictic lakes, by virtue of continuous circulation, exhibit a generally high, uniformly distributed, oxygen content throughout the year. Oligomictic waters, in which circulation seldom occurs, are typically characterized by a condition of hypolimnetic oxygen depletion. In shallow lakes of the colder regions, for example Alaska, oxygen at near saturation appears to be uniformly distributed by circulation from surface to bottom.

The transport of dissolved oxygen throughout a lake is accomplished primarily by currents set up within the lake and by eddy conduction. Lake currents, as we have previously seen, may be set into operation by wind action on the surface and by density differences between layers in the lake. Circulation in the epilimnion during summer distributes oxygen within that zone. Eddy systems thrown up along the interfaces between thermal layers or along stream density currents moving through the lake serve to move substances across boundaries, usually at right angles to the current. During seasonal overturn periods the sinking of upper, dense water masses also delivers oxygen to the underlying regions, thereby accounting for the homogeneous oxygen content during those times of uniform density. Simple diffusion of oxygen molecules plays a minor role, indeed, in oxygenation of natural waters. The process is very slow. As a matter of fact, it has been shown that to raise the oxygen concentration 0.4 mg/l at a depth of 10 m by diffusion from the surface would require over 600 years.

With respect to the utilization and ultimate fate of dissolved oxygen in lakes, many aspects have already been considered. In summary, we might say that the major processes acting to consume oxygen are animal and plant respiration, and organic decomposition. During long periods of cloudiness or where waters may otherwise be shaded, for example underneath broad floating mats of vegetation, organismal respiration and decomposition may rapidly take up oxygen. Where decay bacteria occur in great quantity, as in lakes or streams highly polluted with organic sewage, the water often becomes completely anaerobic.

Although we have been concerned primarily in this section with the maintenance of oxygen in lake waters, the source of the gas, and its distribution and fate, we have had to venture into the subject of productivity. This is a natural consequence because there is an obvious direct relationship between the dissolved oxygen content of natural waters and the amount and rate of energy fixation. Recall that the form of the curve of vertical oxygen distribution indicates a great deal about photosynthesis and decomposition at various levels within a single lake, as well as providing a point of comparison of several lakes. A fuller account of these important aspects of natural waters will be given in Chapter 14.

Redox Potential

It has been known in chemistry for many years that oxygen plays an important role in chemical changes in various kinds of solutions. Only within the past 30 or so years, however, has investigation of certain oxygen-related phenomena been transferred from laboratory beaker and battery jar to natural waters of lakes and seas. One aspect of these laboratory and field investigations has been concerned with oxidation and reduction processes. This line of research has already been fruitful in contributing to our knowledge of chemical limnology and its influence on the activities and distribution of certain organisms.

The term *oxidation* is applied to the process in which oxygen is added to a substance, or in which hydrogen is lost from a compound, or in which an element loses electrons. Conversely, the loss of oxygen, the addition of hydrogen, or the gain of electrons is termed *reduction*. At the elemental level an electron transfer in which *ferrous* iron is oxidized to *ferric* iron may be stated:

$$Fe^{++} \rightleftharpoons Fe^{+++} + e$$

Under certain conditions, the reduced iron may undergo a change back to the oxidized state by a shift in electrons as suggested in the reversibility of the process shown above.

The extent to which a substance can undergo oxidation-reduction processes is dependent upon the concentration of other oxidizing-reducing systems and their products in the solution. Within a given solution, the proportion of oxidized to reduced components of a particular system in relation to other systems constitutes the oxidation-reduction potential, or *redox potential.* In a solution, for example lake water, a system with a given redox potential undergoes reduction and oxidizes a system of lower redox value. It may also cause a reduction to take place in a system of higher redox potential.

In practice, the redox potential is determined by immersing a non-reactive electrode (bright platinum is often used) in a solution. The second electrode, necessary to complete the circuit, is usually the hydrogen electrode. The presence of the electrode sets up an electron flow, the direction of which depends upon the proportion of oxidized to reduced material. Where an excess of the reduced state (Fe^{++}) exists, electrons flow to the electrode and oxidation in the system results. In a solution containing a surplus of the oxidized state (Fe^{+++}), flow from the electrode contributes electrons and reduction occurs. The excess or deficit of electrons (relative) can be measured in volts on a potentiometer as an electromotive force about the electrode. This measure gives the intensity of the force (E_h), either positive or negative, and is the redox potential. A positive E_h reading results from a state tending toward oxidation; a negative E_h indicates a system causing reduction. In addition to *intensity*, the oxidation-reduction dynamics also has the attribute of *capacity*. The capacity of an oxidation-reduction system refers to the ability of a system to undergo a certain amount of oxidation-reduction transformation without an intensity change.

Oxygen in natural waters produces a redox potential which is influenced considerably by temperature and the hydrogen ion concentration (pH). Out of consideration for pH effects, the potential is measured at the prevailing pH and then referred to pH 7, this measurement being called the E_7. Correlations between ranges of E_h at pH 7 and oxidation-reduction reactions in certain systems have been established as follows:

NO_3^- to NO_2^- : 0.45 to 0.40 v
NO_2^- to NH_3 : 0.40 to 0.35 v
Fe^{+++} to Fe^{++} : 0.30 to 0.20 v
$SO_4^{=}$ to $S^{=}$: 0.10 to 0.06 v

At a temperature of 25°C and pH 7, well-aerated lake waters exhibit a redox potential of about 0.5 v. This potential remains relatively steady as long as the oxygen content is above approximately 1 mg/l. In other words, the redox potential, as such, is affected but little by the oxygen concentration.

Generally speaking, the curve of vertical distribution of the E_h in lakes follows that of dissolved oxygen and may be a "mirror image" of the ferrous iron. In stratified lakes exhibiting a somewhat orthograde curve of oxygen, the redox potential gives essentially the same pattern. In small lakes in which the oxygen in the hypolimnion is quite low, therefore giving a clinograde oxygen curve, the E_h usually, but not always, appears in a clinograde distribution. Gradations between these forms are common.

Oxidation-reduction activities at the water-mud interface of the lake bottom bear markedly upon the lake chemistry, particularly in the deep

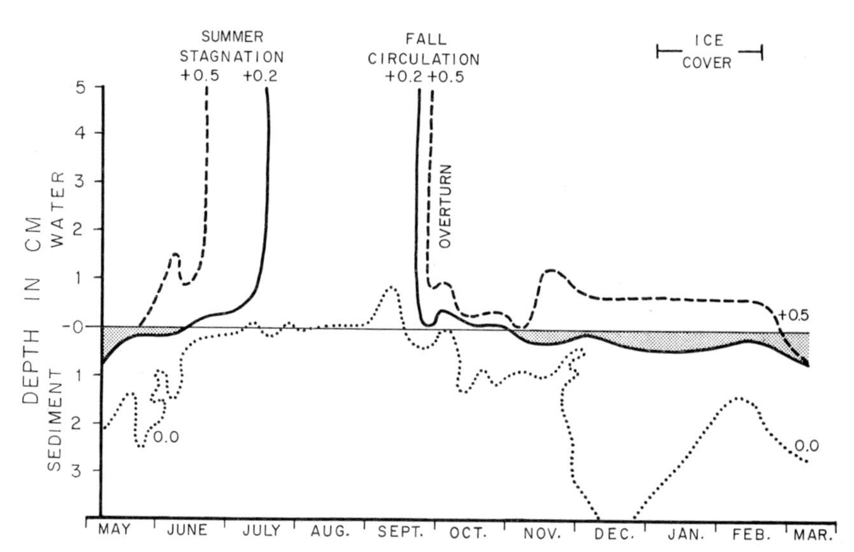

FIGURE 9·5. Seasonal Variation in Redox Potential at Mud-Water Interface at a Depth of 14 m in Esthwaite Water, England. The heavy line represents the isovolt line of E_h of +0.2 v. The stippled area indicates the approximate depth of the oxidized brown mud; the remainder of the sediment is reduced black mud. Depth above and below mud-water interface is given on the vertical axis; time is on the horizontal axis. (After Mortimer, C. H., 1942. "The Exchange of Dissolved Substances Between Mud and Water in Lakes," *J. Ecol.*, Vol. 30, 147-201.)

hypolimnion, and upon the type of organisms present in the shallow part of the sediments. The E_h of mud exposed to oxygenated water varies near 0.5 v. This potential may extend for several millimeters through an oxidized zone of brownish mud at the surface of the bottom sediments, the *oxidized microzone*. Accompanying the decrease in hypolimnetic oxygen with summer stagnation is a diminution in the depth of the oxidized microzone. As the E_h of the interface approaches 0.2 v, the oxidized microzone may disappear (Figure 9·5). Below the microzone, the sediments of deep-water lakes are usually highly reducing in nature, the E_h being near zero volt. An oxidized microzone may be persistent in lakes of size and depth sufficient to exhibit an orthograde oxygen distribution.

The processes responsible for maintaining the integrity of the microzone are not completely understood. It was early suggested that molecular diffusion of oxygen through the mud depended upon the reducing state of the sediment. Turbulence may also be important.

We are agreed that low oxidation-reduction potentials suggest the presence of reducing substances which would, in all probability, utilize such free oxygen as might be brought into the solution. For organisms such as anaerobic bacteria, or other organisms not requiring free oxygen for respiration, a low E_h would pose no problem. Anaerobic bacteria exist where the E_h lies below -0.4 v. Oxygen-dependent plants and animals would be restricted from such a zone, although they may be found inhabiting places where the E_h is as low as -0.2 v. Zonation of certain insect larvae in lakes has been correlated with E_h. The midge *Calopsectra* (*Tanytarsus*) sp. dominated the insect fauna in bottom muds over which the E_h of the water was 0.4 or above. Another midge, *Tendipes* (*Chironomus*) sp., characterized muds in waters in which the E_h was below 0.3 v.

Carbon Dioxide

The very great importance of carbon dioxide as a contributor to the fitness of natural waters as environment derives from essentially three factors. In the first place it serves in a more or less purely chemical sense to "buffer" the environment against rapid shifts in acidity-alkalinity states. In this sense carbon dioxide ameliorates the chemical environment through the ability of the gas to combine with water to form an acid, and to react to give a neutral salt, or a base. We shall give attention to these reactions later. A second contribution of importance by carbon dioxide pertains to regulating biological processes in aquatic communities. Seed germination of some plants, as well as plant growth, is determined by the concentration of carbon dioxide. Various animal processes such as respiration and oxygen transport in blood are related to carbon dioxide. A third and most important contribution by carbon dioxide lies in the fact that it contains carbon. Carbon is one of the most versatile of all elements, due to the possession of four electrons in the outer ring which give great bonding capacity to form a fabulous number of compounds, many of exceeding complexity. The ability to form many compounds is due largely to the asymmetrical nature of the bonds and to the fact that carbon can form chains of atoms almost without limit. This latter characteristic distinguishes carbon from other elements and from inorganic substances, although many inorganic compounds, of course, also contain carbon. Carbon dioxide and water supply the carbon, hydrogen, and oxygen which are major components of protoplasm.

The numerous and varied activities of carbon dioxide in the aquatic ecosystem are made possible primarily by the very high solubility of the gas in natural waters. The solubility of carbon dioxide varies inversely with temperature; at temperatures common in nature, carbon dioxide is much more soluble than oxygen in water. At 20°C and atmospheric pressure of 760 mm Hg, water in equilibrium with atmospheric carbon dioxide contains about 0.88 vol of the gas; only 0.031 vol of oxygen are contained in water under similar conditions. Although air contains some 700 times more oxygen than carbon dioxide (by volume), the proportion in water at equilibrium is more nearly equal, about 4 cc carbon dioxide per liter to 6 cc oxygen per liter. If we consider the amount of carbon dioxide "locked up" in various combined forms, the total amount of the compound in water, particularly the oceans, is much greater than in air.

Carbon dioxide in natural waters is derived from a number of sources. Bacterial decomposition of organic matter in the tropholytic zones, and respiration by animals and plants contribute to the store of carbon dioxide. In the case of plants the greater net contribution is at night, when photosynthesis is not occurring. Ground waters flowing or seeping into lakes and streams may carry carbon dioxide, the amount being determined by the extent of decomposition in the topsoil and, as we shall presently see, by the chemical nature of the underlying rocks. Within the body of water, certain chemical reactions between acids and various compounds of carbonates release carbon dioxide. Finally, the atmosphere directly furnishes some carbon dioxide to natural waters, and rain, as it falls through the atmosphere, dissolves some of the gas and delivers it to lakes and other waters.

The importance of rain in supplying carbon dioxide to inland waters lies in reactions wherein carbon dioxide is maintained and transported in forms other than as a gas. As rain percolates through soil containing carbon dioxide of decomposition, some of the gas becomes dissolved in rain water. The reaction between CO_2 and H_2O results in the formation of carbonic acid (H_2CO_3). If this weak acid encounters carbonate-holding rocks, limestone ($CaCO_3$) for example, the latter dissolves as calcium bicarbonate, or $Ca(HCO_3)_2$. The solution of $Ca(HCO_3)_2$ remains stable only in the presence of a certain amount of *free* or *equilibrium, carbon dioxide*. Free carbon dioxide represents the CO_2 in H_2CO_3 plus that in simple solution. The formation of H_2CO_3 and its dissociations are shown in the reactions:

$$CO_2 + H_2O \rightleftharpoons H_2CO_3 \rightleftharpoons H^+ + HCO_3^- \rightleftharpoons H^+ + CO_3^= \qquad (I)$$

Note in the above reactions that carbon dioxide is contained in two states not previously encountered, i.e., as bicarbonate and carbonate radicals, HCO_3^- and $CO_3^=$, respectively. This carbon dioxide is called the *combined carbon dioxide*. Solution of $CaCO_3$ is dependent upon the addition of CO_2 in an amount greater than that of free carbon dioxide; this additional CO_2 is known as *aggressive carbon dioxide*. The major points to be gained from these considerations include (1) the relations between rain and soil in supplying compounds containing carbon dioxide to natural waters, (2) the chemical reactions by which those compounds are formed, and (3) the occurrence of carbon dioxide in its three forms: free (CO_2 in solution plus that in H_2CO_3), half bound (HCO_3), and bound ($CO_3^{=}$). The direction of reactions involving the forms of carbon dioxide and the very occurrence of these in natural waters are reciprocally related to acid-base relationships in the medium. When carbon dioxide dissolves, the reaction and end products depend upon the nature of the solvent, particularly with relation to the hydrogen-ion concentration. Under acid conditions, the combination is as shown in Eq (I) above in which H_2CO_3 is formed followed by dissociation to $H^+ + HCO_3^-$. Under highly basic conditions the reaction is:

$$\text{Base OH} + H_2CO_3 = H_2O + \text{base}^+ + HCO_3^-$$

Chemical Buffering

The maintenance of near-neutral conditions in mineralized waters is due to *buffering* by chemical systems such as the carbon dioxide-bicarbonate-carbonate complex. Other systems may involve magnesium, sodium, or potassium. In other words, as acid conditions arise, the reaction between acid and base from the bound carbonate, for example, brings about an increase in neutral bicarbonate. Continued acidification releases carbon dioxide and carbonic acid from the bicarbonate accompanied by loss of carbon dioxide from the system. This reaction is the basis for certain limnological techniques for determining the so-called *alkalinity* of natural waters. In these tests a quantity of strong acid is added to water in the presence of a proper indicator. The amount of acid necessary to convert any carbonate or bicarbonate present to free CO_2 is a measure of the HCO_3^- and $CO_3^{=}$ in solution. In this sense, alkalinity refers to the quality and quantity of compounds which bring about a shift in the pH of a solution toward the alkaline side of the pH range. Although not always the case, alkalinity usually reflects the activity of calcium carbonate. Therefore, three forms of alkalinity may be recognized: bicarbonate (determined by the use of methyl orange as indicator, the *M.O. alkalinity*), normal carbonate (*phenolphthalein alkalinity*), and hydroxide. See Welch (1948) or American Public Health Association (1955) for descriptions of methods and interpretations of results. At the other end, increased alkalinity of the solution brings about a reaction between the base and carbonic acid to hold the departure from neutrality to a smaller value than would otherwise be the case if the buffer system were not present.

Definition of pH

The pH is the logarithm of the reciprocal of the hydrogen-ion (or more properly, the hydronium-ion) activity. The pH may be expressed mathematically as follows:

$$pH = \log \frac{1}{(H^+)}$$

where (H^+) is the amount of hydrogen ions in a solution in moles per liter. In a liter of pure water there is 0.0000001 of hydrogen mole/ions (and a corresponding quantity of (OH^-)). The pH of pure (neutral) water is, therefore,

$$pH = \log \frac{1}{0.0000001}, \text{ or } 7$$

Increase in the concentration of H^+ ions results in a lower pH value, or conversely, reduction in the H^+ concentration brings about a higher value. From this relationship, a pH scale has been devised. This scale ranges from pH 0, corresponding to a solution with $(H^+) = 1$, through pH 7 (or neutrality) to pH 14, corresponding to a solution with $(H^+) = 10^{-14}$. From pH 0 to pH 7 solutions are *acid*. From pH 7 to pH 14 reactions are *alkaline*. With respect to the buffering system, we note that only free carbon dioxide is of any import in natural water systems below pH 5, that bicarbonate dominates the range from pH 7 to pH 9, and that the carbonate radical is most important in the range above pH 9.5 or 10. These relationships are shown in Figure 9·6.

The pH range of lakes having some degree of flow through the basin is generally from about 6 to 9. In limestone regions the dissolved carbonates may extend the pH range considerably beyond 9. In basins lacking outlets, evaporation may concentrate alkaline substances, resulting in pH readings of over 12. At the other extreme, accumulation of acids such as sulfuric acid in volcanic lakes gives a pH as low as 1.7.

Occurrence of Carbon Dioxide in Lakes

Having given attention to some of the major aspects of the sources of carbon dioxide in natural waters, as well as to the complex and often critically balanced reactions of the forms of the compound, let us now turn our attention to general considerations of the compound in lakes. From what we have seen of the close relationship between the chemical nature of the drainage basin substrate and the chemistry of waters of the basin, we should expect wide regional variation in carbon dioxide content of lakes. Newly formed lakes in regions of weakly soluble rocks typically contain little carbon dioxide in any form. In these situations the low quantity of carbonates in the substrate results in little of the bound or half-bound forms of carbon dioxide. Similarly, the paucity of soluble minerals as nutrients in biological processes inhibits the development of large biotic populations that would contribute carbon dioxide through respiration, and, of course, decomposition is reduced. Lakes of this type are usually slightly acid, the pH ranging near 6. Lakes of higher acidity (pH 4 to 6) are common in regions of lowlands and bogs. In these waters the free carbon dioxide content is usually quite high, ranging to nearly

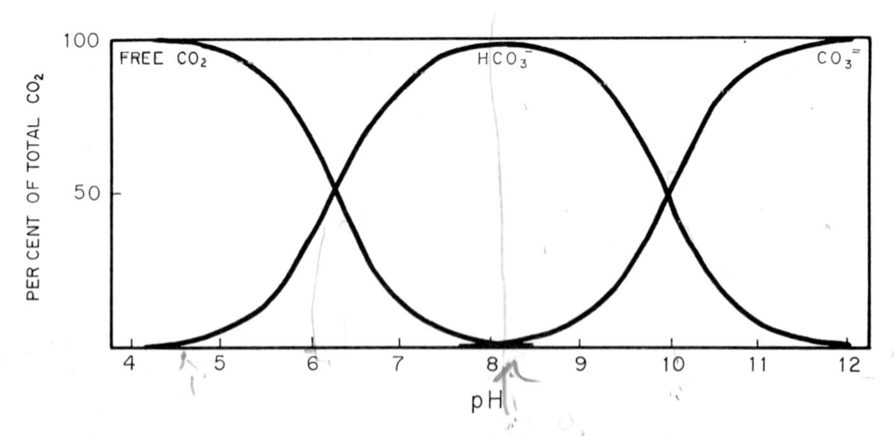

FIGURE 9·6. The Relationship of Hydrogen Ion Concentration to the Percentage of Total Carbon Dioxide in Each of Its Forms in Water. (From Emerson, R. and Green, L., 1938. "Effect of Hydrogen-Ion Concentration on *Chlorella* Photosynthesis," *Plant Physiol.*, Vol. 13, 157-158.)

200 ppm. As would be expected in view of the low pH, the concentration of bound carbon dioxide as carbonate is low, usually less than 9 or 10 ppm. Lakes having these characteristics are termed *soft-water lakes*. A large number of lakes fall into a class in which the pH is circumneutral, and which may be called *medium-water lakes*. Free gaseous carbon dioxide in medium-water lakes varies widely, frequently showing supersaturation relative to the partial pressure of the gas in the atmosphere. These lakes may contain bound carbon dioxide up to 30 or 35 ppm. In regions where the substrate contains easily dissolved minerals, *hard-water lakes* occur. These lakes are characterized by negative values for free carbon dioxide due to withdrawal of bicarbonates at a greater rate than carbonates are precipitated, and by pH values ranging from about 8.5 upwards. Bound carbon dioxide content amounts to over 35 or 40 ppm, often reaching 200 ppm or more. Calcium or magnesium carbonate is often precipitated as *marl*, which will be discussed in Chapter 11. In some highly saline lakes, carbonate concentration may be 8500 ppm or above.

Seasonal Cycle of Carbon Dioxide and pH

During winter in the colder regions ice cover forms over the lake, thereby inhibiting exchange of materials across the air-water interface. Thus the lake becomes essentially "sealed in." Oxidation of organic substances, particularly in the depths, consumes oxygen and increases the carbon dioxide content. This often results in the building up of a gradient of metabolic substances from surface to bottom, especially if there is some transmission of light through the ice permitting photosynthesis in the upper, unfrozen waters. Under such conditions the waters immediately below the ice may contain a considerable quantity of oxygen, sometimes reaching near saturation, and little or no free carbon dioxide. The pH in this zone may show slightly alkaline conditions with low bicarbonate concentration and absence of carbonates. Remember that we should expect to find carbonate only in the absence of free carbon dioxide. In the deep waters, bicarbonates and free carbon dioxide are increased, while oxygen may be nearly or completely depleted; the pH may drop as during summer stratification.

In the warmer regions of the temperate zones, nonfreezing lakes typically contain quantities of carbon dioxide more or less uniformly distributed throughout the waters during winter. This is, of course, due to circulation during the cold season. At this time the phenophthalein alkalinity of surface waters is usually nil, and carbonates are absent.

In holomictic lakes the period of vernal mixing brings about the relatively even distribution of all dissolved materials. During this overturn the pH of the water is uniform from surface to bottom. Free carbon dioxide, together with other gases such as methane and hydrogen sulfide, derived from winter decomposition processes, is lost to the atmosphere. Following the loss of the free form, bound carbon dioxide reappears as carbonate in medium and hard-water lakes as summer stratification begins.

We have already seen that in productive holomictic lakes exhibiting two circulation periods the oxygen concentration diminishes rapidly with depth during summer, giving a clinograde curve. In these lakes the carbon dioxide and carbonate concentration show a general inverse relationship to the oxygen; that is, the concentration of carbon dioxide and bicarbonate increases slightly with depth (Figure 9·3). On the other hand, an orthograde oxygen distribution is usually acompanied by only slight increase, if any, in carbon dioxide. During the summer the epilimnion of medium and hard-water lakes is typically devoid of free carbon dioxide, and contains measurable quantities of carbonates. It has been shown, for example, that in Douglas Lake, Michigan, in July, free carbon dioxide may be absent to a depth of about 14 m, and in this same distance carbonates range from about 10 to 8 ppm. The pH in the upper waters is slightly over 8. At a depth of 20 m, the free carbon dioxide content reaches 11 ppm at a pH of 7.1. As a point of interest, records over a 30-yr period show that the surface alkalinity due to half-bound carbon dioxide varied only from 110 to 128 ppm.

Autumnal circulation brings about uniform physical and chemical conditions throughout the lake. Vertical gradients in density, temperature, dissolved gases, and solids, which became established during summer stagnation, are broken down. The hypolimnetic gases of decomposition are thrown off at the lake surface. The lake's oxygen content is circulated, and with lowering temperatures of autumn an increased supply is dissolved in the water.

Other Lake Gases

Methane is an organic gas (CH_4) widely called "marsh gas," common in many alkaline lakes, ponds, and swamps, particularly during summer stratification. It is produced by bacterial decomposition of organic substances in the tropholytic zone, primarily in the lake bottom. The process involves a multistage breakdown of complex organic material to compounds of simple molecular structure and then decomposition of these, releasing methane and carbon dioxide. It occurs only under anaerobic conditions. In carbohydrates, for example, cellulose is attacked by bacterial enzymes and hydrolyzed to a simple sugar. Anaerobic decomposition of the sugars may result in the formation of hydrogen and methane. The time of highest production appears to be during summer stagnation when the bottom muds have become significantly reducing, that is, when the oxidation potential is low. The process can proceed at relatively low temperatures, about 5°C, because of the tolerance of at least one of the methane-producing bacteria.

In shallow bodies of water such as ponds and swamps, and in the shore zones of lakes, bubbles containing methane and other gases are often seen rising to the surface and erupting. The formation of bubbles is apparently due to insufficient water in the shallow situations to dissolve the gas as it is formed in bottom mud and debris. It follows then that the occurrence of bubbles at the surface over deep water would be unlikely. Under winter conditions methane bubbles may be present in the ice cover. Analysis of the gases in bubbles in a Russian lake (Beloye) revealed a methane proportion of from 74 to about 84 per cent, and 5 to 18 per cent hydrogen. As the bubbles rise to the surface, the hydrogen is lost and the methane volume diminishes to near 24 per cent. In the ascent, nitrogen and a small quantity of oxygen are gained. These transfers of gases into and out of the bubble operate, of course, under the law of partial pressures. Although the full story of methane and its role is not known, it does appear that some of the gas is oxidized by organisms in oxygenated zones of lakes as the bubbles rise. Here we see yet another phase in lake metabolism serving to decrease the oxygen content of deep waters.

Additional gases occur in natural waters and should be mentioned at this point, even though certain of them occur in small quantities. These will be considered in Chapter 10 in relation to dissolved solids. As we have already seen, hydrogen is formed as a decomposition product in the anaerobic zones of bodies of water. It acts, in part, to form methane and also occurs free in bubbles. Free ammonia may, under certain conditions, be present in small quantities in lakes and streams, and elemental nitrogen, derived mainly from the atmosphere, is highly important in lake metabolism; these substances will be taken up in Chapter 10 with nitrates. Hydrogen sulfide, a decomposition product, is frequently present in the hypolimnion of certain lakes during summer; it will be considered in the following chapter along with sulfur.

DISSOLVED GASES IN STREAMS

The behavior and basic relations of gases in streams follow the same fundamental physical and biochemical laws as operate in lentic situations. However, temporal and spatial relationships of dissolved substances, generally, are variously modified and complicated by the very features that characterize streams. In other words, the presence of a current with its inherent turbulence effects, the considerable exchange between stream and surrounding terrain, and the variations in water volume and chemistry associated with climate and drainage basin morphology all serve to make a momentary or long-term picture of stream conditions quite different from a lake. During our time another factor, an "unnatural" one, has become important in relationships in streams. This is pollution, and although the act of ravaging our waters by dumping wastes into them is not to be condoned, we have learned much about biochemical reactions from pollution studies. This topic will be studied more fully in a later chapter.

Oxygen in Streams

There are three primary sources of oxygen in stream water, the contributions of each being far from equal and indeed varying greatly with time of day, season, current velocity and stream morphology, temperature, and biological characteristics.

GROUND WATER AND SURFACE RUNOFF

For most streams these sources are relatively insignificant in supplying oxygen. Water issuing from springs, subterranean channels, or from seepages is typically low in dissolved oxygen, often to the point of being anaerobic. Not only do these ground waters fail to provide oxygen to the spring run, but also the run itself may dilute the oxygen content of a parent stream at the point of junction of the two streams. If, however, subsurface waters flow over broken rocks shortly before reaching the surface, the waters may be near saturation. Similarly, if surface runoff is rapid and vigorous, the water may be high in oxygen content; but sluggish surface drainage seldom contributes great oxygen stores to running waters.

PHOTOSYNTHESIS

In less turbid streams vegetation contributes oxygen to the waters during the day. Rich growths of algae on submerged objects such as rocks or logs, and algae floating free in the water, together with higher plants growing beneath the surface, produce high amounts of oxygen, particularly on cloudless days. During the night and on cloudy days this production may be somewhat balanced by respiratory consumption of the oxygen by plants and animals. In the shallow headwater reaches of a stream a net production of oxygen by photosynthesis contributes to the downstream content. In the lower, more turbid regions of most streams, however, local photosynthesis probably contributes little to the downstream oxygen content. Since turbidity is to a great extent a function of stream discharge and capacity, we can again recognize the importance of stream channel and water-shed features in the chemistry and biology of streams.

PHYSICAL AERATION

The introduction of a large amout of organic substance such as sewage, or debris from swamp or marsh flooding into a stream may bring about a depression of the dissolved oxygen content below the saturation value. The difference between the actual oxygen content and the amount that could be present at saturation is called the *saturation deficit*. This deficit is incurred, of course, through the uptake of oxygen by aerobic decomposition of the organic materials in the stream. Yet, downstream, barring no further immediate depressions, the stream shows evidence of regaining its earlier level of oxygen concentration. The oxygen serving to offset the oxidative loss is absorbed from the atmosphere through reaeration of the stream waters. Reaeration is, therefore, a process by which streams secure oxygen directly from the atmosphere, the gas then entering into the biochemical oxidation reactions in the stream. Within the stream, distribution of the oxygen derived from the atmosphere is accomplished by turbulent transport. The rate at which reoxygenation of a given parcel of water takes place depends upon a number of factors, including temperature, degree of turbulence, depth of the parcel, magnitude of the saturation deficit and, naturally, the momentary oxygen demand by decomposition processes. The importance of temperature lies in its inverse relationship to oxygen solubility, and to its influence on metabolic demands of organisms. Turbulence is a highly variable factor in the oxygenation-deoxygenation relationship, varying from negligible in quiet pools to highly important in riffles and rapids.

How effective is reaeration as a method of oxygenating flowing waters? Figure 9·7 gives data for a 27-mile segment of Holston River below

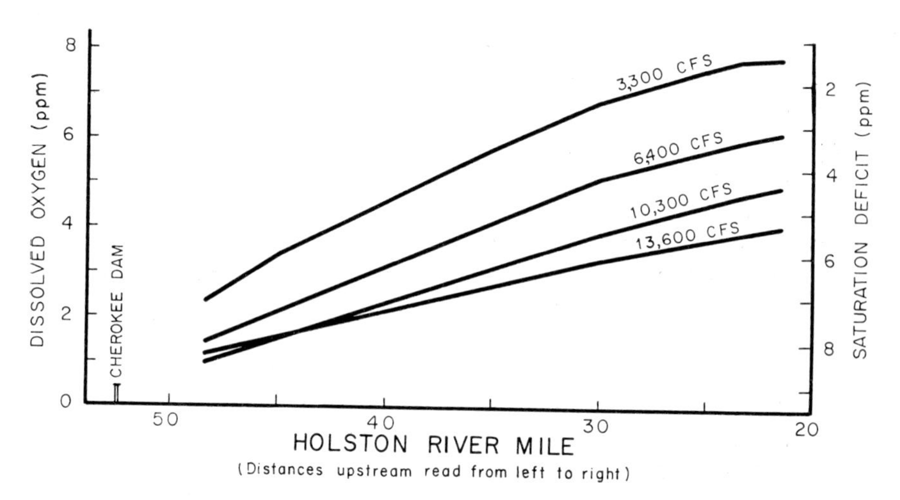

FIGURE 9·7. Rate and Extent of Reaeration as Functions of Discharge Throughout Some 27 Miles of the Lower Holston River Below Cherokee Reservoir, Tennessee. Dissolved oxygen concentration and saturation deficit are given on the vertical axes; distance below dam is shown on the horizontal axis. (From Churchill, M. A., 1958. "Effects of Impoundments on Oxygen Resources," *in* "Oxygen Relationships in Streams," U. S. Public Health Service, pp. 107-129.)

Cherokee Dam, one of the Tennessee Valley Authority impoundments, during a time of uniform flow. Note particularly the relationships between discharge and saturation deficit, and the dissolved oxygen content. The increase in oxygen concentration is greater at the lower discharge than at the higher. The total mass picked up, however, is greater in the higher discharges.

Under natural conditions, the waters of streams typically contain a relatively high concentration of oxygen tending toward saturation. However, a number of factors operate to varying extents to reduce the oxygen content and to contribute to the loss of the gas from streams.

(1) *Turbulence:* We have just witnessed the fact that turbulence plays an important part in the aeration of streams. It follows, therefore, that physical aeration is reduced with decrease in turbulent flow.

In stream areas below riffles and falls, the water is often saturated with oxygen throughout the day and night. In quiet reaches and in streams of low velocity, the waters may be below saturation during the night.

(2) *Respiration of Organisms:* Respiratory activities of plants and animals, and oxidation of organic matter utilize dissolved oxygen of streams. The effects of these processes are more conspicuous at night, being masked during the day by photosynthesis.

(3) *Photosynthesis:* In the more stable zones of streams, submerged plants contribute in a major way to the oxygen content of the water. Consequently, fluctuations in photosynthetic activity will be reflected in the amount of dissolved oxygen present.

(4) *Temperature:* In streams, as in other waters, solubility of oxygen varies inversely with temperature. Thus, raising of water temperature could result in loss of oxygen from streams.

(5) *Atmospheric Pressure:* Since the solubility of oxygen bears a direct relationship with atmospheric pressure, reduction of pressure would bring about a decrease in the amount of dissolved oxygen.

(6) *Inorganic Reactions:* Certain inorganic activities, such as the oxidation of iron, may contribute to the loss of oxygen from polluted streams.

(7) *Inflow of Tributaries:* The introduction of tributary waters of low oxygen content serves to dilute the concentration of oxygen in the receiving streams. This effect is especially noticeable with the entrance of spring or some seepage waters.

The annual cycle of oxygen of streams is closely correlated with temperature conditions. Studies of large rivers and small streams in warm southern regions of moderate temperature regimes and in northern climates of broad temperature fluctuations have shown that the oxygen content of flowing waters is generally highest in winter and lowest in late summer (see Figure 10·3D, Chapter 10). This primary temperature-oxygen relationship may, however, be tempered by a number of factors acting throughout the year. In small, slow streams of northern latitudes, decreased day length and ice and snow cover may serve to inhibit photosynthesis, thereby bringing about a short-term depression of the winter peak of oxygen concentration (Figure 9·8). The vernal decline of oxygen content in slow, clear streams may be attributed to the action of spring floods in removing vegetation. Further, decrease in oxygen content toward late summer, in streams generally, may be due to one or more of several factors. Water temperatures reaching their maxima in late summer hold less of the gas in solution; decreased discharge results in diminished physical mixing and reoxygenation; greater decomposition of summer-produced organic material utilizes some of the available oxygen. On the other hand, a rapid and abundant growth ("bloom") of phytoplankton often increases the oxygen content during the summer. The effect of such a bloom on the oxygen curve for the St. Johns River, Florida, is shown in Figure 9·9; this phenomenon is most likely in streams in which

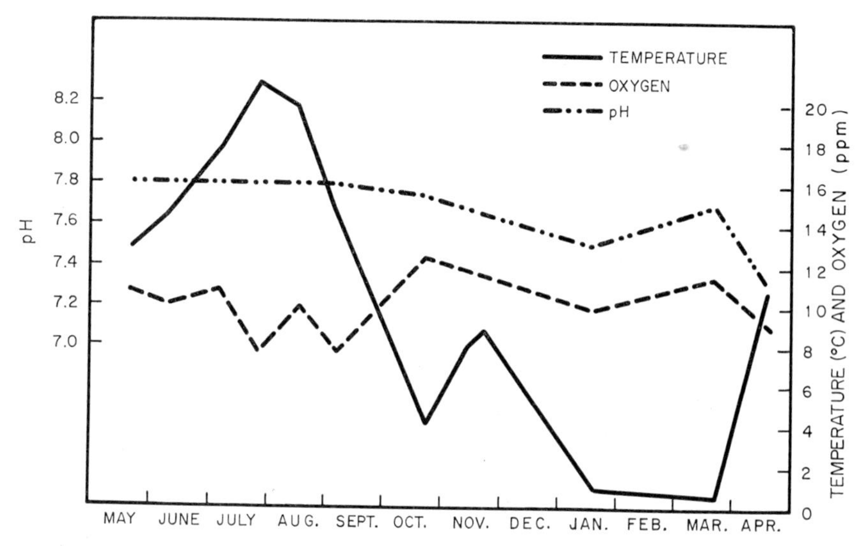

FIGURE 9·8. Seasonal Variation in Temperature, pH, and Dissolved Oxygen Content of Mad River, Ontario, 1930-1931. (After Ricker, W. E., 1934. "An Ecological Classification of Certain Ontario Streams," *Publ. Ontario Fish Research Board, Biol. Ser.*, Vol. 37, 1-114.)

the discharge is sufficiently low to permit the building up of a considerable phytoplankton mass.

A daily oxygen rhythm, the diurnal pulse, is largely a reflection of temperature fluctuations and photosynthesis-respiration relationships. In mountain and hill country the water is typically at between 95 and 105 per cent saturation at all times, so long as there is not a dense growth of rooted aquatic plants. Turbulence is of great importance and easily maintains an average of 100 per cent saturation. In a slow, shallow stream containing a fair abundance of rooted plant growth, daytime photosynthetic production of oxygen exceeds turbulent diffusion into the air and the respiratory consumption of the gas. The result is frequent supersaturation with oxygen during the day, followed by a drop to 100 per cent saturation at night. In winter in the temperate zone the time of maximum daily oxygen production corresponds closely with the time of highest temperature. In summer, oxygen production lags behind rising temperature, often by two or three hours, thus imparting high saturation values even after sundown. Depending upon oxygen demand, the minimum level usually occurs prior to the early-morning temperature low. In streams of this nature the oxygen concentration may be acutely affected by sunlight intensity. In Figure 9·10 the depression in the diurnal curve between about 1300 and 1500 hr is doubtless due to cloudiness shown for

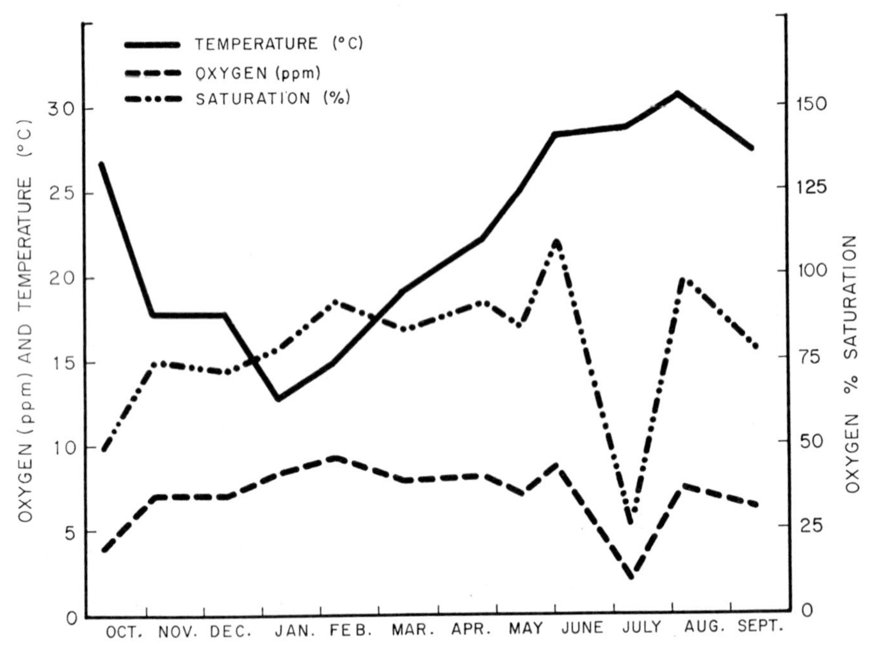

FIGURE 9·9. Seasonal Variation in Temperature, Dissolved Oxygen Content, and Per Cent of Oxygen Saturation of St. Johns River, Florida, 1939-1940. (Data from Pierce, E. L., 1947. "An Annual Cycle of the Plankton and Chemistry of Four Aquatic Habitats in Northern Florida," *Univ. Florida Studies, Biol. Sci. Ser.*, Vol. 4, No. 3.)

the same periods on the light-intensity curve. Such data as these indicate the importance of insolation and photosynthesis in maintaining the oxygen content of slow streams. The considerable diurnal oxygen range in such streams, as for example that in Figure 9·10, also points up the importance of more than a single oxygen determination in a stream study. Obviously an average of several measurements would be much more meaningful than one, and a diel series would most satisfactorily describe the oxygen conditions in this stream type. We have already learned that oxygenation through turbulent mixing is a prominent feature of fast waters and even highly turbid streams of low discharge. The diurnal oxygen pulse of such streams as these is usually less pronounced, since mechanical aeration and oxygen consumption are essentially uniform within a given stream segment.

The distribution of dissolved oxygen throughout the linear extent of a stream is subject to many variables, so much so that generalizations relating to the feature are not readily apparent. As we have seen, the upper reaches of streams are usually well oxygenated, the concentration being a function of turbulence and water temperature. Farther along the

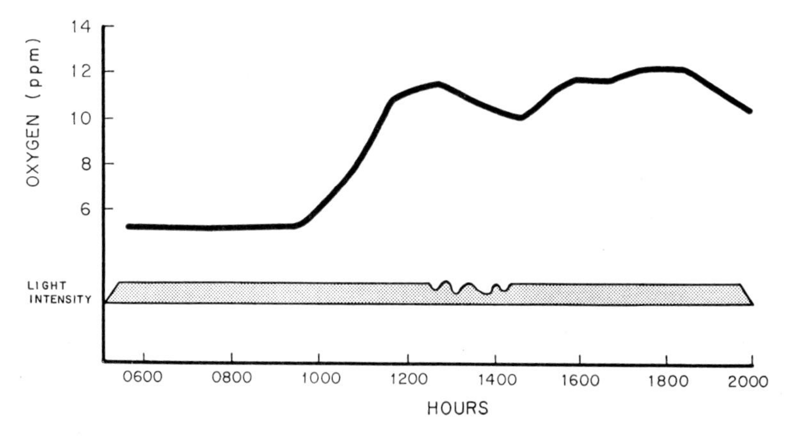

FIGURE 9·10. Oxygen Production in Relation to Light Intensity in Honey Creek, Oklahoma. Note the effect of intermittent cloudiness between 1200 and 1400 hr on dissolved oxygen concentration. (After Hornuff, L., 1957. "A Survey of Four Oklahoma Streams with Reference to Production," *Oklahoma Fish Research Lab. Rept.*, No. 63.)

gradient, the stream becomes slower and rich growths of submerged plants may develop. In this region the oxygen content is influenced less by physical aeration and more by organic oxygen production and respiration. The maximum-minimum relationships here may be the reverse of those found upstream, in that maximum saturation values will be found in the afternoon and the lowest values in the early morning. In the lowlands, increased turbidity and organic decomposition produce a generally low oxygen content. In clear, slow streams supporting an abundant flora, oxygen may be added to the waters along the stream course, thereby resulting in a net increase in downstream segments. The introduction of organic pollutants and inflow from marshes and swamps may serve to decrease the oxygen content at and below the point of inflow of these materials. Thus the amount of dissolved oxygen distributed through the course of a natural stream is strongly determined by channel and flow characteristics, considered in Chapter 3, and by biochemical interactions and processes.

Although our knowledge of reservoir limnology is meager, we know that the effects of the release of impounded waters into streams below the impoundment are varied and important, and worthy of considerably more research in the future. For example, we know that release of water from a reservoir through deep-water intakes during summer stratification introduces poorly oxygenated water and organic materials from the hypolimnion into the stream below the impoundment. Under extreme conditions this actual dilution of the stream and introduction of oxidizable matter could have deleterious effects on the biota. On the other hand, draw-down through shallow intakes from the epilimnion may serve to increase the dissolved oxygen content of the downstream waters.

Carbon Dioxide and pH in Streams

In streams the occurrence and abundance of components of the bicarbonate buffer system and the pH condition are determined primarily by current, biological processes, and the chemical nature of the substrate. The role of current is that of ameliorating the chemical climate of the stream. This is accomplished through mixing and moving concentrations of substances, but usually within a relatively restricted segment of the stream. Over a considerable distance, and depending upon the volume of introduced materials, the chemical composition of stream waters is subject to considerable change. The biological processes acting to influence the nature of the water are those previously considered, i.e., photosynthesis and respiration. In general, the behavior of carbon dioxide in streams is similar to that of oxygen, but with inverse properties. In stream areas of abundant plant growth the concentration of carbon dioxide is minimum during most of the day and maximum in the early morning hours. In sluggish streams a phytoplankton bloom may indeed exhaust the free carbon dioxide supply and obtain further carbon dioxide from bicarbonates. In the lower, sluggish stream course under conditions of high turbidity due to organic suspensoids, high carbon dioxide content occurs in the presence of depressed oxygen concentration, owing to bacterial action.

The chemical composition of mineral-bearing rocks in the stream valley and channel, and also the drainage nature of the valley, may act in a major way to determine the water composition. Under certain conditions these factors may somewhat offset the influences of biological processes. For example, in limestone regions a high concentration of carbon dioxide is not apt to develop in the stream because, as we have already seen, any excess of gas would enter into combination with lime in the substrate to give a carbonate. The direct strong influence of the substrate rests primarily upon the solubility of the buffer substances in rocks. Streams issuing from or flowing over relatively insoluble igneous formations of high silica content are typically soft, because the bicarbonate content is insufficient to buffer pH changes due to accumulation of carbon dioxide. As a result of a shift of the buffer system toward carbonic acid, the pH of such streams will be below neutrality. However, atmosphere-water equilibrium essentially adjusts the carbon dioxide content in such a way that the pH value does not go far toward the acid side, usually coming to balance at about pH 6. If, on the other hand, the stream encounters water from bogs, which commonly occur in areas of siliceous formations, the pH may be further lowered.

In many regions, the Appalachian slopes of North America for example, the fate of soft-water streams is to flow eventually over sedimentary formations in the lower reaches. These formations are normally rich in soluble carbonates, and the addition of carbonate ions serves to shift the pH value toward the basic range of the scale, usually to pH 8 or higher. The increase in pH value is accompanied by a marked rise in total alkalinity, and decrease in carbon dioxide.

Highly acid streams occur primarily on low marshy or swampy terrain, on poorly drained sandy "flatwoods," or under special conditions. These streams are usually stained brownish and support a relatively meager biota. The nature of the acids contained in such streams is not fully known, but frequently considered to be humic, yellow organic, or perhaps sulfuric. In the southern United States, the pH value in sluggish acid streams may range near pH 4, with a concurrent high concentration of free carbon dioxide (25 ppm or more) and low bicarbonate content. Oxygen in these streams is often low, being on the order of 20 to 30 per cent saturation at 17° to 18°C.

Hard-water streams occur commonly as spring runs or surface drainage streams in regions of soluble basic geologic formations. Calcium and magnesium carbonate, often occurring together, are prominent sources of ions which, in solution, contribute to the hardness of streams; however, sulfates, chlorides, and other compounds may also contribute to total hardness. The combination of high carbonate concentration and low carbon dioxide content, characteristic of these streams, apparently constitutes a more favorable environment for plants and animals than do acid conditions—for, generally speaking, hard-water streams support a varied and abundant association of organisms. Most unpolluted major streams exhibit a pH value on the alkaline side of neutrality, and, indeed, tend toward uniformity of composition generally.

Although the relationships of carbon dioxide, bicarbonates, and carbonates will be considered a bit further with respect to dissolved solids in Chapter 10, we might briefly summarize their activities in relation to pH in lakes and streams thus: (1) the pH value varies inversely as the dissolved carbon dioxide concentration, and directly as the bicarbonate concentration; (2) the critical value relating to the presence or absence of free carbon dioxide is pH 8, the free gas being absent above that value; (3) the absence of free carbon dioxide does not limit photosynthesis of certain algae and higher plants, some being adapted for utilization of carbon dioxide from carbonates, usually resulting in very high pH values. Certain of these relationships in a New Jersey stream are shown in Figure 10·3B and C, Chapter 10.

Methane, hydrogen sulfide, and other gases derived from breakdown of organic substance may be present in high concentrations under proper conditions in very slow streams and in stagnant regions of stream pools. In moving water, however, turbulence due to current tends to diffuse and eliminate the gases.

DISSOLVED GASES IN ESTUARIES

The considerable differences between the chemistry of fresh water and that of sea water bring about complex relations of dissolved substances in general within an estuary. The occurrence of dissolved materials relates to the disproportion of dissolved materials found in the waters of the inflowing stream at the upper end of the estuary and in the sea water at the mouth of the estuary. Between these two relatively uniform states there exists a considerable gradient in processes and conditions. Dissolved gases are distributed in an estuary in accordance with turbulence and current factors, biological activities, and salinity and temperature effects. We have previously considered most of these factors and their influence on dissolved gases with respect to lakes and streams; in general, the interrelationships are similar in estuaries. One new factor is quite influential in estuaries; this is salinity, the total amount of dissolved inorganic solids.

Oxygen

In salt water the solubility of oxygen decreases as water temperature and salinity increase. If we visualize cool, fresh (low-salinity) water entering the uppermost reaches of an estuary and grading toward somewhat warmer and more saline sea water at the lower extreme, we can begin to appreciate the adjustments taking place throughout the length of the estuary. Table 9·1 gives saturation (absorption) coefficient values of oxygen in water at various temperatures and salinities. The values show not only that temperature is the most important factor in determining oxygen solubility, but also that salinity manifests considerable influence. We see then that less oxygen can be dissolved in sea water than in fresh water. At saturation at 15°C, a liter of sea water (at 36 ‰ salinity) contains 5.8 cc oxygen per liter of water; a liter of fresh water

TABLE 9·1. SATURATION COEFFICIENT VALUES OF OXYGEN AND CARBON DIOXIDE IN WATER

Concentrations are expressed as ml/l at sea level atmospheric pressure.

(Data from Sverdrup, H. U., Johnson, M. W., and Fleming, R. H., 1942. "The Oceans, Their Physics, Chemistry, and General Biology," Prentice-Hall, Inc., New York, N. Y.)

Salinity (‰)	Temperature (°C)					
	0°		12°		24°	
	O_2	CO_2	O_2	CO_2	O_2	CO_2
0 (fresh water)	49.24	1717	36.75	1118	29.38	782
28.9	40.1	1489	30.6	980	24.8	695
36.1 (sea water)	38.0	1438	29.1	947	23.6	677

at the same temperature holds 10.3 cc.* The importance of these relationships lies not only in the aforementioned possibility of a considerable linear oxygen gradient in a mixing estuary. There is also the possibility of fluctuations associated with stream flood seasons and the inflow of large quantities of fresh water, or with dry seasons when tidal flow of sea water dominates.

In nonmixing estuaries salinity stratification during summer often results in conspicuous differences in dissolved oxygen content in deep and in surface waters. In parts of Chesapeake Bay during summer the oxygen concentration may range from 90 to 100 per cent saturation at the surface, while bottom waters show from 40 to 50 per cent saturation. The surface-bottom differential decreases upstream as the depth decreases. In these shallow zones there is probably more mixing than in the bay proper. The range of oxygen concentration in the bay reflects the activity in the lighted trophogenic zone and the relatively slow replacement of sea water in the lower level. It may well be that even with more rapid replacement the water from the sea might be poor in dissolved oxygen.

Along with vertical differences, oxygen characteristically varies diurnally and seasonally within estuaries. The ranges of such variations differ, depending upon the nature of the freshwater source, the morphology of the basin, and effects of tides. In deep, turbid estuaries lacking the con-

* These values, correlated with Table 9·1, have recently been revised (see Richards and Corwin (1956) for a review of the subject). However, Steen (1958) has found good agreement with the earlier data used above. Since these earlier data have been widely employed in many computations, we have continued their use.

tribution from an abundant bottom flora, diurnal oxygen pulses are apt to be relatively slight. In Chesapeake Bay the range for surface waters was found to be from about 85 per cent of saturation in early morning to about 115 per cent in late afternoon. Shallow, clear estuaries may contain bottom growths of algae on which oxygen bubbles may be seen. In waters such as these the minimum-maximum diurnal range may exceed 200 per cent. The effects of incoming sea water during a tidal cycle may serve to mask the local conditions. Seasonally, the oxygen dynamics in an estuary may be influenced by variations in river discharge, tides, and day length and biological effects. Surface waters in one part of Chesapeake Bay were found to have 143 per cent saturation with oxygen in April, and about 42 per cent in August. Bottom waters in the same area reached 133 per cent saturation in October, and 24 per cent in June.

The Carbon Dioxide System and pH

The solubility of carbon dioxide in estuarine waters is determined primarily by the amount of sea water mixing with fresh water, and secondarily by temperature. The importance of sea water as a solubility factor rests, of course, upon salinity. In Table 9·1 we see that at any temperature the absorption coefficient value of carbon dioxide decreases as the salinity increases. The high solubility of carbon dioxide is due to its chemical reaction with water. Although some of the carbon dioxide in sea water is in the form of the free gas and as carbonic acid, much more is present as bicarbonate and carbonates. This condition results from the fact that sea water contains alkaline radicals in excess of the equivalent acid radicals which, as we have seen, shift the carbon dioxide system toward the carbonate formation, thereby reducing the free carbon dioxide concentration. The presence of the excess bases—boric acid and its borates, carbonic acid and the carbonates—in sea water serves to buffer the water against great changes in pH that might develop from the addition of acids or bases. The pH of sea water at the surface is very stable, usually ranging between pH 8.1 and 8.3.

Since carbon dioxide uptake is greater in the presence of excess base, and since river waters usually contain a lower concentration of excess bases than sea water, we should expect to find a lower content of free carbon dioxide in the mouth of the estuary than in the upper reaches. Similarly, because river waters are seldom buffered, the free carbon dioxide concentration and pH should be more variable in that part of the estuary dominated by stream influences. Streams transporting large quantities of humic material in colloidal suspension are frequently slightly acid. Upon meeting sea water the colloidal particles are coagulated and the pH shifts toward the alkaline side of neutrality. East Bay, Texas, receives considerable runoff from organically rich salt marshes, and it was found that during summer the pH ranged from 6.9 throughout much of the bay to 7.8 near the mouth where it discharges into the Gulf of Mexico. Gulf waters during the same period gave a pH value of 8.0.

In view of what we have learned of oxygen-carbon dioxide-pH relationships generally, we should expect the vertical, diurnal, and seasonal distributions of carbon dioxide and pH to operate acording to certain principles. Although few actual studies have been made of carbon dioxide in estuaries, the pattern of distribution is expected to be the reverse of that of oxygen. By the same rules, pH values should vary inversely as the free carbon dioxide content and directly as the dissolved oxygen concentration.

Chapter 10

DISSOLVED SOLIDS IN NATURAL WATERS

PRACTICALLY ALL OF THE naturally occurring elements of the earth's crust could probably be found in inland waters. Some of these substances would be expected to occur in minute concentrations, however. In addition to the gaseous components considered in the preceding chapter, the more conspicuous materials which are found in varying quantities in natural waters include carbonates, chlorides, sulfates, phosphates, and often nitrates. These anions occur in combination with such metallic cations as calcium, sodium, potassium, magnesium, and iron to form ionizable salts. As a result of availability and high solubility of carbon dioxide, carbonates are usually the most abundant salts in fresh waters. Both the quantitative and qualitative aspects of the chemical composition of inland waters are influenced to a high degree by the geochemistry of the terrain and, in the case of lakes, especially by the form of the basin with respect to inflow and outflow. In general, the inorganic composition of the water of an open lake, that is, one with effluents

TABLE 10·1. MEAN PERCENTAGE COMPOSITION OF NORTH AMERICAN INLAND WATERS

(Data from Clarke, F. W., 1924. "The Composition of River and Lake Waters of the United States," *U. S. Geol. Surv. Profess. Paper*, No. 135.)

Ion	Percentage
$CO_3^{=}$	33.40
$SO_4^{=}$	15.31
Cl^-	7.44
NO_3^-	1.15
Ca^{++}	19.36
Mg^{++}	4.87
Na^+	7.46
K^+	1.77
$(Fe, Al)_2O_3$	0.64
SiO_2	8.60

through which water moves, is a reflection of the nature of the influent waters. The composition of these lakes approaches that shown in Table 10·1. The inorganic composition of closed lakes (or those lacking significant effluents) is greatly modified by precipitation and concentration of salts as determined by evaporation. Lakes near coasts may receive substances from sea spray.

TOTAL DISSOLVED SOLIDS

The total concentration of dissolved substances or minerals in natural waters is a useful parameter in describing the chemical density as a fitness factor, and as a general measure of edaphic relationships that contribute to productivity within the body of water. One measure called *total dissolved solids* is determined by simply evaporating a filtered quantity of water at low temperatures. The dried residue contains both inorganic and organic materials. Ignition of this residue at high temperature eliminates volatile substances, usually organic in nature, and decomposes bicarbonates with the loss of carbon dioxide. The residue following ignition therefore contains the total inorganic solids, the difference between it and the original residue being termed the *loss on ignition.* The amount of residue in each operation and the loss on ignition is expressed as the proportion of the original water sample in terms of parts per thousand (ppt) or parts per million (ppm) depending upon the concentration, or as mass in metric units per liter (mg/l). The *salinity* of fresh waters is defined as the total concentration of the ionic components. Although common in marine sciences, this term has not been widely used in freshwater researches. Because of the relatively small quantities of total ions generally encountered, freshwater salinity is usually expressed in milligrams per liter.

A measure of the total amount of ionized materials in water can be obtained through determination of the electrical conductance of the solution. Commonly called *specific conductance*, this parameter closely approximates the residue in solution and may be correlated with salinity. By definition conductance is the reciprocal of the resistance measured between two electrodes separated by 1 cm and having a cross section of 1 sq cm. The conductivity is usually expressed as micromhos (reciprocal megohms) per centimeter at 25°C (see A.P.H.A., 1955, and Ellis, Westfall, and Ellis, 1946, for details). In general, the range of specific conductance of natural waters should approximate that of total dissolved solids given below. Bunny Lake in the Sierra Nevada, listed in Table 10·2, has a total dissolved solid concentration of 8.2 ppm at mid-depth in July, the specific conductance being 8.9 micromhos. In Hot Lake, Washington, at 3 m depth in August, the total dissolved solids amounts to around 391,800 ppm, and the specific conductance reads 60,440 micromhos.

Most lakes occupying open basins have a total dissolved solid concentration of between 100 and 200 ppm. Evaporation from lakes in closed basins raises the concentration of dissolved solids, in some cases, to over 100,000 ppm. These lakes with extremely high concentrations are considerably more saline than the sea, the latter averaging about 35,000 ppm.

TABLE 10·2. COMPARISON OF CERTAIN IONS IN THE COMPOSITION OF RAIN WITH THOSE OF NATURAL WATERS IN A FLOWING SPRING RUN, A POND, AND THREE LAKES

Quantities are given in parts per million (ppm). See text for details.

(Data from Hutchinson, G. E., 1957. "A Treatise on Limnology, Vol. I, Geography, Physics, and Chemistry," John Wiley & Sons, Inc., New York, N.Y.; Ferguson, G. E., Lingham, C. W., Love, S. K., and Vernon, R. O., 1947. "Springs of Florida." *Bull. Geol. Surv.*, No. 31; Gorham, E., 1957. "Chemical Composition of Nova Scotian Waters," *Limnol. Oceanog.*, Vol. 2, 12-21; Reimers, N., 1958. "Conditions of Existence, Growth, and Longevity of Brook Trout in a Small, High-altitude Lake of the Eastern Sierra Nevada," *California Fish and Game*, Vol. 44, 319-333; Hough, J. L., 1958. "Geology of the Great Lakes," Univ. of Illinois Press, Urbana, Ill.; and Anderson, G. C., 1958. "Some Limnological Features of a Shallow Saline Meromictic Lake," *Limnol. Oceanog.*, Vol. 3, 259-269.)

Ion	Rain	Silver Springs Florida	Purcell's Pond Nova Scotia	Bunny Lake California	Lake Erie	Soap Lake Washington
Cl	0.5	7.8	7.2	0.2	6.6	5467.
SO_4	2.0	34.0	9.5	1.3	9.3	6240.
B	0.01	0.01	not rept.	<0.01	not rept.	not rept.
Na+K	0.43	5.1	5.3	0.9	4.9	13,002.
Mg	0.1	9.6	0.4	0.9	5.7	8.2
Ca	0.1 to 10	68.0	0.8	0.7	23.4	20.6
$N{\cdot}NH_3$	0.5	not rept.	not rept.	0.03	not rept.	not rept.
$N{\cdot}NO_3$	0.2	0.2	not rept.	0.2	not rept.	8.7
pH	4.5	7.8	3.9	5.8 to 6.3	8.2	9.4 to 10
HCO_3	–	201.	nil	6.0	not rept.	5209.

Organic compounds such as the organic states of phosphorus and nitrogen, sugars, acids, and vitamins are known in natural waters. We are very much lacking, however, in our knowledge of the formation and role of many of the organic substances.

Although we shall relate dissolved substances to community metabolism in more detail in later chapters, it is well to keep in mind the idea that the quantity and quality of dissolved solids often determine the variety and abundance of plants and animals in a given aquatic situation. In a most general sense the limiting nature of dissolved solids is essentially two-fold. In the first place the chemical density of the environment of aquatic organisms is a function of the total dissolved solids. According to laws of osmosis and diffusion the water balance, or osmoregulation, of plants and animals is governed by this environmental density factor and the physiological adaptations of the organisms to it. As we shall see later in more detail, osmoregulation is a very salient process in limiting marine and freshwater organisms to their respective habitats. The second way in which dissolved solids influence the nature of the community relates to supply of nutrients and otherwise important materials. The only source of nutritionally important ions available to phytoplankton is the reservoir of matter dissolved in the water. The nature of the animal community is, of course, dependent upon the kinds and quantity of phytoplankton. In another sense certain animals may be directly limited by the availability of a given dissolved substance, animals bearing carbonaceous shells being an example. In highly acid waters some mollusks may be entirely absent, while under mildly acid conditions mollusks may occur but their shells are reduced in thickness as compared with those inhabiting waters of higher pH value.

With these general concepts of the importance of dissolved solids in community function and structure in mind, let us now consider some of the properties and activities of the more abundant and better known inorganic and organic substances in natural waters.

SOLIDS IN SOLUTION IN LAKE WATERS

Calcium and Magnesium

Together, these two alkaline earth metals constitute the most abundant ions in fresh waters. The chemical activity of the two elements is similar, particularly in the formation of carbonate salts, and both may limit biological processes in streams and lakes. Magnesium is an important component of the chlorophyll molecule. Of these two ions, calcium is usually more abundant. In soft waters, or those containing less than 50 ppm of dissolved solids, calcium makes up, on the average, about 48 per cent and magnesium about 14 per cent of the total cations present. In average hard waters the proportion of magnesium to calcium increases, giving approximately 53 per cent of calcium and 34 per cent of magnesium. The increase of these ions in hard waters takes place at the expense of two alkali metals, sodium and potassium, to be considered later. The nature of the lake basin is reflected in the ionic proportions of the waters in open lakes (those with considerable inflow and outflow). In this type of lake, calcium averages about 63 per cent; magnesium, on the other hand, makes up slightly more than 17 per cent of the total cations.

The wide variation in calcium content and the corresponding correlations shown in hardness, stratification, and biological productivity have stimulated many attempts at classification based on this ion. The following scheme was proposed in 1934 by a German limnologist, W. Ohle:

$<$ 10 mg Ca/l	"Poor"
10 to 25 mg Ca/l	"Medium"
$>$ 25 mg Ca/l	"Rich"

Since calcium normally occurs in combination with the carbonate anion, the above classification should correspond generally with the designations of soft and hard waters given previously. Being concerned with the ionic content, both sets of designations are equally applicable to streams and lakes.

The distribution of calcium in stratified lakes generally follows a more or less characteristic pattern. In soft-water lakes of northern Wisconsin, the calcium content ranges from 0.7 to 2.3 mg/l (poor) and during summer stagnation little stratification of calcium exists. Lakes of medium hardness and medium calcium content typically contain a moderately increased calcium concentration in the hypolimnion during stratification. Such has been shown for lakes in southern Wisconsin which ranged from 21.2 to 22.4 mg Ca/l. The hypolimnion of hard-water (rich) lakes characteristically contains a greatly increased load of calcium. In some of these lakes the curve of calcium concentration shows a steady increase of the ion from the surface to the upper region of the hypolimnion, the hypolimnial content being on the order of twice that at the surface and rather uniform throughout the stratum. Under other conditions the epilimnial calcium may be relatively uniform throughout, increasing rapidly with depth through the metalimnion, and with less rapid, but none-the-less increasing change with depth in the hypolimnion. Along with other lake substances calcium is redistributed throughout the lake during vernal and autumnal mixing. An exception to calcium stratification in calcium-rich lakes is seen in data from Cultus Lake, British Columbia. This lake contains 32 mg Ca/l during summer but shows a somewhat uniform vertical distribution of the ion during stratification.

Biologically, soft-water lakes usually contain less living matter per unit area than hard-water lakes. The total estimated biomass of plant substance in certain medium lakes of Wisconsin was found to be three to five times that of poor lakes, while the animal mass, not counting fishes, was as much as triple that of poor lakes. Although the total mass of organisms may be greatest in rich lakes, medium lakes often harbor a greater variety of kinds of plants and animals. In addition to the quantitatively meager mass of organisms, poor lakes generally support a unique assemblage of species.

CALCIUM CARBONATE

We have already seen the interrelationships involving the carbon dioxide-bicarbonate-carbonate equilibrium, water, and the hydrogen ion concentration. If the carbonate formed by dissociation of carbonic acid combines with calcium (or magnesium) ions in the water we arrive at calcium carbonate as follows:

$$CO_3^{=} + Ca^{++} \rightleftharpoons CaCO_3$$

Calcium carbonate, or lime, occurs as a white precipitate. Its formation is dependent upon the loss of carbon dioxide from the carbon dioxide-carbonate system. Recall that the addition of carbon dioxide serves to produce carbonic acid which dissociates and lowers the pH (increases the hydrogen ion concentration), thereby leading to acid conditions. The effect of carbon dioxide removal from the system is to split off CO_2 from HCO_3^-, which brings about the precipitation of $CaCO_3$.

In fresh waters calcium carbonate deposits, generally, are called *marl*, although some may be only accretions of animal shells. More properly, perhaps, marl is the accumulation of calcium carbonate precipitated directly from the water. This precipitation results from the removal of carbon dioxide from the water by physical loss to the atmosphere and by photosynthesis in green plants. It appears well established now that many species of algae and some of the so-called higher plants are capable of precipitating lime in the process of obtaining carbon dioxide for photosynthesis. In some of these plants the carbon dioxide is taken directly from bicarbonate after the free carbon dioxide content has been exhausted. Among the higher plants, *Elodea* (*Anacharis*) and *Potamogeton* often possess dense coatings of lime on their surfaces. In fact, it has been found that 100 kg of fresh *Elodea* can precipitate 2 kg of $CaCO_3$ in a day with 10 hr of sunlight. Green algae (Chlorophyta) such as *Chara* and *Cladophora* under hard-water conditions cause the formation of considerable quantities of marl. *Chara*, or stonewort, feels brittle to the touch and some 30 per cent of its dry weight is lime. *Cladophora* often occurs in lime balls littering the lake bottom. Blue-green algae (Cyanophyta) and bacteria are also capable of forming lime.

Plant activity may not be the only process in the formation of lime deposits in fresh waters. It has been shown that carbon dioxide loss is also related to temperature, hydrogen ion concentration, partial pressure gradient across the water surface, and the nature of other substances in the water. Recently, it has been found that phosphate deficiency may be important in reducing the carbon dioxide content, thereby facilitating lime precipitation.

Calcium and magnesium enter into combination with anions other than carbonate, some of these being found in high concentrations in certain lakes. In the Swiss Alps several small lakes are rich in calcium sulfate ($CaSO_4$), or gypsum, derived from local deposits. Magnesium also combines with sulfate to form epsomite ($MgSO_4 \cdot 7H_2O$), which occurs in high concentration in certain saline lakes. The bottom of Hot Lake, a meromictic lake in Washington, is underlain by a layer of epsomite 4.5 m thick, from which epsom salts have been mined. A layer of gypsum is also found, lying below the epsomite stratum. Unlike the "typical" water we have been emphasizing, this lake lacks carbonates, but the concentration of sulfate in the deep zone (monimolimnion) is over 243,000 mg/l. The same layer contains slightly over 700 mg Ca/l, and nearly 54,000 mg Mg/l. In other types of saline lakes magnesium may occur in combination with chloride as $MgCl_2$. Dolomite, a double carbonate of calcium and magnesium ($CaMg(CO_3)_2$), serves as an important source of both cations in waters in various parts of the world.

THE CALCIUM CYCLE

From our study of lime relationships it is obvious that calcium and carbonates are derived almost entirely from sedimentary rock strata. The ions are dissolved out of these formations and carried in solution in lakes and streams. Biological activity such as shell construction, bone building, and plant precipitation of lime combines and concentrates the ions. Through streams, much of the calcium reaches the sea, where it becomes "locked up" in coral reefs and bottom deposits of animal shells. Here the calcium remains until geologic forces raise the sea-covered deposits and present them to the attacking forces of erosion and solution. The famous white cliffs of Dover, England, and the rich limestone regions of the Alps are examples of now-available sources of calcium in the perpetuation of the cycle.

Sodium and Potassium

Where sodium and potassium occur in low concentrations the proportion of sodium is usually only slightly greater than that of potassium. As the total content of both increases, the concentration of sodium greatly exceeds that of potassium. In average soft waters the equivalent percentage of sodium is second to that of calcium. In hard waters the proportion is less, usually falling below calcium and magnesium. The most common form of sodium in natural waters is as halide ($NaCl$), reaching high concentrations in saline lakes such as Great Salt Lake with its salt content ranging from 24 to 26 per cent. Certain lakes in California, and in other regions of the world, contain sodium in the form of dissolved sodium tetraborate ($Na_2B_4O_7$), or borax. Sodium sulfate (Na_2SO_4) occurs in abundance in some Canadian lakes. In certain of the saline lakes in

Nebraska, potassium accounts for about 23 per cent of the total ionic composition. Because of analytical difficulties, sodium is frequently determined and reported with potassium (Table 10·2). It appears that the limnological importance and behavior of these elements are similar.

Nitrogen and Nitrogen Compounds

The very high importance of nitrogen in aquatic ecosystems rests upon its role in the synthesis and maintenance of protein, which is, along with carbohydrates and fats, a major constituent of living substance. Derived originally from the atmosphere, nitrogen enters into a complex cycle involving plants and animals and several forms of the element. Nitrogenous compounds in natural waters may be derived from outside sources (allochthonous) or may be fixed within the body of water (autochthonous). The former category includes precipitation falling upon the earth carrying its own compounds (such as nitrate and ammonia), surface runoff containing terrestrial compounds of nitrogen (including pollution by human agencies), and inflow of ground waters as springs or seepages. Endogenous nitrogen compounds result from fixation processes carried on by certain bacteria and algae. However, the extent to which "in place" biologically fixed nitrogen contributes to the total supply is not yet known. In other words, we have yet to determine how much of the nitrogen content of a given lake is fixed within the lake by its own organisms and how much is delivered to the lake from the outside.

Once in the system, a great proportion of these compounds is caught up in a cycle of biological assimilation and decomposition, and inorganic processes in the economy of the ecosystem. Depending upon the efficiency of the community, certain quantities of these and other compounds will be lost. Lake, stream, and estuarine communities can lose nutrient materials through outflow, by incorporation of the substances in sediments, and through processes of denitrification releasing elemental nitrogen.

There are at least two possible sources of elemental, or *uncombined* nitrogen (N_2) in natural waters. One reservoir, and very likely the more important, is the atmosphere. The second, but poorly known, source of uncombined nitrogen is that produced by bacterial denitrification of ammonia. This uncombined nitrogen is rather inert; the only organisms capable of using it are certain microorganisms such as the nitrogen-fixing bacteria and algae.

The solubility of molecular nitrogen in fresh waters is related to temperature and pressure, the temperature relationship being an inverse one. Although little study has been made of nitrogen in lakes, it has been established that supersaturation can exist under certain pressure conditions at the air-water interface. During summer stratification, the vertical distribution of uncombined nitrogen follows inversely that of temperature. Mixing, during vernal and autumnal overturn, distributes the nitrogen throughout the lake.

The synthesis of inorganic substances into plant and animal tissues and the metabolic processes of protoplasm produce various compounds containing nitrogen. These *organic nitrogen compounds* include, for example, nitrogen in combination with carbon and other elements, animal and plant protein, and urea and uric acid as animal metabolic wastes. Of the total content of soluble nitrogen in filtered and centrifuged surface waters of lakes, it is probable that 50 per cent or more is in the form of organic nitrogen. Some 60 to 80 per cent of this organic nitrogen is composed of amino compounds such as free amino acids, polypeptides, and proteins. These substances are contained primarily in living plants and animals, and the presence of the compounds in water doubtless reflects metabolic processes of living organisms as well as decomposition of dead bodies. With respect to formation by living organisms, it has been established that many blue-green algae secrete extracellular nitrogenous materials, including polypeptides, amides, and amino acids. The ecological importance of such liberated compounds is not fully known. In the algal species tested, the plants were not able to utilize their own excreted products as nitrogen sources.

The concentration of organic nitrogen may be expected to vary seasonally. Figure 10·1, for example, shows the seasonal distribution of various forms of nitrogen in Lake Mendota, Wisconsin. There is little evidence that much of the organic nitrogen is available as nutrient for plants and animals. The measure of the total organic nitrogen content is, however, a valuable indication of the productivity of the body of water, for certainly most of the substance will ultimately be transformed into states which can enter into production of living matter.

In addition to its occurrence in the uncombined state and in organic compounds, nitrogen is also present in natural waters in the form of *inorganic nitrogen compounds* such as ammonia, nitrite, and nitrate. In most fresh waters the concentrations of these inorganic compounds are relatively slight, but nevertheless very important in determining the productivity of a given community. All of the inorganic forms can be used by most green plants, particularly various algae in their role of primary producers of energy-containing mass that can enter the aquatic food web.

Nitrogen "locked up," so to speak, in organic compounds is returned to the environment through decomposition and, to a lesser extent, by excretion of nitrogenous wastes of animals. The end product of the first stage of oxidative degradation of animal and plant proteins is mainly free ammonia (NH_3). Lesser amounts of ammonium compounds such as the base, ammonium hydroxide, and a salt, ammonium carbonate, are also released. The agents of the decomposition process are microbial organisms, i.e., certain bacteria and fungi. The free ammonia content of natural waters is derived in part from this bacterial decomposition of proteins, and in part from deamination (the removal of an amino group (NH_2) from an amino acid) also involving bacteria.

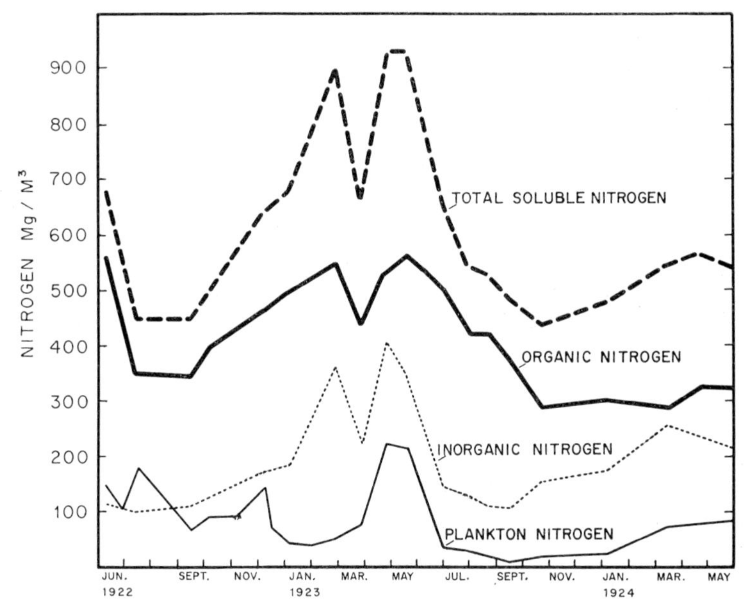

FIGURE 10·1. Seasonal Variation in the Quantity (milligrams per cubic meter) of Various Forms of Nitrogen in Lake Mendota, 1922-1924. (From Domogalla, B. P., Juday, C., and Peterson, W. H., 1925. "The Forms of Nitrogen Found in Certain Lake Waters," *J. Biol. Chem.*, Vol. 63, 269-285.)

Generally, in unpolluted waters ammonia and ammonium compounds occur in relatively small quantities, usually on the order of 1 mg/l or less. With the uptake of oxygen, as in pollution, however, the concentration of ammonia may increase, in extreme cases to 12 mg/l or more. The relationships between ammonia and its basic compound, undissociated ammonium hydroxide (NH_4OH), seem to rest upon the pH of the water. At a temperature of 18°C and pH 6 the proportion of NH_4^+ to NH_4OH is approximately 3000:1, while at pH 8 the proportion is nearer to 30:1. Under highly alkaline conditions the concentration of NH_4OH may reach toxic levels. Free ammonia in concentrations over 2.5 mg/l in neutral or alkaline waters is apt to be harmful to many freshwater species. The common ammonium salt of fresh waters, ammonium carbonate, is usually present in small amounts, but in concentrations of about 20 mg/l or more under alkaline conditions it is also toxic to certain animals.

Although somewhat variable, the ammonia content of the upper, lighted zones of unpolluted lakes is generally low, particularly in stratified lakes in summer. During summer stagnation the hypolimnion of eutrophic lakes may become considerably enriched with ammonia from decomposition in the tropholytic zone. With the advent of autumnal circulation the hypolimnetic ammonia is distributed throughout the lake, thereby increasing the content in the upper strata. Similarly, ammonia produced during winter stagnation is circulated throughout the lake during vernal overturn, producing another seasonally high concentration of the compound. Spring and summer growth of phytoplankton populations in the trophogenic zone, and the resultant increased biochemical demand, bring about a reduction of ammonia during summer and up to the time of fall overturn.

In an intermediate phase of oxidation of nitrogenous organic compounds, ammonia is attacked by certain of the nitrifying bacteria, such as *Nitrosomonas*, which absorb ammonia and release nitrite (NO_2^-) ions in the process. Energy is released in this activity and is used in the synthesis of carbohydrates by the bacteria. Nitrite may also be formed by reduction of nitrate (NO_3^-), and it is probable that more of the nitrite content of most natural waters is derived from this process than from oxidation of ammonia.

Nitrite nitrogen, as it is usually designated, occurs in very minute quantities (if at all) in unpolluted waters. In lakes in which nitrite nitrogen is present, seasonal variation in concentration of the compound appears to follow (except in early winter) that of nitrate from which, as we have seen, the nitrite is probably formed. In Lake Mendota the lowest content was found during late summer, when nitrate was also at a minimum. It is now known that diatoms and certain algae, *Chlorella* for example, are capable of reducing nitrate to nitrite. Since all green plants require nitrate, the supply of this compound may become quite low toward the end of the growing season. It seems reasonable to assume, therefore, that nitrate reduction to nitrite would decline during the same period. During early winter, when nitrate and ammonia concentrations are low, nitrite content may increase; the reason for this is not known. Vertically, the concentration of nitrite is usually maximum between the oxygen-rich trophogenic region with high nitrate content toward its lower boundary, and the oxygen-poor tropholytic zone often high in ammonia. The ecological importance (if any) of small nitrite quantities is not fully understood. Large amounts of the compound usually indicate pollution by sewage.

In what might be considered the final phase in the ecologically important process of decomposition of nitrogenous substance, nitrite is oxidized by nitrifying bacteria to nitrate (NO_3), expressed usually as nitrate nitrogen. It is in this form that nitrogen is most easily taken up by green plants rooted in the substrate or floating in the water. To the bacteria, *Nitrobacter* for example, involved in the process, formation of nitrate is of less importance than the energy released for their own utilization. To the community and its maintenance and productivity, however, nitrate is extremely important as a nutrient in supplying nitrogen for protein synthesis.

Nitrate nitrogen usually occurs in relatively small concentrations in unpolluted fresh waters, the world average being 0.30 ppm. In certain highly saline lakes (those of magnesium sulfate, for example) this form of nitrogen may be entirely absent at times. Lake Superior has a total dissolved solids content of about 60 ppm, making it, in this respect, the lowest of the Great Lakes. Nitrates account for 0.86 per cent of the total in the lake, and this represents the highest concentration of all the Great Lakes. Under the influence of edaphic factors, particularly during flood times, and organic pollution, nitrate nitrogen content may be expected to increase significantly. Under normal conditions, however, the amount of nitrate in solution at a given time is determined by metabolic processes in the body of water, i.e., production and decomposition of organic matter. As we have seen, nitrate is contributed to the ecosystem as a byproduct of bacterial nitrification. The compound is removed from solution through utilization by green plants, and through bacterial denitrification to uncombined nitrogen and reduction to ammonia nitrogen.

Seasonal fluctuations in nitrate nitrogen and other forms in Lake Mendota, Wisconsin, are shown in Figure 10·1. Annual cycles in lakes, generally, may be expected to vary as to time of maximum nitrogen production depending upon latitude, basin morphology and chemical nature of the drainage area substrate, and productivity. Vertical distribution of nitrate is apparently related to lake productivity. In some oligotrophic lakes that have been studied there is little evidence of nitrate stratification; in others the nitrate content of the upper zone decreases during summer from the amount present at spring overturn, with little change in deep waters. In eutrophic lakes, the nitrate concentration is typically decreased in the upper zones by plankton utilization and in the deepest regions by bacterial reduction; the result is a high content near the lower limits of the trophogenic zone.

Phosphorus

In ecological thinking, phosphorus is often considered the most critical single factor in the maintenance of biogeochemical cycles. This extreme importance stems from the fact that phosphorus is vitally necessary in the operation of energy transfer systems of the cell, and that it normally occurs in very small amounts. The latter factor means that there is apt to be a deficiency of the nutrient, and this in turn could lead to inhibition of phytoplankton increase, resulting ultimately in decreased productivity in the system.

Phosphorus is known to occur in several forms, those of greater concern in natural waters being soluble phosphate phosphorus, soluble organic phosphorus, and particulate organic phosphorus of the seston. In water, phosphate may enter into combination with a number of ions but more conspicuously perhaps with iron and with the usually abundant calcium. The pH of the water determines to a great extent the nature of the phosphate compound. Under circumneutral and moderately alkaline conditions calcium phosphate is probably prevalent, while extremely high pH usually results in the formation of sodium phosphate. In acid waters phosphate attraction swings toward iron to form ferric phosphate. Except for precise investigations of the activity of the ion itself or of biological assimilation of the various fractions, it usually suffices in general limnological studies to report total phosphorus, or phosphate.

The concentration of total exchangeable phosphorus in natural waters is determined primarily by (1) basin morphometry as it relates to volume and dilution, and to stratification or water movements; (2) chemical composition of the geological formations of the area as they contribute dissolved phosphate; (3) drainage area features in relation to introduction of organic matter; and (4) organic metabolism within the body of water, and the rate at which phosphorus is lost to sediments. Ground waters and flowing surface waters are typically richer in inorganic phosphate than are surface waters of open lakes, due mainly to less biological demand in proportion to water volume. In most open lakes assimilation by phytoplankton and bacteria serves to reduce the inorganic phosphate content. In closed basins of arid regions evaporation may result in very high concentrations of total phosphate. With respect to geological influence, it should suffice to say that waters in local regions of highly phosphatic substrate contain considerable quantities of the ion. It is doubtless safe to state that all bodies of water that support some plant populations contain a quantity of phosphate, albeit small, in some cases less than 0.001 ppm (frequently expressed as milligrams or micrograms per cubic meter because of the minute quantities normally encountered). Variation in total phosphorus content of selected lakes is shown in Table 10·3. Note especially the difference in mean content of lakes in the highly phosphatic region of Florida and those of other regions. The mean total phosphorus content of most lakes ranges from about 0.010 to 0.030 ppm. Goodenough Lake, shown in the table, is a highly saline lake in an arid region.

The seasonal distribution of phosphorus in lakes is variable, being de-

TABLE 10·3. COMPARISONS OF DISSOLVED PHOSPHORUS IN LAKES OF VARIOUS REGIONS

Concentrations are given in parts per million (ppm).

(Data from Hutchinson, G. E., "A Treatise on Limnology, Vol. I, Geography, Physics, and Chemistry," John Wiley & Sons, Inc., New York, N.Y., after various authors; Odum, H. T., 1953. "Dissolved Phosphorus in Florida Waters," *Rept. Florida Geol. Surv.*, Vol. 9, 1-40.

	Dissolved Phosphorus, ppm	
	Mean	Range
HUMID CLIMATE; EXTERNAL DRAINAGE		
N. E. Wisconsin (Birge and Juday)	0.023	0.008 to 0.140
Connecticut (Deevey)		
Eastern Highland	0.011	0.004 to 0.021
Western Highland	0.013	0.007 to 0.031
Central Lowland	0.020	0.010 to 0.031
Japan (Yoshimura)	0.015	0.004 to 0.044
Austrian Alps (Ruttner)	0.020	0.000 to 0.046
Sweden (Lohammar)		
Uplands	0.038	0.002 to 0.162
North	0.024	0.007 to 0.064
Florida (Odum)		
Phosphate regions	0.290	0.100 to 0.660
Other regions	0.038	0.000 to 0.197
ARID CLIMATE; SALINE LAKES, INTERNAL DRAINAGE		
California		
Owen's Lake	0.078	. . .
British Columbia		
Goodenough Lake	0.208	. . .

termined to a great degree by basin form, chemical composition of the surrounding terrain, land use, behavior of other substances in the particular lake, and the annual cycles of mixing. In relatively open basins receiving considerable influx of surface water, phosphorus concentration is often regulated by stream discharge, particularly following high rainfall. The Maumee River drainage of Ohio annually contributes on the order of 96 metric tons of soluble phosphate to the western basin of Lake Erie. Of this total, 68 per cent is carried in seasonally during January and February, the time of greatest stream discharge. The Maumee load is apparently derived as leachings from the highly agricultural region, for other streams of the area transport proportionately less phosphorus. In lakes of more self-contained dynamics seasonal variations in total content and vertical distribution of phosphorus are rather complexly related to other compounds and mixing cycles. In eutrophic lakes during summer stratification the phosphorus content in the hypolimnion increases significantly following oxygen depletion. The factors involved in this increase are not fully understood. One investigator has shown that this phosphorus is apparently released from a ferric iron-phosphorus complex which is insoluble in the presence of oxygen; in the absence of oxygen the ferric compound is reduced to a soluble ferrous form, thus liberating phosphorus. With lake overturn and the reintroduction of oxygen the insoluble ferric phosphate is again formed and distributed throughout the lake. From another source has come evidence that in the presence of oxygen phosphorus is adsorbed on basic iron compounds in the oxidized microzone of the mud. Removal of oxygen brings about a reaction in which the ferric ion is reduced and phosphorus is released. While it is possible that both mechanisms operate, it seems certain that the presence or absence of oxygen is a critical factor. Soluble inorganic phosphorus in the upper waters of eutrophic lakes is usually low throughout the year, becoming depleted at times during summer. The concentration of total phosphorus (mainly the organic form) may increase in late summer.

THE CYCLE OF RADIOACTIVE PHOSPHORUS IN LAKES

We are all aware of the astounding array of new tools of everyday living and instruments of research made possible by the twentieth-century exploitation of atomic energy. Some of the materials and techniques have already entered into limnological investigations, and many more are destined for future usefulness. Prior to the "radioactive era" our knowledge of phosphorus activity in natural waters was based on field and laboratory analyses and measurements of seasonal variations in concentration of the several forms of phosphorus, and the distribution of the substances in natural waters. Now through the use of radioactive phosphorus (P^{32}) we know quite a bit about the rate and processes of phosphorus circulation in at least certain types of natural waters. Even so, our knowledge at this stage is still meager, and much is yet to be learned. Actually, radioactive phosphorus is only one of many such materials presently available for use in new and exciting research in the dynamics of natural communities, be they aquatic or terrestrial. Recently, isotopes of calcium (Ca^{45}), strontium (Sr^{90}), and yttrium (Y^{90}) have been utilized in researches bearing on biological limnology.

Of the many questions asked in investigations of nutrient cycles, those of utmost importance concern the source of the nutrient, the forms in which the nutrient occurs throughout the cycle, the manner of transformation of the fractions, and the rate at which the nutrient is circulated. In the case of phosphorus, for example, the major forms such as we have just considered have been known for some time. How are these forms produced and at what rate? We also know that when phosphorus fertilizer is dumped into a lake most of the compound disappears rapidly, the disappearance sometimes being acompanied by increased plant production and sometimes not. How do we account for this? Among other things, we are aware of the mud-water differential in phosphorus concentration, but as indicated previously we are not sure of the causes, or the rate of turnover of phosphorus from these and other sources.

If a quantity of radioactive phosphorus (P^{32}) is diluted to a desired strength of radioactivity (usually as counts per minute) and added to a body of water or used in laboratory experiments, the isotope enters into the system under study. The fate of the introduced P^{32} can then be determined as radioactivity of inorganic and organic phosphorus in living and nonliving components of water and mud. The rate at which the isotope is transferred within the system may be measured also. Let us now review some of the findings from experiments designed along those lines.

In the first place, it was found that radiophosphorus, when added to a lake surface, decreased at a high rate as had been observed for fertilizer. Within a few days between 70 and 90 per cent of the P^{32} was lost from the epilimnion. This led to the idea that phosphorus in lakes is held in a kind of repository from which the nutrient is withdrawn to maintain a steady-state equilibrium. In other words, there probably exists a sort of reciprocal exchange relationship between phosphorus in the water and its contents, and phosphorus in solids in contact with the epilimnion (recall that during stratification there can be little mixing of bottom-produced substance in the epilimnion). This state of equilibrium leads to the question of rate of exchange, or turnover time, between the two phases. Turnover time, in this case, refers to the rate at which atoms of phosphorus move from one phase to another, as for example from the lake water to the bottom mud. More precisely, it is the time during which as many atoms are transferred through a phase as there are atoms in that particular phase. The use of tracers permits recognition of atoms of a given phase. (Further details relating to concepts and methods are found in publications of Coffin *et al.*, Harris, Hayes, and Rigler cited in the Bibliography.)

A cycle of phosphorus, including turnover times of the components, is shown in Figure 10·2. The exchange rate between mud surface and water, at the bottom of the figure, was determined from laboratory experiments rather than from studies of a lake–in which the rates are apt to be rather variable. Exchange rates in the water phase are probably constant within lakes generally. Of particular interest is the rate at which the introduced inorganic P^{32} is taken up. Note that there is considerable competition among several factors for inorganic phosphate. Laboratory experiments indicate that over 95 per cent of the original radiophosphorus may be taken up within 20 min by phytoplankton and bacteria, the turnover time of the inorganic fraction in the epilimnion being on the order of 5 min. Certain phytoplankton can convert inorganic PO_4 to the organic state in less than 1 min. As shown at the lower right (and this is probably the same substance as at lower left) about 4 per cent of the planktonic cells that take up inorganic PO_4 are sedimented to the bottom each day. In-

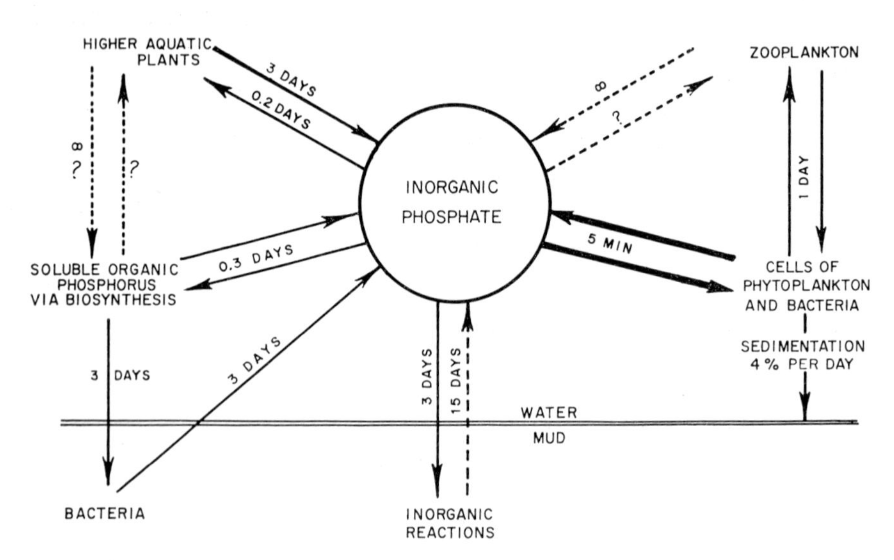

FIGURE 10·2. Phosphorus Relationships in a Lake. Turnover times for the various equilibria are shown. The bold solid lines indicate the first reaction with floating cells with time shown in minutes. All other times are given in days. The lighter solid lines are reactions at intermediate rates, on the order of two or three times slower than the initial reaction. The dashed line shows the return from mud by inorganic release, a much slower turnover. The dotted lines at upper left indicate reactions too slow to measure, being infinitely slow by comparison with the others. (After Hayes, F. R. and Phillips, J. E., 1958. "Lake Water and Sediment, IV. Radiophosphorous Equilibrium with Mud, Plants and Bacteria Under Oxidized Conditions," *Limnol. Oceanog.*, Vol. 3, 459-475.)

organic reactions involving the uptake of inorganic phosphate through the water-mud interface are given at lower center, these reactions proceeding when bacterial activity is inhibited. It is here seen that the rate of return of the phosphate to the water (15 days) is much slower than the turnover from the water (3 days).

The activity of bacteria in phosphorus utilization revealed in these studies is of great ecological significance. It was found, for example, that these microorganisms take up large quantities of inorganic phosphate either by assimilation into their own bodies or by conversion to the organic fraction, thus making the nutrient unavailable for use by green plants. In the absence of bacteria, the rate of uptake of P^{32} by algae and rooted plants is very high. It appears that in the competition bacteria get a large share of the available inorganic phosphorus. This in turn could seriously limit production of animal mass in a community, although we do not know to what extent these autotrophic bacteria are utilized by consumer organisms in the food web. The problem of bacteria *versus* phytoplankton is difficult to assess, however, for we know very little about the phosphorus requirements in the maintenance of natural populations of the various algal species. Many studies have been made of this subject of plant food requirements, but in most cases laboratory-determined minimal concentrations necessary for growth have been higher than the average content of most natural waters. On the other hand, it has been shown that an excess of phosphorus inhibits plankton growth in laboratory cultures.

Iron

Iron is of particular interest because of its importance as a vital element in respiratory pigments of many animals, and because of its part in many chemical reactions in water. We have already seen various aspects of iron in relation to oxidation-reduction systems, and its reactions with carbonate and phosphate compounds. Whether or not these reactions enter into prominence in particular bodies of water depends upon the presence of the compounds other than iron and the momentary form in which the iron occurs. Iron is found widely in nature, usually as either bivalent Fe^{++} or trivalent Fe^{+++}. The bivalent, ferrous, state is soluble, but only under anaerobic conditions. In the presence of oxygen the trivalent, ferric, form is present as a colloidal complex in combination with other inorganic ions and simple decomposition products. With oxygen depletion, as for example in the hypolimnion in summer, the ferric form is reduced to ferrous, the latter going into solution. As a result of the breakdown of the ferric complex the concentration of silicate, phosphate, bicarbonate, or iron is often increased, depending, of course, upon the original chemical nature of the waters. It might be pointed out here that the mere absence of oxygen will not bring about the transformation from ferric to ferrous iron. The depletion of oxygen is a result of organic decomposition which also forms organic compounds that reduce the ferric iron. In some instances, that of ferrous bicarbonate for example, high carbon dioxide content and near-neutral or acid conditions are necessary in addition to absence of oxygen and presence of reducing substances of organic origin.

The vertical distribution of iron in lakes is primarily a composite picture of several forms of the element influenced by the solubility factors just considered. In view of the chemical conditions prevailing in the epilimnetic regions of most lakes the iron content, maintained by several forms, is quite low, usually less than 0.2 ppm. In the deeper zones the iron concentration is a function of oxygen content and redox potential. Where the hypolimnion is enriched with oxygen, as in oligotrophic lakes, the iron is normally contained as the nonsoluble ferric complex. If oxygen becomes deficient at the mud-water interface, ferrous iron may go into solution. In eutrophic lakes the hypolimnion typically contains iron in solution during the later stages of summer stagnation. Even at this time the content is normally low, generally a few parts per million.

One of the forms of iron which appears to be most readily available to phytoplankton is ferric hydroxide ($Fe(OH)_3$). This form is the usual product of the oxidation of ferrous iron in waters containing dissolved ferrous salts. It normally occurs as an amorphous mass of a ferruginous organic complex. In the utilization of the organic component of the complex by bacteria, ferric hydroxide is precipitated as a by-product. This process also leads to the formation of lake ochre, the "parent" of iron ore. In the filamentous bacteria of the genus *Leptothrix*, iron hydroxide is left in sheaths surrounding the microbe as it metabolizes the organic material. The species of *Gallionella* excrete ribbon-shaped strands of ferric hydroxide that attach to objects in the water. This and such forms as *Ochrobium*, *Siderocapsa*, and others probably derive energy from the oxidation of ferrous iron.

Sulfur, Silica, and Other Elements

SULFUR

The most frequently encountered forms of sulfur in fresh waters are as the anion sulfate ($SO_4^{=}$) in combination with the common cations, and as hydrogen sulfide (H_2S). Sulfate enters bodies of water with rain, and through solution of sulfate compounds in sedimentary geologic formations in the drainage area. With respect to the latter source we should add that sulfate is not highly soluble. Surface waters are generally low in sulfate, except in regions locally rich in the ion, and in closed lake basins where evaporation raises the concentration. In some geographic regions rain may be the major source of the ion. As shown in Table 10·1, the average percentage content of sulfate in North American waters is about 15 per cent. Although this quantity places the ion second in abundance, it is also less than half that of carbonate. Some idea of the range of sulfate concentration in natural waters can be gained from Table 10·2. The low sulfate content of Bunny Lake is probably characteristic of youthful lakes (geologically speaking) formed in regions of weakly soluble terrain. Purcell's Pond (Table 10·2) is in an area of granitic rocks which normally contain little sulfur in proportion to the alkali and alkaline earth metals. However, this lake has a large peat bog along one side, and, since *Sphagnum* is known to accumulate sulfur from the atmosphere and release it as sulfuric acid, this contribution would augment that from rainfall. Nevertheless, the content of 9.5 mg SO_4/l for this pond is quite small compared with that of Hot Lake, Washington, which amounts to 103,680 mg SO_4/l at the surface and 243,552 mg/l at the bottom.

Sulfate is ecologically important in natural waters in several ways. It is apparently necessary for plant growth; short supply of the material can inhibit the development of phytoplankton populations and, therefore, production. Sulfur is important in protein metabolism and is supplied to the organisms originally as sulfate. As we shall presently see, under anaerobic conditions sulfate is utilized in chemosynthetic processes of sulfur bacteria, the sulfate being reduced to hydrogen sulfide. This in turn may result in the liberation of phosphates and other nutrients held in ferric complexes.

Seasonal pulses in sulfate concentration have been demonstrated in certain lakes. In some, the sulfate reaches a high in spring followed by decreasing content to a low in autumn. Bicarbonate variation in the same pond behaves in the reverse fashion. It has been suggested that the sulfate decrease is due to reduction to sulfide which becomes taken up in the bottom mud; the bicarbonate fluctuation apparently results partly from carbon dioxide removal by photosynthesis and partly from sulfuric acid activity in the winter oxidation of sulfide.

The presence of hydrogen sulfide and some aspects of its occurrence have been considered in Chapter 9. In relation to the sulfur cycle, however, we should here note the activity of bacteria in maintenance of the cycle. The reactions involving H_2S are essentially between oxidation and reduction. Hydrogen sulfide can be oxidized by a number of so-called "colorless sulfur bacteria," among which *Beggiatoa* is probably the best known. The species of this genus are large filamentous forms often said to be more closely related to blue-green algae than to bacteria. Under normal conditions the cells of the filaments include granules of elemental sulfur accumulated only when both oxygen and H_2S are present. When H_2S becomes exhausted from the environment, the sulfur in the granules is metabolized by the organism and sulfuric acid is released. The cellular oxidation exhibited by these bacteria apparently proceeds as follows:

$$2H_2S + O_2 \longrightarrow 2S + 2H_2O + \text{energy}$$
$$2S + 2H_2O + 3O_2 \longrightarrow 2SO_4^{=} + 4H^+ + \text{energy}$$

Another group of bacteria, particularly the genus *Thiobacillus*, is obligately chemotrophic and can oxidize elemental sulfur to release sulfuric acid. Certain other bacteria, including the euphoniously named *Desulfovibrio desulphuricans*, can reduce sulfate (and other sulfur compounds), resulting in the release of H_2S. With this activity the cycle of consumption and production of H_2S is completed.

SILICA

Over 60 per cent of the rocks and soils at the earth's crust is silicon dioxide (SiO_2). In view of this proportion we should expect to find the compound in most unpolluted waters. As is the case with many other substances, silica is most abundant in sedimentary rocks and therefore generally occurs in higher concentrations in waters in regions of such rocks. Even within a given region, however, the silica content of waters may be quite variable. For example, in Florida, an area in which sedimentary soils and rocks predominate, the concentration of SiO_2 ranges from less than 1 mg/l to over 30 mg/l. Unlike many other minerals, silica does not appear to be important in the composition of animal or plant protoplasm. It does occur, however, in the "shells," or frustules, of certain algae (the diatoms), in cysts of yellow-brown algae, and in skeletal spicules of some sponges. Population development of such diatoms as *Asterionella*, *Melosira*, and *Tabellaria* (see Chapter 11) is limited, at least partially, by concentrations of from 0.5 to 0.8 mg SiO_2/l. The utilization of silica by one or more of these organisms takes up the mineral from lakes and streams, and often during blooms of diatoms the quantity removed may be great.

OTHER ELEMENTS

In addition to those elements previously considered, natural waters may contain others in various proportions. Manganese is commonly found, and though less abundant, its role is similar to that of iron. Aluminum, zinc, and copper are usually found in natural waters in varying quantities. Traces of molybdenum, gallium, and nickel have been detected on occasion. Uranium appears to be common in small amounts in most lakes and streams, and man may be a strong environmental factor influencing future concentrations of this element.

Organic Substances

Organic matter is present in solution in water, and as fractions of plankton and organic debris, the seston, is suspended in water. Organic substances are also found, often abundantly, in bottom sediments. The total content of these materials in lakes is derived essentially from two sources: (1) incoming waters and wave action at the shore, which introduce allochthonous matter into the lake to account for a portion of the total—the nature of the substances brought into the lake from the outside being quite varied, ranging from leaf litter of the shoreline to upland soil leachings and organic pollution; (2) autochthonous matter produced in the lake by living organisms and by decomposition of plant and animal bodies. The relative contribution of each of these sources to the lake total is of considerable importance in maintaining productivity. It would, therefore, be of interest to measure each, but at present we do not have precise, readily usable methods for doing so. Hutchinson (1957) has reviewed and commented upon several techniques.

Data on the organic composition of a large number of lakes in Wisconsin have shown that the total content and the proportions of various fractions of the total organic concentration vary widely in lakes. In organically poor lakes the carbon content ranges from 1 to 2 mg per liter, and, of the total organic content, carbohydrates comprise slightly over 73 per cent and proteins about 24 per cent. The richer lakes contain up to about 26 mg carbon per liter, with nearly 90 per cent of the organic matter being composed of carbohydrates and about 10 per cent of proteins. It appears that the protein content essentially decreases with increasing total organic content. In lakes poor in organic matter (little autochthonous substance) the seston organic material averages about 16 per cent of the total. In richer lakes seston matter comprises about 4 per cent of the total organic content. The data from these studies indicate that the particulate organic matter of lakes increases with increase in dissolved organic substance up to a point; beyond this point the seston content does not seem to be influenced by additional dissolved matter. This relationship would seemingly point to the fact that autochthonous organic materials do not contribute to the productivity of lakes. In fact, it has been shown that the protein content generally is indeed low in the allochthonous portion of lake substances.

Lake color may be correlated as a rule with the source of organic matter. Allochthonous material derived from peat or shallow-water plant debris is mainly responsible for the dark-brown color of bogs. In such waters the protein fraction of the total organic content is low in proportion to carbon. In the waters of lakes receiving little allochthonous material the protein component is high in proportion to carbon, the autochthonous matter being formed mainly by plankton decay. These waters are relatively uncolored. Color and source of organic matter are apparently associated with plankton production also. In highly stained waters which contain considerable quantities of allochthonous substances, plankton may comprise as little as 3 per cent of the total organic substance. In unstained lakes receiving little allochthonous materials, over one fourth of the organic matter may be plankton.

The total organic content of most "average" lakes is as much as three or four times that of open seas. Sea water generally contains about 2 mg carbon and about 0.2 mg organic nitrogen per liter. These differences are of interest, but probably bear little on the composition of estuaries.

Most estuarine basins receive considerable quantities of allochthonous materials from stream inflow and drainage from surrounding marshes or other land features.

An impressive array of organic compounds has been isolated, identified, or otherwise recognized in fresh and marine waters, bottom sediments, and soils (see Vallentyne (1957a) for a comprehensive review of the subject). At present we know very little about the role of most of the organic compounds in the over-all ecology of freshwater or marine communities. There is sound evidence that certain bacteria can utilize dissolved organic substances as an energy source. Beyond this it has not been established that multicellular animals make use of such materials, even though a considerable controversy over the matter has existed for some 50 years.

Fats, in the form of simpler molecules of fatty acids, are known from freshwater and marine seston and water. Acids and certain waxes have been recovered from sediments. Some attention has already been given to organic nitrogen in natural waters. Other compounds such as free amino acids, tryptophan, glycine, glutamic acid, tyrosine, and others, are known from water, seston, and sediments. These have been recovered upon hydrolysis. Although studied but little, some of these amino substances may be important to organisms which are unable to synthesize certain of the acids, the organisms obtaining the items directly from the environment. Hydrolysis of seston and sediment components reveals a number of sugars and sugar-like compounds, the best known being glucose and galactose. In estuaries a compound having the qualities of a reducing sugar has been shown to influence the water-pumping rate of oysters.

Yellow carotenoid substances of fluorescent qualities have been extracted from shallow marine waters, and from fresh waters. The marine compound may not, however, be the same as yellow acids of lakes.

Fresh and marine waters contain biologically active compounds such as vitamin B_{12}. The seston of both environments has yielded also vitamin A and vitamin D. Thiamin (B_1), biotin (H), and niacin (B_6) are known from lakes, the last-mentioned being present in amounts up to 0.89 mg/m^3. Enzymes are present in natural waters and probably enter into various biological reactions in the environment.

It is now known that some organisms, particularly algae, produce metabolites that serve in various ways to inhibit, stimulate, or generally regulate the development of natural populations of other algae. The chemical nature of the particular metabolites is not known, although the possibility of hormone-like substances and fatty acids has been suggested. The effects of these and the more specifically toxic compounds will be considered further in Chapter 13, in relation to populations.

DISSOLVED SOLIDS IN STREAMS

The reactions and behavior of the various ions and organic substances in streams are governed by chemical and biological processes and conditions such as we have considered in the preceding pages. The main differences between lake and stream characteristics with respect to dissolved solids usually pertain to relative concentration, composition, and longitudinal distribution of the substances. Seasonal variation in rainfall and surface runoff, and the geochemical nature of the drainage basin strongly influence the composition of waters of small streams, thereby imparting considerable individuality to streams even within a restricted region. Variability becomes the key word in any attempt to categorize small streams. This variability is seen not only when comparing individual streams; it is also observed along the gradient of a particular stream, being markedly influenced by edaphic factors, human endeavors, and channel morphology. Waters of large streams, on the other hand, typically exhibit general uniformity of composition, so much so that a quantitative expression of average content becomes meaningful. The mean composition, as percentage proportions, of the major cations of river water has been given as: calcium 63.5, magnesium 17.4, sodium 15.7, and potassium 3.4. Note that the order of prominence here differs slightly from that of North American waters, generally, as given in Table 10·1. In fact, the average composition of rivers is quite similar to that of open lakes, as might be expected.

The calcium content of surface waters varies with pH, substrate composition, and temperature. Acid swamp streams and streams on igneous rocks contain little or no calcium, but streams in regions of sedimentary geologic features often contain high proportions of the ion. Springs issuing from limestone frequently contain on the order of 70 ppm of calcium. This concentration is similar to that of many large North American streams; throughout much of its length the Missouri River contains, on the average, about 60 ppm of calcium. Much of the ion in large streams is carried to the sea, where it enters into the calcium cycle.

As in lakes, there also appears to be a correlation in streams between quality and quantity of algae, and calcium carbonate concentration. Under proper conditions calcium in streams may be precipitated as the monocarbonate by lime-encrusting plants, and some is taken up by shell-building animals. The magnesium content of flowing waters is also highly variable, ranging from trace quantities to nearly 500 ppm in warm mineral waters. Although calcium and magnesium react similarly, particularly in entering into carbonate combinations, their proportions in freshwater streams vary from about 2:1 to 10:1 or more, respectively.

Nitrogen compounds normally exhibit conspicuous seasonal fluctuations and pronounced variations along the gradient in small streams. The amount of nitrate nitrogen and nitrite nitrogen is influenced to a great degree by surface runoff and associated stream level and discharge (see Figure 10·3, A and E). Nitrogen in ammonium compounds is released into streams through decomposition of organic debris. In unpolluted streams the concentration is small, normally less than 1 ppm. Pollution raises the concentration of ammonium compounds, and this, within limits, increases biological productivity. In excess, certain compounds of ammonium can be toxic to stream organisms. High concentrations of nitrogen occur during times of winter and spring flooding, when plant populations are minimal. Interestingly, the floods that serve to bring in nitrogen may also scour algae from rocks in the stream bed, thus minimizing consumption during the temporary increase in nitrogen. We might add that rapid decrease in nitrogen follows closely behind the flush of flood water. In streams in more stable watersheds, the nitrogen content may be lowered in spring and summer by plant utilization, with continued decrease to late summer. The major source of nitrogen compounds in many streams is pollution by drainage from agricultural areas and from sewage. These factors contribute large quantities of the substances regionally and have noticeable effects on stream communities.

The phosphate compounds of streams are derived from biological and chemical processes along the stream course. During summer the concentration of inorganic phosphates may increase somewhat, due to biological activity. Increased surface runoff contributes to the phosphorus content of streams by introducing allochthonous phosphorus-containing substances (compare A and E in Figure 10·3). Although there is considerable longitudinal and seasonal fluctuation in content, there does not appear to be any great depletion of stream phosphorus such as we have noted in surface waters of lakes during the growing season. This is due partly to mixing and partly to the normally greater proportion of water to phytoplankton. In fact, as we shall see later, there is little, if any, true plankton in streams of fair velocity.

Iron occurs naturally in streams, although in relatively small proportions. In unpolluted flowing waters this ion usually takes the ferric form because of continual aeration and the presence of oxygen. Where organic decomposition is great, oxygen depletion may result in the transformation to the ferrous state and the precipitation of iron hydroxide. Similarly, in stagnant stream pools, particularly following subsidence of flood conditions, iron bacteria may grow rapidly and form masses of ferruginous substance in the pools.

Sulfur, as sulfate or hydrogen sulfide, occurs in varying quantities depending upon the composition of the substrate, the area of the catchment

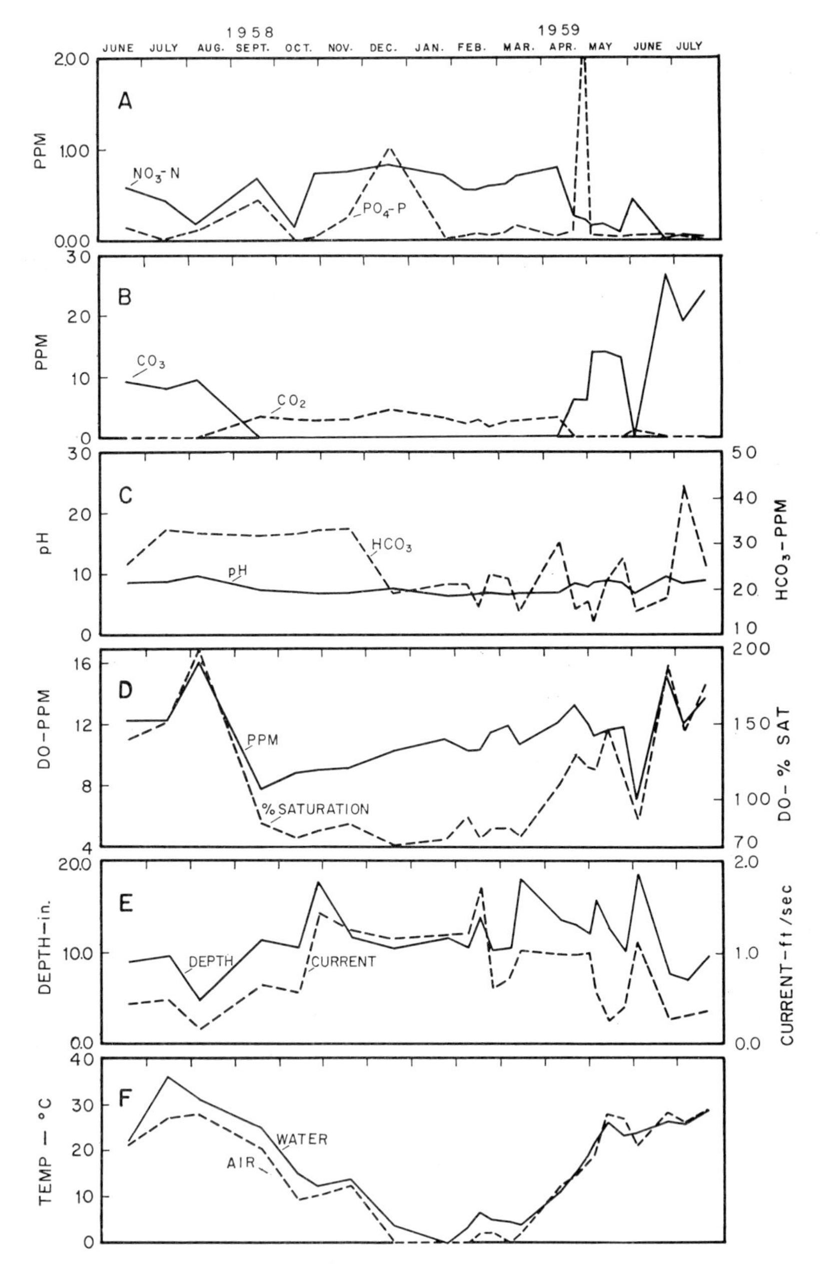

FIGURE 10·3. Seasonal Variations in Certain Physical and Chemical Aspects of a Segment of Stony Brook, a Small Stream on the Piedmont of New Jersey, Near Princeton. The graphs of the various data are presented in one figure in order to show correlations more clearly (dissolved oxygen content is designated "DO"). (From Woods, W. J., 1960. "An Ecological Study of Stony Brook, New Jersey" (unpublished dissertation), Rutgers Univ., New Brunswick, N.J.)

basin (its relation to rainfall), and the course of the stream. Obviously, the importance of the substrate rests on the proportion and solubility of minerals present. The stream course determines to a great extent the occurrence of hydrogen sulfide, for, if the course permits slow flow through organically rich swamps or marshes, the gas may be formed by decomposition or, as we have seen, by the reduction of sulfate.

A number of additional ions, including those discussed in the preceding section, are present in streams. Their abundance is usually associated with land forms and chemistry, and the ecological role of many of them is poorly known. The importance of climate, geochemistry, and physiography in stream composition has not been studied sufficiently. We can see some of the interrelationships in a few summarizing statements relating to certain of the major ions.

(1) The waters of rivers whose headwaters lie in semiarid plains are generally high in sodium, sulfate, and chloride, but low in calcium and carbonate.
(2) Streams draining youthful granite mountains typically contain high concentrations of silica, and the total dissolved solid content is low.
(3) Valley streams in granitic mountain regions show an increase in total dissolved solids.
(4) Rivers draining extensive plains in humid, temperate regions are characterized by high sulfate and carbonate concentrations.

DISSOLVED SOLIDS IN ESTUARINE WATERS

The chemical composition and activity and the distribution of ions in an estuary represent the end results of the meeting of normally dissimilar waters, those of streams and sea. We have just seen something of the variety in chemical composition exhibited by streams, and in any case an entering stream is almost certain to differ from the relatively uniform nature of sea water. Table 10·4 gives the percentage composition of ions in two streams that form estuaries on the North American coast, and of sea water. Even discounting the influences exerted by channel and basin morphology, tides, and currents, it is apparent that the chemical features of the two estuaries should differ.

Of the cations in sea water, sodium is by far the most abundant. Streams typically contain more calcium than any other element, and calcium and magnesium together constitute a greater fraction than does sodium. In streams the anion sulfate characteristically occurs in greater proportion than does chloride, and in many cases carbonate does also, but compare the carbonate and sulfate percentages for the Delaware and the Rio Grande in Table 10·4. Offshore and away from stream influences, sea water exhibits a nearly constant content. Sodium predominates among

TABLE 10·4. PERCENTAGE COMPOSITION OF THE MAJOR IONS OF TWO STREAMS AND SEA WATER

(Data from Clarke, F. W., 1924. "The Composition of River and Lake Waters of the United States," *U.S. Geol. Surv., Prof. Paper*, No. 135; Harvey, H. W., 1957. "The Chemistry and Fertility of Sea Waters," Cambridge University Press, Cambridge.)

Ion	Delaware River at Lambertville, N. J.	Rio Grande at Laredo, Texas	Sea Water
Na	6.70	14.78	30.4
K	1.46	.85	1.1
Ca	17.49	13.73	1.16
Mg	4.81	3.03	3.7
Cl	4.23	21.65	55.2
SO_4	17.49	30.10	7.7
CO_3	32.95	11.55	$+HCO_3$ 0.35

the cations of sea water, making up nearly one third of the total salts content, and chloride is the most abundant anion, comprising over one half of the total ionic composition. Thus the chemical composition of any estuary must represent the dual contribution from fresh and salt water, in addition to which local drainage may play a part.

SALINITY

Numerous references have been made to salinity in preceding sections. These references have shown the influence of this factor on various estuarine parameters such as density and stratification, temperature, and dissolved gases. Let us now give attention to a more precise definition and to consideration of salinity in its more or less true light, that of its chemical aspects.

The ions shown in Table 10·4, together with strontium, bromide, and boric acid, contribute over 99 per cent of the total salts in solution in full-strength sea water. The weight in grams of these salts dissolved in 1 kg of sea water is termed the *salinity*. In the process of drying sea water in order to weigh the salts, carbonates are decomposed to oxide, and bromine and some quantity of chlorine are liberated; these transformations and losses are taken into consideration in the definition. Salinity is usually expressed as *parts per thousand* (‰), although *per cent* (%) and *grams per kilogram* are also used. Since the salts of the metallic ions, the conservative elements, occur in very uniform proportions, it is possible to determine the salinity by measuring the chloride content, or *chlorinity*, and applying it in the relationship:

$$\text{Salinity ‰} = 0.030 + 1.8050 \times \text{chlorinity ‰}$$

Chlorinity is rather easily and accurately measured by silver nitrate titration, using potassium chromate as an indicator. Salinity can be determined by electrical conductivity, and from measurements of density obtained by the use of hydrometers.

The salinity of open seas generally ranges between about 33 and 38‰, with the average being near 35‰. Since the average salinity of soft fresh water is 0.065‰ and of hard fresh water 0.30‰, it is apparent that high estuarine salinities are derived almost wholly from sea water, while the diluting effect of the influent serves to reduce the concentration of dissolved salts. In general, the proportion of dissolved salts in estuaries resembles that of sea water, while the total concentration is variable along the axis of the estuary. In some instances, however, the concentration of inflowing fresh waters may be such as to modify the normal ionic relationships in estuaries. Where this occurs the result is usually an increase in the ratios of carbonate and sulfate to chloride, and of calcium to sodium over those of average sea water. The momentary salinity may be regarded as a function of the quantity and quality of inflowing and outflowing waters, rainfall, and evaporation. Since these factors may vary with season (in some instances rather drastically), the general structure of the estuary also shifts. Therefore, attempts to fit estuaries into schemes of classification are often difficult. We can, however, recognize that certain estuaries are typically, or on the average, more or less saline than others. For example, the average salinity of the estuary-like Laguna Madre of Texas nearly always exceeds that of sea water. The waters of the James River estuary of Chesapeake Bay, on the other hand, grade from fresh to a salinity of about 17‰ near the mouth. On the basis of salinity characteristics such as mean and range, estuaries can be classified under various systems. One of these, the 1958 "Venice System," classifies marine waters according to certain approximate limits as follows:

Zone	Salinity, ‰
Hyperhaline	> 40
Euhaline	40 to 30
Mixohaline	(40) 30 to 0.5
Mixoeuhaline	> 30 but $<$ adjacent euhaline sea
(Mixo-) polyhaline	30 to 18
(Mixo-) mesohaline	18 to 5
(Mixo-) oligohaline	5 to 0.5
Limnetic (fresh water)	< 0.5

According to this scheme the James River would be classified on the basis of salinity range as mixohaline, while its mean salinity of about 10‰ permits a designation of mixo-mesohaline. Laguna Madre would obviously be termed hyperhaline. The salinity range of Great South Bay (Figure 4·2) indicates mixo-polyhalinity.

Salinity is a very critical factor in the distribution and maintenance of many organisms in the estuary. To these plants and animals mean salinity is probably of less importance than the rate and magnitude of seasonal and tidal fluctuations in salinity. A bottom-dweller in the upper part of the Tees estuary (Figure 10·4) may be subjected to a change of 12 or 13‰ during a winter tidal cycle, and seasonal variations in other

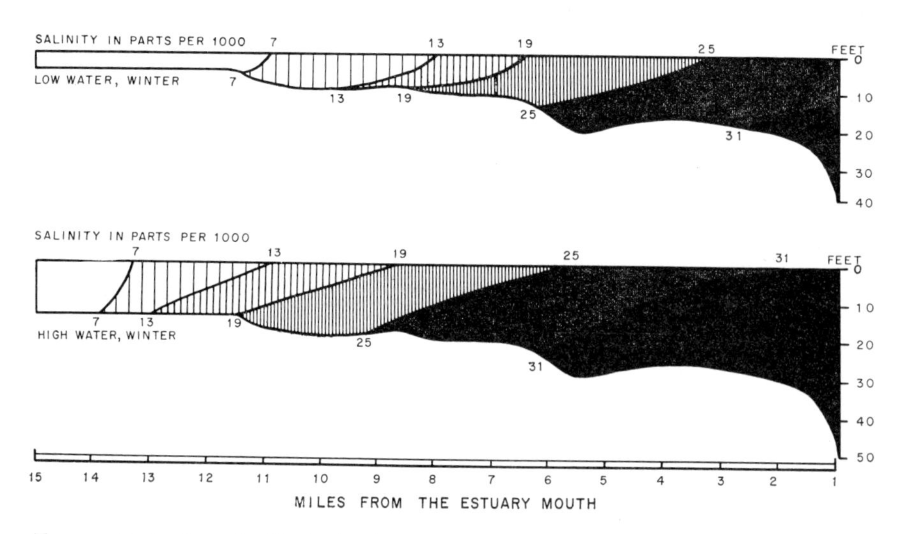

FIGURE 10·4. Longitudinal Section of the Tees Estuary, England, Showing Salinity Gradient at High and Low Tide in Winter. (From Macan and Worthington, 1951. "Life in Lakes and Rivers," *New Naturalist Series*, Wm. Collins Sons & Co. Ltd., London, after Alexander, Southgate, and Bassindale.)

estuaries can be much greater. The ability of organisms to withstand such changes is primarily related to their osmoregulatory adaptations, which will be considered in Chapter 13.

The horizontal and vertical distribution of saline waters in an estuary is closely related to the form of the basin and channel (Chapter 4), to tides and currents imposed by sea and stream (Chapter 8), to relative water volume contributed by the two sources, and to evaporation. These processes operate to varying extents to create a diversity of often complicated estuarine patterns. Generally, however, we may recognize a *positive estuary* in which the influx of fresh water is sufficient to maintain mixing and a pattern of increasing salinity toward the mouth of the estuary, usually in the mixohaline range. This type of estuary is characterized by low oxygen concentration in the deeper waters and considerable organic material in the bottom sediments. Vertical distribution of salinity in normal estuaries may range from top to bottom homogeneity, through oblique layering, to complete vertical stratification, depending to a great degree upon depth of the basin, morphology of the bottom and especially the mouth, and relative temperatures and magnitude of currents in the conflicting waters. Great South Bay is a shallow, mixing estuary-lagoon in which the vertical salinity is nearly uniform, and in which the horizontal pattern is one of increasing salinity toward the mouth (Figure 4·2). The positions of lines connecting masses of similar salinity (isohalines) indicate deflection of inflowing currents of high salinity toward the left side of the estuary (facing toward the mouth), and outflowing masses of low salinity toward the right side. This configuration results from rotational effects of the earth (Coriolis forces) in the Northern Hemisphere. The profile of salinity distribution in the estuary of the River Tees, England (Figure 10·4), shows a vertical stratification maintained during complete tidal cycles. As depicted in the figure, the entire water mass shifts vertically and slowly horizontally with tides. As the surface layers of fresh water move seaward, salt water flows upstream. The average time for all masses to move down the estuary is about six days during the dry season, and between two and three days during the wet period in winter.

It appears from the two examples of positive estuaries that salt water may flow into an estuary either as a sheet below a freshwater layer, or as something of a current moving along one side of the basin. This is essentially true, except that patterns of inflowing sea water may be complicated by movements ranging from the advection of a large salt-water wedge to eddy diffusion resulting from shallow-water tidal currents.

The second major type of estuary based upon salinity is the *negative estuary*. In arid regions, in particular, evaporation from an estuary may exceed the inflow of fresh water. This usually results in increased salinity in the upper part of the basin, especially if the mouth of the estuary is restricted by shoreline features that inhibit tidal flow. The resultant pattern of salinity distribution is therefore reversed with respect to that of positive estuaries. Negative estuaries (and lagoons) are typically hyperhaline, but possess a moderate oxygen concentration at depths. Bottom muds are generally poor in organic content. The Laguna Madre, previously mentioned, is an example of a negative lagoon. High salinity, sometimes up to 90‰, occurs in the upper segment. The salinity distribution pattern changes often, due to highly saline (more dense) masses being blown about in the lagoon by shifting winds.

We have given considerable attention to the more conspicuous and general reactions and relationships of the major ions in fresh water. These factors and processes have been studied very little in estuaries. However, as in freshwater research, much is known about certain phases of the chemistry of sea water. The chemical activities of calcium, magnesium, sodium, and potassium in salt water are basically the same as in fresh water, being tempered somewhat by salinity and by variation in relative proportion, and bearing certain relations to chlorinity which can be modified by the introduction of fresh water. These aspects of marine chemistry, along with fuller treatment of salinity, are found in Harvey (1957) and Sverdrup, Johnson, and Fleming (1942). In the estuary, therefore, we might expect the essential processes to be complicated by the meeting of chemically variable fresh water with relatively uniform sea water over an often considerable linear distance.

Similarly, the activities of nutrient elements and processes of nutrient cycles in estuaries are not greatly different from freshwater dynamics. The organisms involved, such as green plants, animals, and bacteria, may differ in name, but they perform similar ecological functions in both fresh water and estuaries. Estuaries may contain greater concentrations of certain nutrients than the adjoining sea, due to the introduction of the substances from terrestrial sources. Phosphorus, for example, may be in higher concentration in the upper reaches of an estuary than near the mouth. The element has also been found in the deeper, unlighted and anaerobic regions and in surface waters at times when photosynthetic activity is low; this corresponds to what we have found in fresh waters. Similarly, the content of nitrogen (in its various forms) and silica is often greater in estuaries than in sea surface waters. With respect to most of these, the estuarine concentration typically increases linearly toward that of the contributing fresh water. Surface-bottom differentials in the concentration of nitrite, nitrate, phosphate, silica, and others have been observed in stratified estuaries. In some, the concentration of silica is highest in fresh waters flowing over the surface, while the greater content of nitrite and nitrate is in the deeper water. The seasonal cycles of nitrate, phosphorus, and silica, as nutrients, in estuaries approximate those of lakes and the sea insofar as they are known. The pattern is essentially one of thorough mixing in winter due to uniform temperatures and winds, followed by spring and summer growth of phytoplankton, and depletion, or near-depletion, of nutrients in the upper regions during summer stratification (if present). During summer there may be little circulation of nutrients through the thermocline. Mixing and replenishment of the nutrient supply in autumn complete the cycle.

PART IV

Organisms in the Environment

THUS FAR IN OUR introduction to the fundamental features and processes in inland waters we have directed our attention to certain major aspects which we have conveniently termed the physical and chemical factors and processes. Our concern for these has derived from two basic points of view. In the first place, we are interested in the relationships between natural waters and each of the physicochemical factors for its own sake. Remember, however, how closely most of these factors are interrelated to other factors, many of them involving biological entities and processes. Our second concern for the physical and chemical features has been from the standpoint of their individual and collective contributions to the fitness of the environment. This concept relates, of course, to the surroundings of living organisms; and this brings us to the next major feature of natural waters, namely: the plant and animal inhabitants.

No organism lives in its environment without in some way having an effect upon it. We have seen a number of such effects reflected in the concentration of respiratory gases, the maintenance of nutrient cycles, and in other ways. Similarly, the environment exerts considerable influence over the kinds, abundance, and distribution of organisms inhabiting a given set of conditions. Thus, no organism lives unto itself in the economy of a natural body of water (and we shall pursue this idea much further in Part V); nor does any organism exist but that it bears some phylogenetic or evolutionary relationship to others. If these concepts are to be related to the great number of species in various aquatic communities and if the total of the interactions is to be understood, it is apparent that we must recognize the organisms and their characteristics. This job of recognizing and describing species belongs to the highly important biological fields of taxonomy and systematics.

Our purpose here in Part IV is essentially two-fold. In the first place we wish to know something of the types of organisms, their roles, and distribution in natural waters. Secondly, a knowledge of the characteristics of the organisms should make recognition of the kinds easier when we meet them in their habitats. To anyone who has had courses in general botany and zoology, the major phyla will be familiar. Our emphasis will, therefore, be on some of the less well-known, but taxonomically and ecologically important, representatives of those phyla, and on the phyla not normally stressed in introductory courses but which are, nevertheless, important in freshwater and estuarine communities. Similarly, illustrations will be devoted to forms frequently encountered in nature but not emphasized in elementary biology.

We cannot, of course, expect to account for all known species, even in North American waters; there are too many. We have, however, listed certain of the more recent and comprehensive treatments of aquatic plants and animals, at the end of each chapter in this Part. Monographic works on many of the animal or plant groups are available and are given in the bibliographic listings of the books named in our chapters.

Chapter 11

PROTISTS AND PLANTS IN INLAND WATERS AND ESTUARIES

AQUATIC COMMUNITIES SUPPORT a great variety of bacteria and other plant-like protists and plants. Most of the commonly recognized phyla are represented in fresh waters and estuaries. These segments of the biological world include the all-important populations which decompose organic material to the elemental state, and the photosynthetic species which are capable of fixing solar energy in protoplasm, thus forming the basis of aquatic food relationships. The small algae are particularly important in this latter aspect. Larger algae and the so-called "higher plants" often provide cover for populations of animals in aquatic communities. All green and colorless plants of natural waters contribute significantly to cycles of nutrients and respiratory gases in the ecosystem.

THE PROTISTA

In recognizing this group of organisms we break somewhat with a large segment of zoological and botanical tradition which has attempted to classify all living things as either plant or animal. The tripartite concept of protists, plants, and animals in the living world is not new; in fact, it is nearly 100 years old. As new knowledge is gained it is becoming more and more apparent that bacteria, slime molds, and flagellates (and perhaps the protozoa of Chapter 12) have more in common with each other than with what we might call "true" plants and animals. Three major features unite these four groups: (1) they are quite old and probably represent ancestral types of modern, true plants or animals; (2) many of them show both plant and animal characteristics, thereby making classification difficult; and (3) they are not clearly related to modern forms, therefore, a separate category seems advisable. Even now, however, there is lack of agreement as to what should be included (or not included) in this and other taxonomic categories. This is natural, because in the phylogeny and nature of living things there are no sharp lines of distinction.

PHYLUM SCHIZOMYCOPHYTA (BACTERIA)

Bacteria are extremely diverse organisms with respect to form, habitat, and function in the economy of inland waters. They range in size from about 0.5 microns (micron, abbreviated μ, equals 0.001 mm) to over 50μ. Mostly however, the size is less than 10μ. True bacteria occur in two primary forms: the spherical types, called *cocci*, and the cylindrical types, which include the *bacilli* and *vibrios*. Many secrete sheaths or stalks, thereby increasing the variety of body forms. Some of the true bacteria are capable of "swimming" locomotion. Another group, the Myxobacteria, distinct from true bacteria, move by gliding.

Within the aquatic environment bacteria inhabit the water phase, the bottom muds, and, of course, live on and in plants, animals, and detritus. As we have already seen, some exist under completely anaerobic conditions (obligate anaerobes), others require oxygen (obligate aerobes), while still others function equally well whether oxygen is present or not (facultative forms). In terms of deriving energy and maintaining their own metabolism and synthesis, a few bacteria are capable of photosynthesis and are said to be *autotrophic*. Chemosynthetic bacteria obtain energy from inorganic substances. Most bacteria, however, are *heterotrophic* in that energy is obtained from environmental sources such as organic compounds in the surroundings.

The contributions of a number of bacteria to community maintenance have been considered in preceding pages. We refer to those forms so instrumental in deriving carbon (as carbon dioxide, ammonia nitrogen, phosphate phosphorus) and sulfur (as hydrogen sulfide) from complex organic substances. Additionally, bacterial activity can reduce certain of those compounds to methane, molecular nitrogen, hydrogen sulfide, or nitrites. We see here some of the important activities of certain kinds of bacteria in the ecology of natural waters. The development and maintenance of the various bacterial populations is, of course, dependent upon the availability of raw materials in a particular body of water. For example, a whole host of bacteria, the sulfur bacteria, is found only in waters containing hydrogen sulfide. Some of these, the green sulfur bacteria, do not normally appear until the hydrogen sulfide concentration exceeds about 50 ppm. Similarly, the bacterial assemblage in polluted waters is determined primarily by the chemical nature of the pollutant.

PHYLUM MASTIGOPHORA (FLAGELLATES)

As a whole, this phylum includes a decidedly heterogeneous assortment of organisms. Most of the group possess flagella, or whip-like structures, that serve for locomotion (up to 300μ/sec), although some lose the structure in early development and move in amoeboid fashion. Among the many mastigophorans, almost every known type of nutrition is to be found. Some are holophytic, carrying on photosynthesis; others are holozoic, ingesting particulate food; many are saprophytic, absorbing decay products through the cell surfaces; and quite a number are parasitic. Certain flagellates are capable of alternating modes of nutrition; *Euglena*, for example, can be photosynthetic or saprophytic. Within this phylum three classes are recognized.

CLASS PHYTOMASTIGINA

This class includes a number of flagellates which are photosynthetic, some which are nonphotosynthetic, and others which may be either. The lumping of these into one class is based largely on biochemical and morphological similarities. Ecologically, the contributions of the members of this class are varied. Photosynthetic forms such as *Euglena*, *Cryptomonas*, and others are important in the cycles of respiratory gases. Some, in the role of saprophytes, aid in the breakdown of complex organic compounds. Many serve as food for larger organisms and thereby enter into food webs in the community. In the sea and estuaries, photosynthetic members of this class and the dinoflagellates, to be considered next, share with certain of the true algae some fundamental roles as producers in energy cycles.

CLASS DINOFLAGELLATA

This class includes forms which are holozoic, holophytic, and saprophytic. Most of the dinoflagellates are unicells, some possessing covering plates of cellulose while others are naked. *Peridinium* and *Ceratium* (Figure 11·1) are examples of the armored types. Both of these genera, as well as others, are represented in both salt and fresh waters. As shown in the figures, these dinoflagellates typically have one or more flagella lying in a groove about the body. The photosynthetic dinoflagellates possess chlorophyll *a*, chlorophyll *c*, carotene, four xanthophylls, and other yellow-brown pigments. Bioluminescence occurs in many of the organisms, and these are often responsible for "luminescent waters" of the sea and estuaries. Under certain conditions, the details of which are not fully known, the population of one or another of the flagellates in estuaries or open sea may increase rapidly, giving rise to a "bloom" which colors the water purplish or reddish-brown. *Gymnodinium breve* is responsible for the recent "red tides" that produce great mortality of fishes in the Gulf of Mexico. Other species of *Gymnodinium* may also discolor estuarine waters. In fresh waters, blooms of *Peridinium* or a host of others impart "fishy" odors to reservoirs or lakes. *Ceratium* in abundance produces a particularly obnoxious odor.

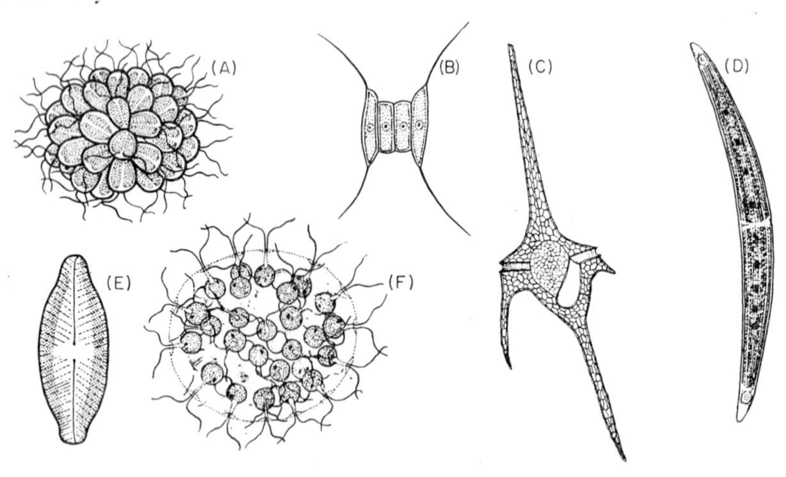

FIGURE 11·1. Some Representative Protists and Algae of Lakes and Ponds. (A) *Synura uvella* Ehrenberg, × 200. (B) *Scenedesmus quadricauda* (Turp.), × 450. (C) *Ceratium hirundinella* (O.F.M.) Schrank, × 250. (D) *Closterium* sp., × 125. (E) *Navicula* sp., × 675. (F) *Eudorina elegans* Ehrenberg, × 250. See text for classification. ((C) from Smith, G. M., 1950. "The Freshwater Algae of the United States," McGraw-Hill Book Co., Inc., New York, N.Y.)

CLASS ZOOMASTIGINA

This class is characterized by flagellated forms which are free-living or which have entered into various symbiotic relations, such as saprophytism or parasitism. Because of their nutrition, these are often placed with Protozoa. Some of these are pathogenic, others are obligate mutuals in the gut of cellulose-ingesting animals. The ecology of the free-living flagellates is not well known. In laboratory "wild" cultures, they typically represent a successional stage in a series of population changes, and in natural waters often show seasonal cycles of abundance.

PLANTS

In contrast to the vague phylogenetic relationships and the primitive features of the plant-like protists, the lineages, structure, and biochemistry of true or "advanced," plants are rather clear and indicate separation of these into a category, the Plant Kingdom. However, much controversy over this matter has prevailed for many years, and even today the classification of plants, animals, and "in-between" forms at the unicellular level is not settled. Reference to almost any textbook of zoology or botany will amplify this point. In the classification adopted here, plants include algae, fungi, bryophytes, and tracheophytes. All of these have representatives in fresh water; bryophytes apparently have not become adapted to marine existence and are, therefore, of limited distribution in estuaries.

Algae

Algae are found in natural waters in an impressive array of shapes, sizes, biochemical characteristics, and ecological roles. Most of the vegetation in the sea consists of algae, for few tracheophytes (no ferns at all) inhabit saline waters. In fresh waters a great variety of algae and tracheophytes (including some ferns) are found. Classification of algae is at present based upon chemical composition of food storage substances and of the cell wall, and upon the quality of pigments present. The algae and the plant-like protists constitute the major segment of the aquatic "pastures," the level in which radiant energy is fixed in protoplasm and transferred to non-autotrophic organisms ranging from zooplankton to fish to man.

Phylum Cyanophyta (Blue-Green "Algae")

The reference to "algae" in the name of these organisms indicates that the designation is used loosely, for they are often considered to be more closely related to bacteria than to true algae. Furthermore, many of them are not blue-green in color, but rather various shades of blue, yellow, red, and green. Each body, or cell, is encased in a gelatinous sheath secreted by the organism. Some of the blue-greens are filamentous, the forms being made up of numerous cells attached end to end; others are simply masses, or colonies, of cells. The cells contain a number of pigments, including chlorophyll, carotene, phycocyanin, and phycoerythrin, which impart certain colors to the cells. Also present in the cell are false vacuoles, various granules, and the diffused nucleus. The pigments of blue-greens are not contained in discrete bodies, the plastids, as in true green plants, but are diffused throughout the cells. Some of these organisms, such as the Nostocaceae, are able to fix free nitrogen in a fashion similar to bacteria. Reproduction is wholly asexual by cellular fission.

Cyanophytes most frequently occur in masses, either floating or attached to some object in the water. They are mainly inhabitants of fresh waters, but many are found in estuaries and in the sea. "Pond scum" or "water blooms" often seen in ponds and small lakes are sometimes formed by *Anabaena*, a filamentous form resembling a string of beads, or *Coelosphaerium*, a spherical mass of cells, or by other blue-greens. The color of the Red Sea is said to be caused by the presence of a red-pigmented cyanophyte, *Trichodesmium erythraeum*, which is a small filamentous species. The pigment phycocyanin, so common in these forms, is water-soluble, and release of it into a lake through disintegration of a water bloom may color the water. Although not necessarily "indicators" of pollution, blue-greens often thrive under such conditions, *Lyngbya* and *Oscillatoria* being notable examples.

PHYLUM CHLOROPHYTA (GREEN ALGAE)

These plants occur widely wherever there is water. Many are found in moist soil and on tree trunks, and even in ice. They are especially

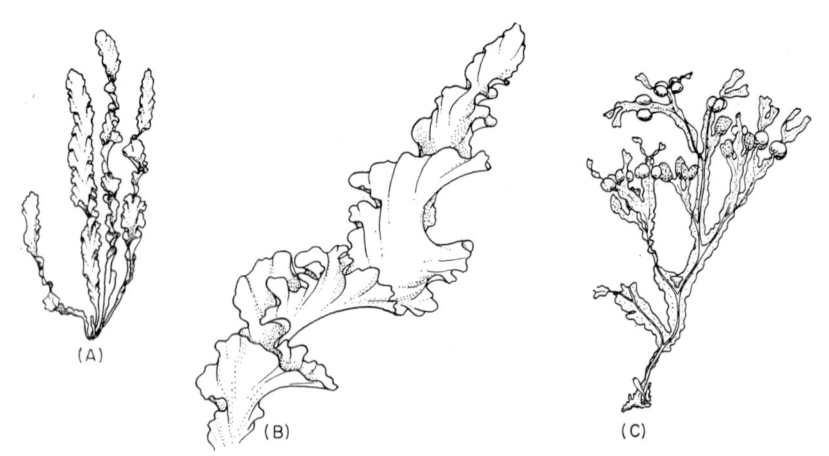

FIGURE 11·2. Green Algae of Estuaries and Marine Environment. (A) *Enteromorpha linza* (Linnaeus) J. Agardh. (B) *Ulva lactuca* L. (C) *Fucus vesiculosus* L. (Redrawn from Taylor, 1937. "Marine Algae of the Northeastern Coast of North America," Univ. of Michigan Press, Ann Arbor, Mich.)

common in a variety of body forms. Some of the greens are highly conspicuous, sheet-like colonies, particularly those such as *Ulva* and *Enteromorpha* (Figure 11·2) inhabiting estuaries and the sea. The thallus, or body, of other greens may appear as a continuous filament or as a partitioned, or septate, branching filament (*Cladophora* in Figure 11·3).

Single-celled forms such as *Chlorella* are sometimes common in lakes and ponds, often in numbers sufficient to color the water. *Chlorella* is also interesting in that it occurs in the cell body of some protozoans, *Stentor* for example, and in the tissues of certain coelenterates. The green hydra, *Chlorohydra viridissima*, owes its color to the presence of zoochlorellae in the animal tissues. It is presumed that the alga-hydra relationship is a symbiotic one in which there is exchange of nutrients and respiratory gases. The details of the association are not fully understood, however, for both members can exist independently. Colonial green algae, such as *Volvox* and *Eudorina* (Figure 11·1) are often common in lakes and ponds.

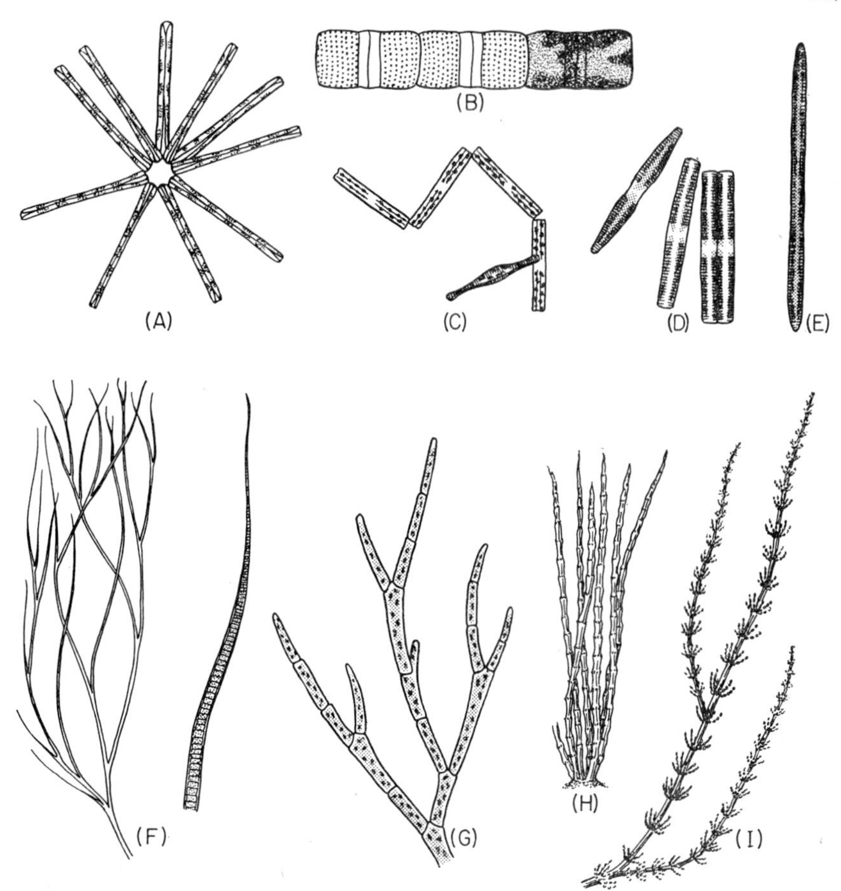

FIGURE 11·3. (Upper Row): Algae of Streams. (A) *Asterionella formosa* Hass, × 320. (B) *Melosira granulata* (Ehr.) Ralfs, × 660. (C) *Tabellaria fenestrata* (Lyngb.) Kuetzing, × 250. (D) *Fragilaria capucina* Desmazieres, × 330. (E) *Synedra ulna* (Nitzch) Ehr., × 350.

(Lower Row): Benthic Algae of Streams. (F) *Compsopogon coeruleus* Mont., nat. size. (G) *Cladophora crispata* (Roth) Kuetzing, × 80. (H) *Lemanea annulata* Kuetzing, nat. size. (I) *Batrachospermum vagum* (Roth) Ag., × 15. (Figures (F), (H), and (I) from Wolle, F. A., 1887. "Freshwater Algae of the United States," Bethlehem, Pa.)

Desmids are green algae which occur as single cells or sometimes as small colonies and filaments of minute cells. These often exhibit intricately beautiful patterns of body form. Desmids may be found on objects in the lake or freely floating as important constituents of phytoplankton. *Cosmarium* and *Closterium* (Figure 11·1) are common forms in lake or pond plankton.

Stoneworts, of which *Chara* is an example, are often placed in the Chlorophyta. These large algae exhibit a form and habit similar to tracheophytes in possessing whorls of branches along a stem-like filament and holdfasts to anchor the plants in the substrate. The common name derives from the brittle, stony texture of the algae caused by deposits of lime frequently found on the plant surface in highly calcareous streams or lakes. *Nitella*, similar in form to *Chara*, often thrives in soft or acid waters. Under such conditions, *Nitella* lacks the hard texture.

Chlorophytes range in size from microscopic cells and filaments to large colonies several feet long and quite broad. This group typically contains chlorophyll *a* and chlorophyll *b*, together with various carotenoid and xanthophyll pigments contained in chloroplasts. Some green algae may not appear green due to the masking of chlorophyll by one or more of the accessory pigments. Manufactured food is stored principally as starch, this feature making the iodine starch test a useful aid in identification of the group. Cellulose is present in the cell walls.

PHYLUM CHRYSOPHYTA (YELLOW-GREEN OR YELLOW-BROWN ALGAE AND DIATOMS)

This is a rather heterogeneous assemblage of algae containing chlorophyll *a* and chlorophyll *c*, beta carotene, and xanthophylls including fucoxanthin. Starch is absent, the food substances being stored primarily as other carbohydrates or as oils. Silica, rather than cellulose, is the major component of the cell walls.

Among the chrysophytes, diatoms are probably of greater importance in the energy cycle of natural waters. These algae are frequently considered among the foremost of the photosynthetic producers, especially in the sea, and to a great extent in the fast regions of upland streams. Diatoms are typically unicellular, although in some forms the individuals may be aggregated or variously shaped colonies such as the radiating colony of *Asterionella* or the chain-like colony of *Tabellaria* or *Melosira* as shown in Figure 11·3.

A most unique characteristic of diatoms is the enclosure of the cell in two siliceous shells, or frustules, which fit together in overlapping, or "pill-box," fashion. The frustules are frequently decorated with highly ornate and precise sculpture consisting of fine striations, pits, and shallow depressions. Reproduction is normally vegetative, and during the process the frustules separate; each new cell forms a new frustule to fit inside the original one. Thus part of a population is constantly being reduced in size. This diminution of size apparently continues until a certain minimum is reached, whereupon a form of sexual reproduction occurs involving the formation of an auxospore by two individuals. During this process the organism may grow and form new and larger frustules.

Diatoms are generally microscopic, but some reach a size of nearly 200μ. They occur abundantly as floating forms in plankton and on submerged objects such as stones, larger plants, and animals. In streams, the slick nature of stones may be due to dense diatom populations and their secretions. The "plankton" of streams is derived almost wholly from scouring of the substrate by stream action, and this plankton (tychoplankton) consists in large part of diatoms. It often serves as the principal food source for many small stream animals.

PHYLUM PHAEOPHYTA (BROWN ALGAE)

Of the thousand-odd known species of this group, only a few inhabit fresh water. One kind, *Heribaudiella*, forms relatively large disk-shaped crusts of dark color on stones in rapidly flowing upland streams. Most of the brown algae are adapted to existence in salt water, the distribution of many extending into estuaries until limited by decreasing salinity. These algae are the familiar large kelps and rockweeds found in coastal and intertidal waters. All are robust, multicellular plants. *Fucus* (Figure 11·2) is a common rockweed of cooler climates. The giant *Macrocystis* may grow to over 80 m in length. Most of the browns are attached to the substrate by holdfasts. One species of *Sargassum*, usually harboring a unique community of plants and animals, is a free-floating plant of the open seas. The floating of this form is made possible by morphological adaptations in the form of air-filled tissues which add buoyancy. In coastal waters and estuaries, brown algae in abundance provide numerous habitats in which populations of animals and of other plants develop and constitute a vigorous and productive community.

The pigments of phaeophytes include chlorophyll (*a* and *c*), carotene, and xanthophylls including fucoxanthin. The last-mentioned pigment predominates to give the characteristic color. Food reserve is maintained as a sugar, including mannitol, or as laminarin, the latter being unique to the brown algae.

PHYLUM RHODOPHYTA (RED ALGAE)

The members of this phylum are characteristically marine. A few forms, including *Batrachospermum* and *Lemanea*, shown in Figure 11·3, inhabit fresh water, mainly streams. *Lemanea* sometimes occurs in the upper reaches of estuaries where it receives washings in saline water. *Compsopogon* (Figure 11·3) is also a stream form, but it may be found growing on mangrove plants in estuarine or coastal waters. The marine species, about 3000 presently known, are widely distributed in the more temperate seas. All red algae are multicellular and exhibit considerable variety of form and coloration. One group, the genus *Gelidium*, is important as a major commercial source of agar. Most of the rhodophytes live at depths not normally encountered in estuaries. They are, therefore, somewhat beyond our consideration.

The red algae pigments include chlorophyll *a* and chlorophyll *d*, xanthophyll, carotene, phycocyanin, and phycoerythrin, the last two being chemically different from pigments of the same names in cyanophytes. Food reserve in the reds is in the form of floridean starch, which does not react with iodine in the standard starch test. The presence of the pigment phycoerythrin apparently accounts for the ability of these algae to exist at depths greater than most other plants. This pigment efficiently absorbs light in the spectrum's blue range which, as we have seen in Chapter 6, is usually transmitted more deeply than others in water. Another interesting feature of this group is the presence of nonmotile gametes, the meeting of sex cells being dependent upon water movements. It is almost universal that sexual reproduction in aquatic organisms involves gametes capable of self-propulsion.

Fungi

PHYLUM MYCOPHYTA

The fungi are plants lacking chlorophyll. The group includes, generally, such well-known forms as mushrooms, yeasts, and molds. Representatives of this group are essentially saprophytic or parasitic and are found everywhere that living matter exists. Out of a considerable array of kinds only relatively few are found in the aquatic environment. These, however, engage in the same processes as the nonaquatic fungi, i.e., aiding in the decay of organic substance and, in some instances, causing disease.

A majority of the aquatic fungi are classified in the class Phycomycetes, the tubular fungi. The fundamental structural unit of these fungi is the hypha, a slender filament of cytoplasm containing many nuclei and lacking cross walls. A mass of hyphae constitutes a mycelium, which may be loosely organized, as in bread mold, or compactly bound, as in a toadstool. Asexual reproduction is by flagellated spores borne in a sporangium. Sexual reproduction involving gametes also occurs. The fungus *Saprolegnia*, called "water mold," is a typical example of an aquatic phycomycete. It commonly causes infection of small fishes and appears as a whitish, fuzzy mass on the animal. Other molds, very similar in appearance, attack fish eggs during development, and in the natural habitat or in hatcheries may be of serious consequence.

A small number of sac fungi (Ascomycetes) occur in natural waters. These usually resemble miniature mushrooms (which they are not) on decaying plants. Many sac fungi cause diseases, and others, such as the famous *Penicillium*, produce antibiotic substances. The importance of these antibiotic materials in the natural environment is not known. As it relates to human welfare, the field of exploration for new compounds from aquatic fungi is as yet relatively untouched.

Bryophytes

PHYLUM BRYOPHYTA (LIVERWORTS AND MOSSES)

The bryophytes are relatively small plants lacking the flowers and specialized water-conducting tissues characteristic of "higher" plants. Water may be retained in the tissues of these plants for considerable periods of time, however, due to presence of numerous empty cells scattered throughout the plant body. The life cycle of bryophytes typically consists of two phases: the leafy, green gametophyte that produces motile gametes, and the usually brownish, nonleafy sporophyte generation that produces spores.

As pointed out previously, bryophytes do not occur in marine waters. About 45 genera of these plants occur in or near fresh waters of North America. Of these, about 12 are liverworts (class Hepaticae), which are small, flattened green plants. Some of these lack distinct development of stems and leaves; others possess these structures. *Riccia* is a slender, branched liverwort usually floating individually just beneath the surface of ponds, canals, and ditches. Often common in the same environment, *Ricciocarpus* is a notched, semicircular form about 1 cm in diameter, with black rhizoids on the undersurface. *Jungermannia* is a leafy liverwort found in streams and slow waters.

The class Musci includes the true mosses, those bryophytes with distinct stems and leaves. Peat moss, *Sphagnum*, is probably the best-known of the mosses. It is widespread and under proper conditions forms extensive bogs. *Fontinalis* and a number of other true mosses often occur in large waving masses, particularly in spring runs and mountain streams. These are typically long, sinuous plants with thick-set leaves arising in various fashions along the stem. Interesting assemblages of small plants and animals usually inhabit masses of these mosses.

Vascular Plants

PHYLUM TRACHEOPHYTA

As the name implies, these plants are characterized by the possession of conducting tissues (xylem and phloem) for the transport of materials throughout the plant body. Most of the tracheophytes exhibit structural specialization into true roots, stems, and leaves. The development of these organs is associated ecologically with life in a terrestrial habitat. Aquatic vascular plants comprise only a minority of the group.

The lower tracheophytes are not tolerant of salt concentrations normally found in sea water. Their distribution and importance in estuaries is, therefore, limited. Even in fresh waters the number of species of these lower vascular plants is relatively small. Water horse-tail (*Equisetum*) and the grass-like quillwort (*Isoetes*) are frequently common along the shores of fresh waters and sometimes grow submerged in a lake or stream. Among the ferns, the water shamrock (*Marsilea*) inhabits shallow zones; its leaves, consisting of four broad leaflets, arise from rhizomes growing in the substrate. Water fern (*Azolla*), shown in Figure 11·4, is

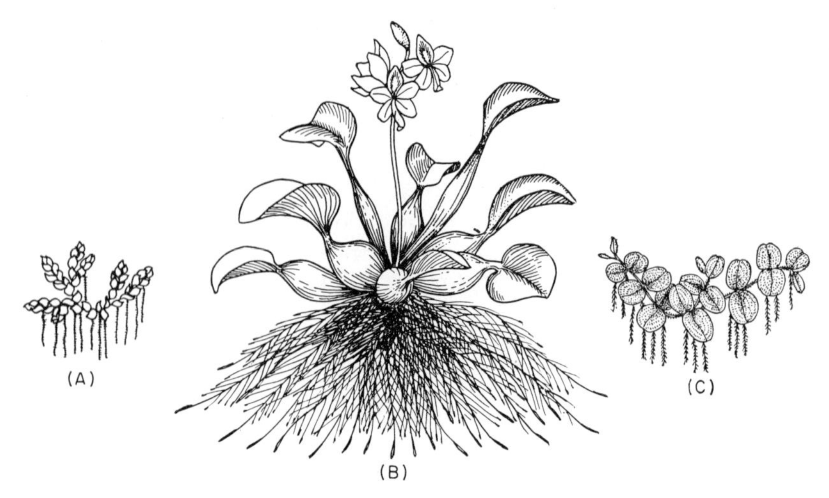

FIGURE 11·4. Floating Plants of Fresh Waters. Fern-like *Azolla* (A), the flowering water hyacinth, *Eichornia crassipes* (Mart.) Solms., (B), and *Salvinia*, (C).

a small, floating fern with overlapping leaves and fine roots hanging from the undersurface. The leaves, usually no more than 5 mm long, are green during summer, but turn reddish in the fall. Where abundant, these plants may completely obscure the water surface, the shading effect thereby inhibiting production in the water below.

Class Angiospermae (Flowering Seed Plants). The angiosperms include the plants which have, through the course of evolution, become quite successfully adapted to life on the land. The characteristics which distinguish these forms and also make possible the terrestrial habit include: (1) the presence of specialized parts, the sporophylls, which enclose seeds, (2) the seed itself which protects and nourishes the plant embryo, (3) the possession of flowers and fruits, and (4) the development of an elaborate system of tissues for conduction of food and water and for support; the continuous xylem vessels of angiosperms represent the ultimate in conductive structures of tracheophytes generally.

Of some 200,000 known species of flowering plants, relatively few grow as "true" aquatics in fresh waters, and only about 30 species inhabit salt water. This distribution, however, presents an interesting illustration of the part played by natural waters in the evolution of organisms and their invasions of "new" habitats. The angiosperms first appeared in the early Mesozoic. Some time during subsequent years some of these land-adapted plants invaded fresh waters, carrying with them the features necessary for terrestrial existence. In the water, adaptive radiation has resulted in many species occupying a variety of habitats and exhibiting considerable morphological diversity. From two widely adapted families of freshwater angiosperms, the Najadaceae, and the Hydrocharitaceae, arose the present-day marine angiosperms.* Their evolution was probably one of slow adaptation to increased salt concentration accompanied by migration down freshwater streams through the estuary and into the sea.

The recognition of a category to be labeled "true aquatic plants" demands a certain amount of latitude and qualification, for within the world of plants we find a very broad spectrum of adaptation to water. The adaptations relate to water requirement and water tolerance, and range from strongly xerophytic, through mesophytic, to hydrophytic forms. We may define "aquatic plants" as those whose seeds germinate in either the water phase or the substrate of a body of water and which must spend part of their life cycle in water. This ecological grouping includes plants which grow completely submerged (except when flowering) as well as a variety of emergent types. This general idea is shown in Figure 11·5. In the United States about 50 families of flowering nonwoody plants may be considered primarily aquatic.

The family Najadaceae is said to be the largest of the families of aquatic plants. In the United States this group includes about 13 genera representing nearly 60 species, some of which are estuarine and marine. The predominantly marine forms are *Zostera*, *Phyllospadix*, *Cymodocea*, *Thalassia*, and *Halodule*. *Zostera*, or eelgrass, is the most widely distributed marine angiosperm, occurring in coastal waters and estuaries throughout a vast portion of the cooler regions of the northern hemisphere. It grows commonly in shallow zones on sandy mud bottom. During 1931-1932, great mortality of eelgrass was experienced over extensive areas, particularly on the Atlantic coast of North America. This was serious because the plant was an important food of some waterfowl. Since about 1940, however, there has been considerable return of the plant. The cause of the mortality has never been fully explained. *Phyllospadix* inhabits coastal embayments and rocky shores of the Pacific coast of

* The meaning and validity of this generalization depend upon the point of view of systematics. Some authorities relegate several of the marine angiosperms to separate and dictinct families. Nevertheless, the basic idea of evolution of the marine forms from freshwater ancestors remains sound.

North America. *Cymodocea* and *Holodule* are restricted to the warm coastal waters of southern and tropical zones. *Thalassia* is common in these regions. Broad, shallow "flats" of sandy mud and the mouths of small estuaries on the northern Gulf Coast of Florida support abundant growths of both of these plants, together with another genus, widgeon

FIGURE 11·5. Some Examples of Flowering Plants of Fresh Waters. *Myriophyllum* (A) and *Utricularia* (B) are essentially submersed plants. The various species of *Sagittaria* (C) and *Potamogeton* (D) exhibit diverse adaptations, some possessing floating leaves, some being emergent (C), and others completely submersed.

grass (*Ruppia*), the latter being also common in alkaline waters throughout most of the United States. Two genera of pondweeds, *Potamogeton* and *Najas*, are essentially freshwater inhabitants, although a few species do extend into brackish waters. In the United States, these two genera include over 80 per cent of the species of Najadaceae. One or another of the species occurs in nearly all types of fresh waters. These plants are mostly submersed, some with floating leaves, exhibiting a variety of leaf form ranging from linear to broadly ovate. Ecologically, the pondweeds are of great importance in the cycles of nutrients and respiratory gases, and in often providing very dense habitats which supply food and shelter to numerous small organisms living on and among the plants. Many of the potamogetons serve as a major item of food for ducks and geese.

The water plantain family (Alismaceae) contains about 50 species of emergent marsh plants and submersed aquatics. In most of these the leaves possess long slender petioles with flattened blades. Two of the genera, *Alisma* and *Sagittaria,* are widely distributed in a variety of waters, some species of the latter entering the upper regions of estuaries. The conspicuous flowers of *Sagittaria* often add color to marsh communities. Also noteworthy here is the family Hydrocharitaceae, mentioned previously. This group contains the well-known genus *Anacharis* (*Elodea*) so popular with aquarists. One species was introduced into the United States from South America; another species, native to North America, was introduced into Europe and became quite a pest by clogging streams and canals. Hydrocharitaceae also includes two marine genera, *Halophila* and *Thalassia,* of warm waters, and *Vallisneria,* a form with long ribbon-shaped leaves, widely consumed by waterfowl.

Many grasses (Gramineae), sedges (Cyperaceae), and rushes (Juncaceae) typically inhabit marshes and shore zones; these include manna grass, cut grass, reed (*Phragmites*), cord grass, wild rice, sedge, saw grass of the Florida Everglades, spike rush, true rush, and others. These plants normally grow in the very shallow waters, but usually in profusion, and upon death and decay contribute to the richness of the body of water.

A number of angiosperms have become adapted for floating on the water surface. Their role in community activities is interesting in that the roots of the plants, dangling as they do in the water beneath the plant, extract nutrients from the water phase in competition with phytoplankton. The leaves, borne above the water, contribute little to the cycle of respiratory gases in the water. As the plants die, nutrients are returned to the water. Where abundant, these floating forms often shade the water from sunlight, thereby inhibiting production. *Pistia,* the growth form of which resembles garden lettuce, is common in the Gulf Coast states. The family Lemnaceae includes the duckweeds, which are minute plants of flattened or spherical body form. *Lemna* and *Spirodela* are duckweeds widespread in the United States, often in association with the liverwort, *Ricciocarpus,* and the fern, *Azolla,* described previously. Another floating species, but of the family of pickerelweeds, the water hyacinth (*Piaropus,* or *Eichornia*), shown in Figure 11·4, was introduced into the United States from South America during the latter part of the 19th century. Because of its very showy flower, it has been introduced widely. In favorable areas, however, the plant quickly becomes a pest—

navigationally, economically, and, insofar as recreational fishing is concerned, ecologically.

Various members of the water lily family are familiar to almost everyone. To this family belong the white water lily (*Nymphaea*), the yellow water lily (*Nuphar*), and the lotus (*Nelumbo*). In many lakes and streams patches of water lilies are favorite nesting areas for certain sunfishes. *Brasenia,* a small member of the water lily family, is found chiefly in acid waters in the eastern half of the United States; the stem and undersurface of the leaf of *Brasenia* are thickly coated with a gelatin-like substance.

Of the various habits and adaptations exhibited by aquatic plants, one more is worthy of mention; this is *Utricularia* (Figure 11·5). Most species of this plant are free-floating forms lacking roots. They may float near the surface of the water or hang near the bottom. In some, the leaves may function as roots in anchoring or absorption. Most of the species possess small bladders, developed from leaf segments, which have trap doors that can be sprung to capture minute aquatic organisms.

General References to Aquatic Plants of the United States

Fassett, N. C., 1957. "A Manual of Aquatic Plants," Univ. of Wisconsin Press, Madison, Wis.

Muenscher, W. C., 1944. "Aquatic Plants of the United States," Comstock Pub. Associates, Ithaca, N.Y.

Smith, A. M., 1950. "The Freshwater Algae of the United States," McGraw-Hill Book Co., New York, N.Y.

Tiffany, L. H. and Britton, M. E., 1952. "The Algae of Illinois," University of Chicago Press, Chicago, Ill.

Chapter 12

THE ANIMAL INHABITANTS OF INLAND WATERS AND ESTUARIES

In contrast to the ten or so phyla of plants usually recognized, the number of animal groups of the same level approaches 30 to 33. All of these phyla are represented in natural waters by free-living forms or by parasites inhabiting aquatic organisms, or both. The number of species known to spend some part or all of their lives in an aquatic environment is staggering. It is estimated that some 30,000 known insects are aquatic in some way; and these are restricted to fresh water (except for a very few). Approximately 4800 species of echinoderms are known; these are entirely marine. Another estimate gives 8500 as the number of known invertebrate animals, exclusive of protozoans, in the fresh waters of the United States. We have emphasized the *known* in these statements, for in many groups only a fraction of the existing species have been scientifically "described."

The first three phyla considered here are sometimes classified as classes in a phylum Protozoa. Some authors add to this phylum the Mastigophora, or at least the nonphotosynthetic forms. In still another scheme, all of the unicellular organisms are treated as Protista. For reasons stated previously, the classification adopted herein places the flagellated unicells with protists, and on the basis of strong differences in locomotor adaptations of the nonflagellated forms recognizes three phyla of single-celled animals.

Phylum Sarcodina (Pseudopod-bearing Unicells)

In addition to the well-known *Ameba,* a free-living sarcodine, and *Entamoeba,* a pathogenic parasite, this phylum includes a great number and variety of forms inhabiting freshwater or estuarine environments. Pseudopods, the distinguishing feature of the group, are used for feeding and locomotion. Many sarcodines are naked species; others possess shells composed of various substances.

Some of the materials of which sarcodine shells are composed contain ions that are important in biological and chemical processes; thus these animals enter prominently into biogeochemical cycles. In the sea, and

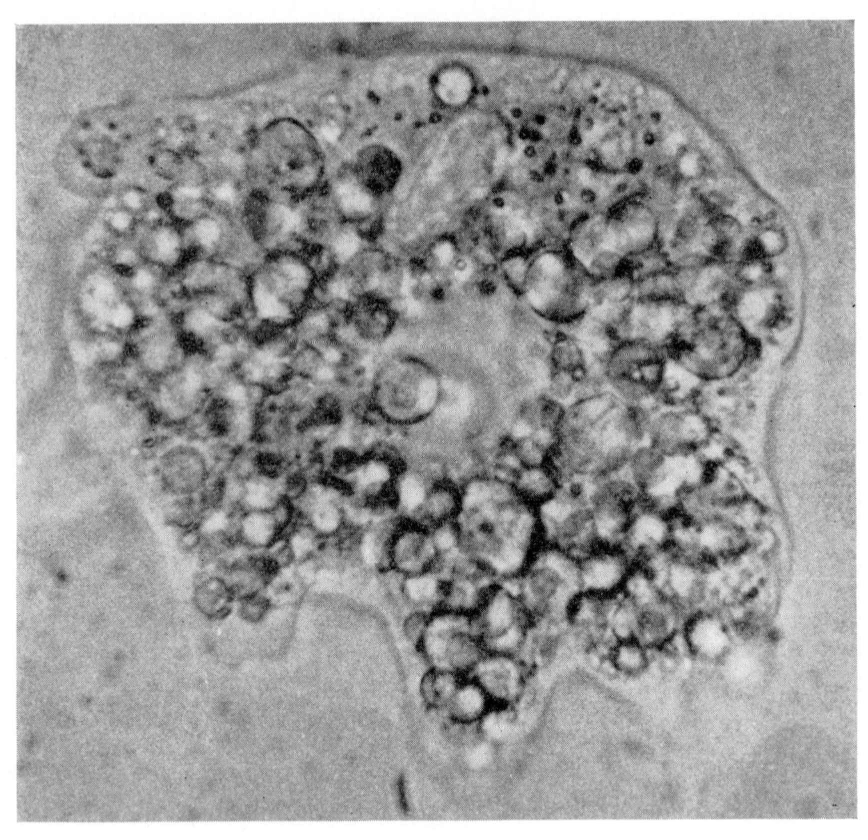

FIGURE 12·1. *Hydramoeba hydroxena* (Entz), a Sarcodine Predator on Hydra. Just to the right of the clear center region can be seen a freshly ingested cnidoblast of a hydra. (Photograph by Helen Forrest.)

to a small extent in fresh water, sarcodines of the order Foraminifera secrete coiled and spiralled, snail-like shells composed predominantly of calcium carbonate. Deposition of these shells in ocean sediments essentially "locks up" considerable quantities of carbon and calcium. The magnitude of such deposition can be appreciated from the knowledge that the famous White Cliffs of England are of marine origin and are composed mainly of "foram" shells. The sarcodine order Radiolaria is composed entirely of marine forms. These animals secrete variously shaped shells in which silica is abundant. In fresh waters, *Difflugia* (order Lobosa) builds urn-shaped coverings of sand grains cemented by a chitinous substance. *Arcella* secretes a hemispherical shell with a single opening (foramen) in the center of the flattened surface; the pseudopods extend through the foramen. Some of the sarcodine shells contain a bubble of gas, apparently given off by the animal, which is probably an adaptation for floating. The composition of the coverings of the shelled sarcodines varies considerably from mucilaginous materials to the above-mentioned chitinous substance.

The Heliozoa are sarcodines usually enclosed in gelatinous sheaths, and characterized by the possession of numerous long, stiff radiating projections of cytoplasm. The animals, among which *Actinophrys* and *Actinosphaerium* are typical examples, resemble sunbursts. Heliozoa are primarily freshwater forms.

In the activities of the community, sarcodines feed upon organic detritus, bacteria, small algae, and even other unicells. This they do by taking the food item into cytoplasmic vacuoles.

Hydramoeba hydroxena is a large sarcodine that preys upon hydra, a coelenterate animal. The ameba, shown in Figure 12·1, attacks the hydra, loosens and devours the cells, and eventually kills the prey. Sarcodines, in turn, serve as food for larger organisms.

Phylum Ciliophora (Ciliated Unicells)

This group is characterized by the possession of numerous hair-like projections, or cilia, which in some forms serve almost wholly for locomotion, and in others for creating currents which deliver food to the ciliate. These animals are the largest of the unicells and also exhibit the greatest complexity of structure. A variety of morphological types are known. The "slipper-shaped," actively swimming *Paramecium* is generally the best known. *Stentor* is a horn-shaped form that frequently attaches to some object in the water; when swimming it is somewhat globular. *Lacrymaria* has a serpent-like form when extended and moving. A number of genera, such as *Epistylis* and *Carchesium*, are sessile, colonial forms in which the cell bodies are borne on stalks. The family Tintinnidae is a marine group of vase-shaped ciliates bearing feathery projections about the oral region.

Although the body form of ciliates exhibits considerable diversity, the internal structures are highly developed as permanent organelles. The ciliates typically possess a mouth-like opening which leads into a cytopharynx, contractile vacuoles, neurofibrils which coordinate, contractile fibers for movement, and locomotor structures. A form of sexual reproduction involving exchange of nuclear substance as two animals conjoin is characteristic of ciliates. In all, the structure and organization of ciliates represent a considerable evolutionary "advance" over the protists and the sarcodines. To us, the structural complexity suggests that these are not "simple" organisms.

Phylum Sporozoa

This phylum includes unicellular nonmotile organisms all of which are parasitic, many being pathogenic. In our province of interrelationships of free-living inhabitants of natural waters, sporozoans are of little interest.

Phylum Porifera (Sponges)

The bodies of these animals consist of loosely organized masses of more or less independent cells. Some of the cells are specialized for the production of spicules, which are variously shaped structures that serve primarily as skeletal elements. The chemical composition of spicules is used as a basis for classification of sponges, some being made up primarily of calcium carbonate ("chalk sponges"), some of silica ("glass sponges," including the freshwater forms), and others of protein compounds, with or without inorganic spicules ("horny sponges"). The basic structure of sponges involves a somewhat porous or sac-like body, with numerous pores or canals ramifying throughout the tissues. Many variations in form and color exist, especially among marine species. Food is obtained from water circulated through the pores or canal system. Although the larvae may be motile, the adults of all sponges are sessile, and most are attached to the bottom or to some object in the water.

Nearly all sponges are marine. They are found in all seas in both shallow waters and at depths greater than 400 m. Most species are rather narrowly restricted by temperature and salinity. Marine species vary in size from small crustose masses to large vase-like or globular forms over 2 m high. Not many animals eat sponges, but the poriferan does frequently harbor an assemblage of algae, worms, insects, and crustacea, and fishes often nibble at the sponge in seeking the more tasty items.

One family of sponges (Spongillidae) occurs in unpolluted fresh waters. These are characterized by the formation of specialized reproductive structures, known as gemmules, which sometimes contain unique spicules. Although highly variable in form and size, the freshwater sponges do not resemble their more elaborately fashioned marine relatives. Most freshwater species are inconspicuous, green or dull-colored flat masses encrusting objects in lakes or streams. Among the 150 or so species, considerable variation in adaptation to environmental conditions is shown. Some sponges inhabit soft, or acid, situations; others are found in hard waters. In spite of the requirement of silica for spicule construction, many species exist in waters of remarkably low silica content.

Phylum Coelenterata

This phylum consists of a variety of simply constructed animals, the greatest number and most conspicuous of which are marine and estuarine. Included in the group, typically under three classes, are sea anemones and corals (Anthozoa), jellyfishes (Scyphozoa), and freshwater and marine hydroids (Hydrozoa). These animals are constructed of two relatively well organized layers of tissue in the form of a vase. The

Figure 12·2. A Small, Submersed Bit of Twig Inhabited by *Hydra americana* Hyman. (Photograph by Helen Forrest.)

single opening is usually bordered by tentacles, which aid in feeding. The cavity of the animal is the digestive chamber. A unique aspect of coelenterates is seen in the presence of "stinging cells" (nematocysts) variously modified to eject filaments which ensnare or penetrate and poison prey, or protect the animal itself. The larvae of coelenterates are ciliated and motile; the adult stage may be as a polyp attached to a substrate, or as a free-swimming (or drifting) medusa. Alternation of these forms is typical in the life cycle of hydrozoans such as the marine and estuarine *Obelia*, and the freshwater *Craspedacusta*.

Only a very few hydrozoan coelenterates have become adapted to the freshwater environment. These have evolved from marine stocks using the estuary as a route to freshwater habitats. No scyphozoans or anthozoans have apparently invaded fresh water, even though the phylum is a very ancient one and has, in the marine world, become highly diversified. In fresh waters, the hydras, well-known to almost anyone who has had general biology, are probably the most widespread. Of these, however, only about a dozen species are known in the United States. Two species, *Hydra oligactis,* the brown hydra, which ranges from several western states to New England, and *Hydra americana,* of the eastern

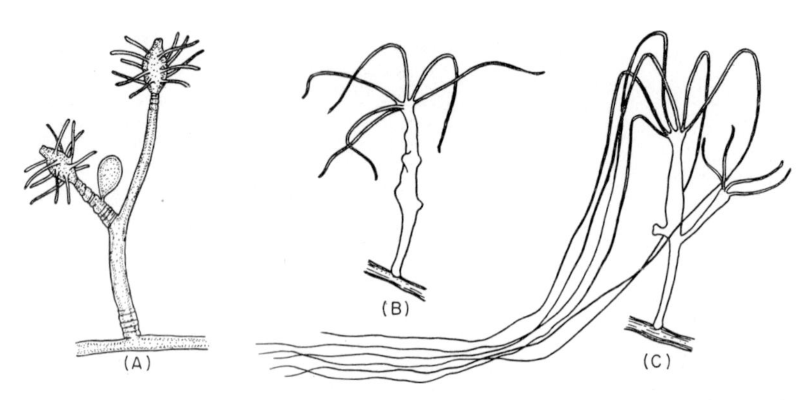

FIGURE 12·3. Freshwater Coelenterates. (A) *Cordylophora lacustris* Allman. (B) *Hydra americana* Hyman. (After Hyman, L. H., 1929. "Taxonomic Studies on the Hydras of North America. I. General Remarks and Descriptions of *Hydra americana,*" *Trans. Am. Microscop. Soc.,* Vol. 48, 242-255.) (C) *Hydra oligactis* (Pallas), the brown Hydra. (After Hyman, L. H., 1930. *Ibid.* II. "The Characters of *Pelmatohydra oligactis* (Pallas)," *Trans. Am. Microscop. Soc.,* Vol. 49, 322-333.)

United States, are shown in Figures 12·2 and 12·3. *Chlorohydra viridissima,* rather widely distributed in North America, and *Chlorohydra hadleyi,* of the northeastern United States, are of ecological interest, for the tissues of these forms normally contain large populations of an alga. This unicell, usually *Chlorella,* imparts a rich green color to the coelenterate. As pointed out in our consideration of algae, this relationship may not be an obligate one, for the algal cells can be removed from the hydra by glycerin treatment and the alga continues to exist. Similarly, the hydra can survive without the presence of the alga. The other hydrozoans of fresh water include a single species of "jelly fish" (*Craspedacusta sowerbii*), which appears to have been introduced into North America from South America or the West Indies. Its distribution in North America is not fully known, for it is sporadic in occurrence, sometimes not being seen for years in a locality in which it is known to have once been present.

The medusa of *Craspedacusta* (Figure 12·4) is generally about 15 mm in diameter and bears numerous tentacles. The hydroid is small and often difficult to find, because it is usually covered with silt or other debris. One species of a colonial polyp, *Cordylophora lacustris* (Figure 12·3), is widely distributed in fresh waters and in low-salinity regions of estuaries. Two types of polyps occur in the colony, a globular reproductive member which produces sperm or eggs, and a feeding polyp bearing tentacles. *Protohydra leuckarti* is a small hydra-like hydroid no larger than about 5 mm, and lacking tentacles. It typically inhabits estuaries, being known

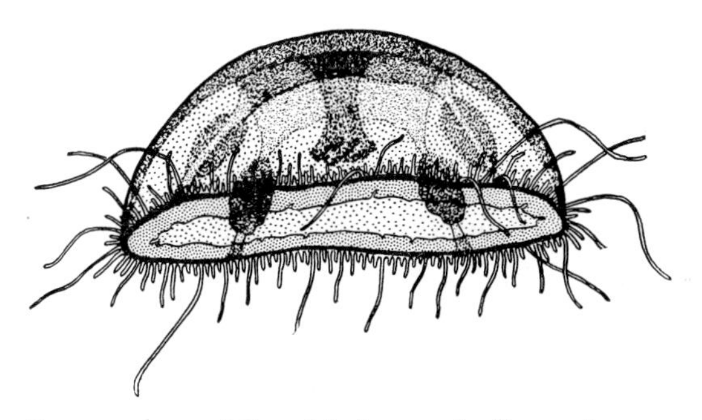

FIGURE 12·4. The Medusa of *Craspedacusta sowerbii* Lankester, the freshwater jellyfish.

at present from localities along the northern Atlantic coasts of North America and Europe.

Nearly 10,000 species of coelenterates are known from marine waters. Many of these fall within our estuarine province to varying extents, being limited by currents and salinity. These include a number of colonial hydroids, *Cordylophora* for example, and some anthozoans, particularly the soft-bodied sea anemones. Medusoid jellyfishes, such as the sea nettle (*Dactylometra quinquecirrha*) and the white sea jelly (*Aurelia aurita*), invade estuaries, sometimes being observed in salinity as low as 16‰. The anemone *Aiptasia* and coral (*Astrangia*) occur in the more saline zones of some estuaries, in Texas, for example. The upstream distribution of sessile, attached hydroids and anthozoans (and probably other animals as well) in estuaries is generally limited more by water level and salinity fluctuations during tidal cycles than by actual salinity. Given a comparatively "quiet" estuary, we should expect further penetration of the more strictly marine species.

In the over-all economy of natural waters, coelenterates are, in the main, active predators. Many of the anemones are highly carnivorous, feeding upon fishes that venture too close to the coelenterate's tentacles. In the laboratory, hydras are regularly fed planktonic crustacea; under natural conditions, however, such organisms are seldom found in the digestive cavity.

PHYLUM CTENOPHORA (COMB JELLIES)

Of the nearly 100 known species of this group, all are primarily marine. Some, however, do penetrate estuaries. In certain Texas bays, *Beroë* occurs abundantly during summer months in a salinity of 10 to 12‰. Ctenophores are generally small, nearly transparent, ovoid animals containing a large amount of gelatinous material in the body layers. They possess a pair of tentacles, which lack stinging cells, and eight rows of "comb plates," extending from pole to pole, which are used in locomotion. A number of species exhibit bioluminescence.

PHYLUM PLATYHELMINTHES (FLATWORMS)

This phylum includes a large number of free-living and parasitic species characterized by dorso-ventrally flattened bodies showing greater differentiation and specialization of parts than do the preceding groups. The basic plan of construction is similar to that of the coelenterates in that a single opening leads to a digestive cavity. Three classes are usually recognized, and at least certain representatives of each are familiar to students of introductory biology. These groups include the free-living flatworms of the class Turbellaria, and the flukes and tapeworms, the last two being parasitic. Although parasitism is of great interest and importance for its own sake as well as in the health of a community, our interest here is in the animals which range freely in aquatic situations. Therefore, emphasis will be placed mainly on the turbellarians.

Turbellarians are flattened or cylindrically shaped forms found widely in marine and freshwater habitats. These animals "creep" over bottoms and submersed objects by means of cilia and subtle body undulations. Some marine forms, such as *Leptodoplana,* may reach a size of 40 mm; others are microscopic. Dorsal coloration of turbellarians is highly varied, but usually dark and often changeable to blend with the habitat. The ventral surface is typically creamy or grayish. Some species are green, due to the presence of zoochlorellae. Turbellarians are characteristically carnivorous, feeding on small organisms by means of a highly extensible pharynx, which is extruded through a ventrally located mouth.

Five orders have usually been recognized in the class Turbellaria, although in recent years there has been a tendency to promote some of the subordinate groups to higher rank. The Acoela are marine, usually inhabiting intertidal zones. These flatworms lack a hollow gut and muscular pharynx. Most acoels are distinctly flattened, and some, such as *Convoluta,* contain algal cells. The Polycladida are marine forms with thin, leaf-like bodies. The gastrovascular cavity is highly branched and a sucker is present on some species. One group of polyclads is of especial biological interest in that a planktonic, ciliated larval stage is included in the life history of the species. This would seemingly be of adaptive value in aiding in dispersal of these types. The three remaining orders, Tricladida, Alloeocoela, and Rhabdocoela, include both freshwater and marine forms, and some which are euryhaline, or widely salt-tolerant. In the triclads, the gastrovascular cavity is hollow and three-branched, one branch passing anteriorly and two directed posterolaterally. This group includes the more familiar "planarians," among which *Dugesia tigrina* is probably the most widespread in the United States. Triclads, generally, are stream-adapted forms. Some, however, such as *Procotyla fluviatilis*, are found in a variety of habitats ranging from streams and ponds to brackish waters. The Alloeocoela and Rhabdocoela possess a nonbranched intestine, the two groups being distinguished on the basis of pharynx construction. These forms are typically more common in standing waters than in streams. One rhabdocoel species, *Gyratrix hermaphroditus*, is known from freshwater, estuarine, and marine habitats. Another rhabdocoel, *Microstomum*, feeds on hydra and in the process obtains stinging cells which the worm uses to its own advantage. The bodies of *Gyratrix hermaphroditus* and other worms such as *Stenostomum* and *Catenula* are composed of chains of segment-like zooids associated with asexual reproduction. Sexual reproduction occurs in many turbellarians, often followed by the formation of a cocoon around the eggs. The cocoon is stalked and attached to a submersed object in the habitat.

Phylum Nemertea (Proboscis Worms)

Proboscis worms are unsegmented, slender, often highly colored animals of fresh and salt waters. Nemertea bear a general resemblance to flatworms. Unlike the flatworms, however, these animals possess a true digestive tract, with mouth and anus. The most unique feature of the group is a protrusible proboscis, which is extended to grab prey. Some species are equipped with a hypodermic-like "stylet" and glands which secrete a toxin into the puncture made by the stylet. The Nemertea of fresh waters are relatively small, seldom attaining a length greater than 20 mm, and are greatly contractile. At present, only one species is known for the United States. This is *Prostoma rubrum*, which inhabits plant surfaces and bottom debris from coast to coast, but which is seldom collected.

About 500 species of nemerteans are known from marine communities. Many are common inhabitants of shells, rocks, and bottom muds of estuaries. Others are free-swimming components of plankton. The marine and estuarine species of proboscis worms exhibit greater size variations than their freshwater kin. *Cerebratulus lacteus* is probably the largest of the American shallow-water forms, often reaching a width of 25 mm and a length of over 6 m. A European species may be over 30 m in length.

As suggested by the food-capturing adaptations, nemerteans are carnivorous. They feed upon a variety of organisms and, in turn, enter into the diet of various worms, fishes, and other larger predators. Special habitats such as oyster reefs, mussel assemblages, and algal mats characteristically support proboscis worms as conspicuous members of the local communities.

Phylum Nematoda (Roundworms)

Nematodes are small, cylindrical worms with tapered ends, a chitinous cuticle, and straight digestive tract from mouth to anus. Locomotion is by vigorous, whip-like thrashing of the body. Although a great majority of these animals is parasitic, the free-living forms occur wherever organic substances are present, and usually in tremendous numbers. A single spadeful of humus soil or cupful of lake, stream, or estuary bottom will contain thousands of individuals. In spite of their great abundance, aquatic nematodes have been studied but little.

These worms, in general, play an important role in the turnover of organic matter in aquatic communities. Many cause disease and ultimately death of plants and animals. The diversified feeding adaptations of the free-living forms include detritus utilization, predation upon other living animals, and plant consumption. With respect to habitat restriction, only a few genera, such as *Chronogaster* and *Tripyla*, are found exclusively in fresh waters. The genus *Haplectus* has representatives in fresh, brackish, and marine habitats. Of ecological and evolutionary interest is the fact that a peculiarly estuarine, or brackish-water fauna of nematodes apparently has not developed.

Phylum Nematomorpha (Horsehair Worms)

As indicated by their common name, these animals are long (up to 1 m) and slender, thereby resembling horsehair. In fresh waters, the adults are usually most common in summer in pools or slow-moving brooks. The marine forms typically swim in a rapid writhing fashion near the surface. The developmental stages are spent as parasites of certain arthropods or mollusks. The adults do not feed.

Phylum Gastrotricha

The gastrotrichs are microscopic animals ranging in length to somewhat over 500μ. All are more or less ciliated ventrally, and some possess conspicuous spinous processes; others are more or less naked, with but a few spines. The internal morphology of some has been likened to that of rotifers (see below), and others have been compared with nematodes. Locomotion is produced by the beating of cilia and the action of spines.

Of some 400 known species, most are freshwater forms. However, little work has been done on this group and it would not, therefore, be proper to say that gastrotrichs are predominantly of one environment or an-

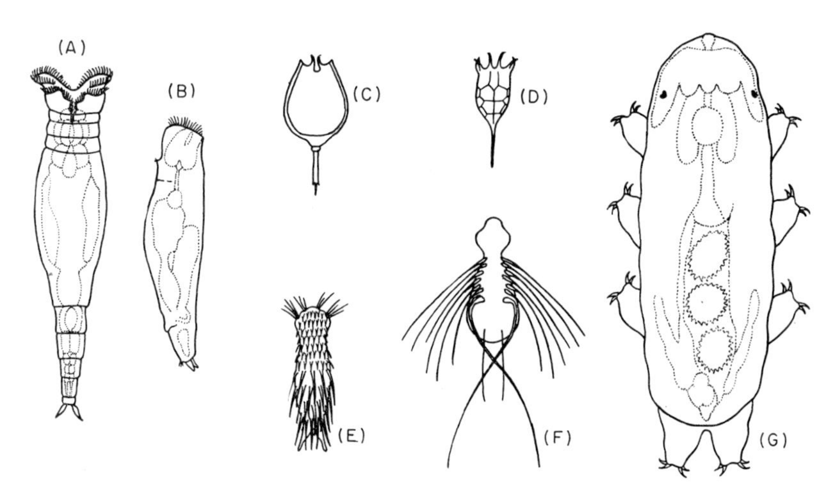

FIGURE 12·5. Rotifers (A) to (D), Gastrotrichs (E) to (F), and a Tardigrade (G) of Fresh Waters. (A) *Philodina*. (B) *Proales*. (C) *Monostyla*. (D) *Keratella*. (E) *Chaetonotus*. (F) *Dasydytes*. (G) *Macrobiotus*. ((A) and (B) after Edmondson, W. T., Ed., 1959. "Fresh-Water Biology," John Wiley & Sons, Inc., New York, N.Y.; (C) after Harrington, H. K. and Myers, F. J., 1926. "The Rotifer Fauna of Wisconsin. III. A Revision of the Genera *Lecane* and *Monostyla*," *Trans. Wisconsin Acad. Sci.*, Vol. 22, 315-423; (D) after Ahlstrom, E. H., 1943. "A Revision of the Rotatorian Genus *Keratella* with Descriptions of Three New Species and Five New Varieties," *Bull. Am. Mus. Nat. Hist.*, Vol. 80, 411-457; (E) and (F) after Brunson, R. B., 1950. "An Introduction to the Taxonomy of the Gastrotricha with a Study of Eighteen Species from Michigan," *Trans. Am. Microscop. Soc.*, Vol. 69, 325-352; (G) from Pennak, R. W., 1953. "Freshwater Invertebrates of the United States," Ronald Press Co., New York, N.Y.)

other. The genus *Dasyatis* (Figure 12·5) is one of a half dozen genera restricted to fresh water. A few genera, including *Chaetonotus* (Figure 12·5), have representatives in both fresh and marine habitats. Within their environment, gastrotrichs are usually closely associated with debris or submersed objects. Some are occasionally taken with plankton. Quiet pools, sphagnum bogs, and shallow zones of lakes are most productive habitats. Few gastrotrichs have been found in rapidly flowing streams,

The food of gastrotrichs consists mainly of smaller organisms such as bacteria, algae, and protozoans.

Reproduction and sexual development among gastrotrichs is of more than usual interest. A number of marine species are hermaphroditic, the opening of the female structures being posterior to that of the male, both being situated on the ventral surface of the animal. Males are unknown in freshwater forms, the females being produced parthenogenetically, i.e., without fertilization.

Phylum Rotifera (Rotifers)

The Rotifera are generally microscopic animals possessing conspicuous organ-systems and characterized by an anterior corona of cilia surrounding a mouth. The cilia aid in feeding and swimming. Posteriorly, most rotifers possess toe-like appendages and glands which secrete a cementing substance, these adaptations serving to anchor the animal or aid in a creeping locomotion. In many species the outer body covering, or cuticle, is stiffened to form an armor-like coat, the lorica. The extent of development and the form of the lorica vary greatly and are useful taxonomic features.

Sexuality and reproduction in rotifers are similar to those of gastrotrichs, this similarity suggesting close phylogenetic relationships of the two groups. In certain of the rotifers, males are unknown—the populations consisting entirely of parthenogenetic females. In other groups, certain types of females produce eggs which develop into degenerate males whose sole purpose appears to be that of fertilizing "special" eggs. These eggs are highly resistant to adverse environmental conditions and thus represent an evolutionary adaptation of considerable survival value.

The distribution and ecology of rotifers also have interesting evolutionary implications. Of the more than 1500 species, over 90 per cent are freshwater inhabitants. This suggests a freshwater origin of the group. Rotifers have undergone considerable evolution and adaptive radiation, resulting not only in morphological differentiation (Figure 12·5), but also in their occupancy of a great variety of habitats and their filling of a number of roles in the economy of fresh waters. Many species are cosmopolitan, their occurrence being nearly world-wide in suitable habitats. Within a body of water, various species are found in the sediments of shallow shore-zones and in the bottom deposits of deep water. Many are typically planktonic; others inhabit masses of vegetation, certain species showing close affinities for particular plants. A few rotifers are ectoparasitic, and some have entered into other symbiotic relationships. Feeding habits are similarly varied; some species are carnivorous, others utilize the sap of algal cells, and many are detritus feeders or wholly omnivorous.

PHYLUM BRYOZOA (MOSS ANIMALS)

Bryozoans are found in both fresh and salt water, but predominantly in the latter. The common name is derived from the fact that these forms typically occur as colonies encrusting objects in the water. In fresh waters, *Pectinatella* grows as a large gelatinous mass, while *Plumatella* and *Fredricella* (Figure 12·6) colonies are loosely developed and mat-like. *Membranipora,* a cosmopolitan estuarine form, grows as calcareous encrusting mats on submersed structures. Many marine bryozoans, *Bugula* for example, resemble large ferns; others may be coral-like, possessing a rigid

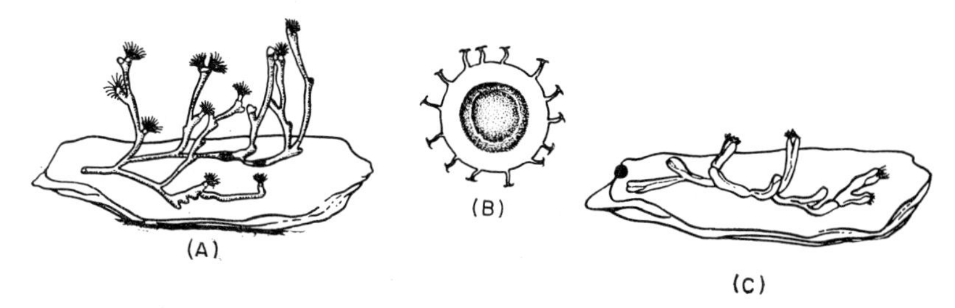

FIGURE 12·6. Common Bryozoans of Fresh Waters. (A) *Fredricella sultana* (Blumenbach). (B) the statoblast of *Pectinatella magnifica* Leidy. (C) *Plumatella repens* L. (All after Rogick, M. D., 1935. "Studies on Freshwater Bryozoa. II. The Bryozoa of Lake Erie," *Trans. Am. Microscop. Soc.*, Vol. 54, 245-263.)

framework of carbonates or chitinous substance. The basic unit of the colony is the individual animal, or zooid. The mouth of a zooid is bordered by ciliated tentacles which create currents and bring in food. The digestive tract of these animals is a tube bent upon itself, with the result that the mouth and anus lie close together. Sexual reproduction by hermaphroditic individuals occurs for a short time each year. Asexual reproduction by budding contributes to the increase in size of the colonies. In addition to budding, certain freshwater species produce asexually formed bodies (statoblasts) containing germinative tissue. These bodies are constructed of a tough chitin-like material and are very resistant to desiccation. One type, called a spinoblast, is produced by *Pectinatella* and is shown in Figure 12·6. Other statoblasts may lack spines or hooks and are termed floatoblasts. Both spinoblasts and floatoblasts are free-floating and often taken in plankton collections. A third type, such as that produced by *Fredericella* and *Plumatella,* is attached to the colony; these are sessoblasts.

Freshwater bryozoans occur in a variety of habitats, but seldom in polluted waters. Highly turbid and acid waters also inhibit colony development. Summer is the favorable period for growth of these animals. During this season, bryozoans commonly develop on submersed objects in shallow shore zones of lakes and ponds, and in both slow and rapid waters in streams. Estuarine forms are particularly attracted to pilings, mollusk shells, and rocks. Many species, especially those forming statoblasts, are found world-wide in favorable habitats. Such widespread distribution is due in part to transport of the statoblasts by many agents, including waterfowl.

At times, some of the bryozoans enter prominently into human economy. Large colonies of *Pectinatella* may grow on intake pipes and screens of water-supply plants and in time clog the pipes, demanding manual removal. In estuaries and marine communities, bryozoans contribute to fouling of boat hulls and pier pilings; efforts to control this condition also involve labor and expense.

Phylum Chaetognatha (Arrowworms)

In terms of numbers of species, this is a relatively small phylum; about 30 species, all marine, are known. At certain times, however, the abundance of individuals places the group high in importance in energy relationships in marine waters, for seasonally at given localities these animals form a major item in the diet of fishes. Chaetognaths are slender, transparent animals with lateral and caudal fins, usually no longer than about 40 mm. The head bears several curved spines which serve to seize prey consisting of plankton, small fishes, and, frequently, other chaetognaths. Being restricted to generally high salinity, these forms probably inhabit positive estuaries no farther than the lowermost reaches. The genus *Sagitta* is widespread, and probably the best known of the chaetognaths.

Phylum Annelida (Segmented Worms)

Annelids are typically cylindrical, elongate animals with serially segmented bodies, bearing a mouth at one end and an anus at the other. Although the terrestrial earthworm is the most familiar annelid, the greatest variety and abundance of species are found in estuaries and the marine environment. Of four recognized classes, only three need receive our attention in this synopsis; these are: Polychaeta, the sandworms; Oligochaeta, the earthworms; and Hirudinea, the leeches.

CLASS POLYCHAETA

This class consists mostly of marine animals characterized by the presence of serially arranged pairs of fleshy appendages (parapodia) usually bearing numerous stiff bristles (setae). The surface of the parapodia serves as a gaseous exchange region in "breathing"; the beating of the appendages moves water across the surfaces and also aids in locomotion. Although mostly small, some species of the genus *Nereis* may attain 50 cm in length. Many polychaetes inhabit burrows in the substrate, others build tubes in which the animal lives permanently, and some species are free-swimming. In the estuary and the sea, a number of species typically inhabit shells of living mollusks. Sexual reproduction predominates in this group, and the sexes are separate. Breeding takes place seasonally, and in many species is marked by precisely timed periodicity. The life history of polychaetes involves an unsegmented larva (the trochophore) which metamorphoses into an adult.

Of the more than 3000 known species of polychaetes, less than 20 are typically freshwater forms. A number of families are represented in fresh waters, suggesting their separate evolution and invasion of a different environment. In North America, less than a dozen species, mostly of the family Nereidae, inhabit fresh waters. Most of these are euryhaline species distributed in estuaries and in their freshwater tributaries. *Neanthes succinea* is found in estuarine and nearly fresh environments on both coasts of temperate North America. The freshwater genus *Manayunka* (family Sabellidae) has representatives in streams in Pennsylvania and New Jersey, and in Lake Erie and Lake Superior. Notorious as marine fouling organisms, the family Serpulidae is represented in fresh waters by a species found in a California lake and in streams entering the western Gulf of Mexico. Of special interest is the occurrence of a sabellid species in Lake Baikal, Russia. This is a very ancient lake and is over a thousand miles from salt water.

Polychaetes are characteristically more abundant in estuaries than in freshwater communities. In the zone of intermediate salinity in the Elbe Estuary polychaetes were found at a density of 25 per square meter. This is to be compared with zero for the freshwater zone and 19,000/m^2 in the seawater region of the same stream. Polychaetes contribute to the metabolism of the estuarine community in several ways. Being active feeders, in large numbers they consume considerable quantities of organic substance. Many of them are integral items in the diet of larger organisms; certain species of fishes, for example, consume great stores of polychaetes. The continual workings of the burrowing forms turn over much material of the sediments. These and other animals join with various plants to produce dense local associations contributing to productivity generally.

CLASS OLIGOCHAETA

The characteristics of the class Oligochaeta, as typified by the earthworm, are familiar to all and need not be considered here. The structure of aquatic oligochaetes deviates little from that of earthworms. In contrast to polychaete reproduction, however, some of the aquatic oligochaetes generally reproduce by asexual budding; this is true of the

Naididae and Aeolosomatidae. The other oligochaetes engage in sexual reproduction involving cross-fertilization between two hermaphroditic individuals. Direct development from eggs and embryos contained in cocoons attached to submersed objects follows.

Also in contrast to polychaete adaptations, nearly all of the aquatic oligochaetes are inhabitants of fresh waters. Freshwater species occur widely in all types of aquatic conditions, but the greatest abundance is usually associated with organically rich substrates. The family Tubificidae is distinguished by a number of species adapted to existence in estuaries and in near-anaerobic sediments in deep lakes. *Tubifex tubifex,* a red worm, found world-wide, reaches greatest population densities in highly polluted waters. A number of tubificid species are tolerant of the low oxygen concentrations present during winter stagnation in lakes. Many oligochaetes, such as *Tubifex* and the naid genera *Dero* and *Nais,* construct mud tubes for habitation. The construction of tubes and the burrowing by other forms contribute to the overturn of sediments, indicating that one ecological role of aquatic worms is similar to that of their terrestrial counterparts.

CLASS HIRUDINEA

This class includes the leeches, found primarily in fresh waters, although a few marine and terrestrial species are known. These annelids are flattened, and usually possess sucking devices which aid in locomotion and in attachment while eating. Some members of this group are ectoparasites, others feed on detritus, and still others are carnivorous. Leeches reach maximum abundance in standing waters containing large amounts of debris.

The phylum Annelida presents an interesting picture of evolution and ecological and physiological adaptation with respect to the freshwater-estuary-marine series. As we have seen, the density per unit area of individuals of marine species of polychaetes is much higher than that of estuarine forms. Conversely, the density of oligochaetes decreases from fresh water through the estuary to the marine. It would appear that physiological adaptations of each group to environmental salinity have been slow to change. This could be the entire answer. On the other hand, ecological adaptations may be important. Even if large numbers of species should migrate, they might be unable to successfully colonize because their particular roles and habitats might be already filled. The answer to the problem is not fully known. Certainly laboratory experiments have shown that many present-day forms can not transgress the salinity barrier, but the tests tell us little with respect to the possibility of change over the many millions of years that these forms have been in existence.

Phylum Tardigrada (Water Bears)

This is a relatively small phylum of animals of "uncertain systematic position." Some authors include the forms with the arthropods, others consider the water bears to be more closely related to annelids, and still others simply place the animals in a separate phylum. Tardigrades are small, typically less than 500μ in length. As shown in Figure 12·5, these are "well-formed" organisms possessing a distinct head, usually with eyes, four pairs of legs with claws, and a segmented body. Internally, a brain and ventral nerve cords coordinate locomotion and other functions. The digestive and reproductive systems appear well formed, but respiratory and circulatory systems are absent. Water bears develop from eggs laid by a female and fertilized beneath a shedding cuticle by males. Cuticles are shed several times during the life of the animal. In many species, individuals are capable of contracting into a small mass inside the cuticle and surviving many years of adverse conditions of dryness and low temperature.

About 350 species are known to inhabit aquatic and semiaquatic habitats; the latter includes damp masses of algae, mosses, and sand. Only a few species are marine.

Phylum Arthropoda (Joint-legged Animals)

This phylum is said to be the most successful (in terms of evolution and adaptation to numerous environments) and the largest (in respect to numbers of species). Representatives are found everywhere that life can be supported, and rather more than a million species are known. The classes with aquatic representatives are: Crustacea, Insecta, and Arachnida. Our survey of the arthropods here must be brief and quite general, with attention being directed toward the less familiar but ecologically important types. For a deeper appreciation for the great variety of these animals, consult the several references suggested at the end of this chapter.

CLASS CRUSTACEA

The larger members of this class, such as the crabs, lobsters, shrimp, crayfish, and barnacles, are quite familiar. Less commonplace, however, by virtue of their size are the myriads of crustacean species which occur in the plankton of fresh and marine waters. The tremendous importance of these animals rests upon their indispensable position in the food web and energy relationships in lakes, streams, estuaries, and the sea. An important concept to bear in mind is that for practically every major type found in fresh waters there is an ecological counterpart, often similar in appearance, inhabiting the sea; many of these meet in the estuary.

Four of the crustacean orders are sometimes called the branchiopods.

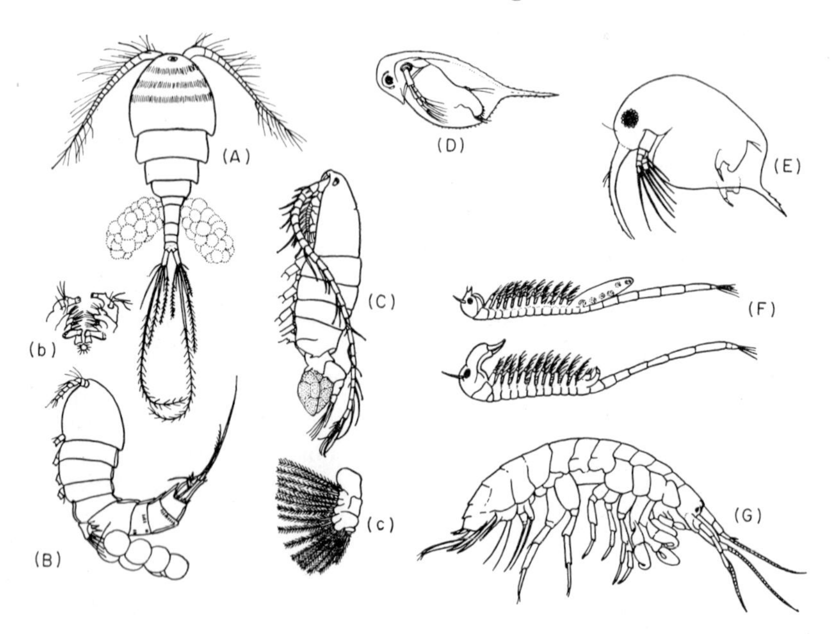

FIGURE 12·7. Some Representative Freshwater Crustaceans. (A) Cyclopoid copepod, *Macrocyclops ater* (Herrick) with two egg sacs. (B) Harpacticoid copepod, *Moraria virginiana* Carter, with single egg sac and antenna approximately the same length as first body segment. (C) Calanoid copepod, *Diaptomus birgei* Marsh, with one egg sac and antenna about same length as body. (D) and (E) Cladocerans, *Daphnia* and *Bosmina*, respectively. (F) Fairy shrimp, *Branchinecta;* female above, male below. (G) Amphipod, *Gammarus*. At (b), scraping mouthparts of harpacticoid copepod; and at (c), filtering mouthparts of calanoid copepod. ((A) through (F) after Coker, R. E., 1954. "Lakes, Ponds, and Streams," Univ. of North Carolina Press, Chapel Hill, N.C.)

These include superficially different groups of small animals. Three of these, the fairy shrimps, tadpole shrimps, and clam shrimps, are entirely freshwater forms typically found in small ponds or temporary puddles and pools; in summer even a water-filled cow-track may harbor a population of one or more of these. The eggs are quite resistant and are disseminated by a number of agents, but primarily by wind. The fairy shrimps (Figure 12·7) generally are elongate forms, usually about 20 to 30 mm in length, which swim slowly and gracefully on their backs. One species of fairy shrimp, *Artemia salina*, the brine shrimp, is found throughout the world; in the United States it occurs in abundance in Great Salt Lake. The clam shrimps give the appearance of being fairy shrimps bent and enclosed between bivalve shells. Some of these attain a length of about 16 mm, although most are smaller. The tadpole shrimp is characterized by an arched carapace over much of the body; these branchiopods have not been found east of the Mississippi River.

The fourth group of the branchiopods includes the water fleas of the order Cladocera (Figure 12·7). These are small (up to 3 mm in length) crustaceans found in all types of fresh waters; a few species occur in the estuary and the sea. Over 130 species are known in the United States. The animal is contained within a folded bivalve carapace. In a number of species the posterior of the carapace is produced as a long spine. The surface of the carapace may exhibit fine etchings, or *striae*. A large compound eye is conspicuous on the head. Also in the head region is a pair of antennae used in locomotion. Under favorable environmental conditions cladoceran populations consist mostly of females produced parthenogenetically from eggs carried in the upper part of the carapace. As unfavorable conditions arise, some of the eggs develop into males and, simultaneously, a different type of female is also produced. When she is fertilized, her carapace thickens and forms a protective *ephippium* about one or two eggs. With the following molt, the ephippium is shed and becomes sealed. Thus, the embryo is able to withstand the unfavorable times, and upon return of proper environment the developing animal(s) are released. This adaptation serves to ensure repopulation of waters following periods of drought or other unfavorable circumstances.

A most interesting and noteworthy aspect of cladoceran biology involves seasonal changes in the body form of several species (Figure 12·8) in some geographical areas. During the warm months, *Bosmina*

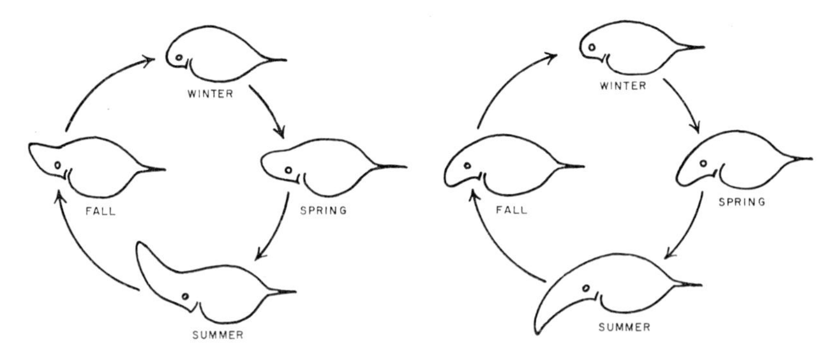

FIGURE 12·8. Rather Extreme and Conventionalized Cyclomorphosis Exhibited by Different Races of *Daphnia* from Large Lakes. The winter generations of the two races are nearly identical; individuals of summer generations were, at one time, thought to be different genera from the winter forms. (From Coker, R. E., 1954. "Lakes, Ponds, and Streams," Univ. of North Carolina Press, Chapel Hill, N.C.)

coregoni becomes noticeably humpbacked and the rostrum greatly elongates; *Daphnia cucullata, D. longispina,* and *D. retrocurva* are marked by the development of a conspicuous "helmet" in summer. Termed *cyclomorphosis,* this phenomenon has not been fully explained. It was early thought that such enlargements served to increase surface area as an aid in flotation during times of lowered viscosity of water. Nutrition, temperature during development, and various physiological conditions have also been suggested to account for cyclomorphosis. It has recently been found that circulation of water might influence the development of helmets in *Daphnia cucullata,* the suggestion being that the produced helmet is an adaptation to resistance of water current.

Within a given lake, certain groups of cladocerans exhibit affinities for a particular set of environmental conditions. Nearly all species of the family Chydoridae are found in stable embayments, or "backwaters." Such genera as *Bosmina, Diaphanosoma,* and *Daphnia* are typically found in the limnetic, or open-water, region of the lake. Others are associated with the mud zone, and a number of forms inhabit various intermediate zones.

Cladocerans are worthy of much attention because of their tremendous importance in food and energy relationships, especially in lakes and ponds. All carnivorous fishes pass through an early growth stage in which zooplankton is the major source of food. From one third to more than one half of the zooplankton in the vegetated littoral zone, a favorite haunt of sunfishes and perch, consists of cladocerans. Upon reaching larger size and becoming piscivorous, large fishes are more or less dependent upon the plankters, for the small fishes taken as food usually feed regularly upon the microscopic forms.

The order Ostracoda includes small pea-shaped, bivalved crustaceans generally about 1 mm in length. Of the 2000 known species of ostracods, most are marine and estuarine. The valves of many species are often colored and patterned. Some of both the freshwater and marine species are cosmopolitan and found in a great variety of habitats. In both major environments ostracods are usually most abundant near the surface of bottom muds or among dense vegetation. The development of ostracods following hatching includes a larva known as a *nauplius,* which is often abundant in plankton. Although generally of less importance than other microcrustaceans as food for fishes, ostracods are consumed by the gizzard shad (*Dorosoma cepedianum*), an important forage species in waters of the southern United States.

The crustacean order Copepoda embraces an ecologically important group of small animals widely distributed in freshwater, estuarine, and marine environments. Copepods are mostly less than about 2 mm in length, and, of the 6000 or more species, a majority is marine. A few forms are parasitic. The free-living species fall into three suborders showing relatively distinct body forms, living habits, locomotion, and ecological roles. These groups are: Calanoida, Cyclopoida, and Harpacticoida (Figure 12·7).

Calanoid copepods are generally characterized by long antennae of 23 to 25 segments extending the length of the animal, and by an elongate *metasome* (anterior body) of generally similar segments. Eggs are carried in one or two closely appended sacs. All of the calanoid families except one have representatives in fresh and marine waters. The genus *Eurytemora* includes several euryhaline species, some of which are the dominant calanoids of estuaries of certain regions. *Acartia* is primarily a marine genus, but some species occur abundantly in estuaries. The family Diaptomidae includes the most common calanoids of standing fresh waters. Calanoids are typically planktonic in the limnetic regions of lakes, *Diaptomus* being especially abundant under certain conditions. Calanoids are also distinguished by their feeding habits in that they are primarily "filter-feeders," consuming small organisms and detritus delivered by currents set up by the antennae. Locomotion in these copepods is typically by swimming.

Cyclopoid copepods possess short antennae (of 6 to 17 segments) seldom more than about one half the length of the metasome; the metasome is generally compressed longitudinally and tapered posteriorly. Eggs are characteristically carried in two sacs attached laterally. Species of *Halicyclops* inhabit mesohaline waters. *Cyclops* is distributed widely in fresh waters. Within a given body of water, cyclopoid copepods are mainly inhabitants of the littoral, although one or two species may be abundant in the limnetic. These forms possess mouth parts adapted for seizing and biting small organisms. Locomotion in cyclopoids is essentially of a leaping nature involving both antennae and legs.

In harpacticoid copepods the antennae are quite short, consisting of no more than nine segments, and seldom longer than the cephalothorax. The body is nearly cylindrical, with little differentiation between anterior and posterior regions. Nearly all genera of North American harpacticoids are represented in fresh, estuarine, and marine waters. In fresh waters, *Canthocamptus* and *Bryocamptus* are probably the most common, being found frequently in bottom debris in both shallow and deep zones. Several genera, including *Tachidus*, *Nitocra*, and *Mesochra*, are notably estuarine, some forms being widely distributed. *Nitocra spinipes*, for example, is known from estuaries of northern Europe and North America, being reported from such widely distant localities as Hudson Bay, Mexico, and Alaska. In all bodies of water, harpacticoids show extremely close affinities for the littoral zone and its cover, being especially common in masses of vegetation, even above water and on beaches. Interstitial waters of beach sands may often harbor these copepods.

Emphasis must be given to the tremendously important place occupied by copepods, generally, in the economy of natural waters. These crustaceans occur in great numbers and consume great quantities of phytoplankton and detritus. Under certain conditions the density of limnetic copepods in lakes may exceed 1000 organisms per liter. As much as 85 to 90 per cent of estuarine zooplankton may consist of cyclopoids. In turn, plankton-feeding fishes ingest copepods; a single gizzard shad may contain hundreds of the crustaceans. The great fishing areas of the world are usually in plankton-rich regions, and the zooplankton is largely composed of copepods.

During the course of evolution a few copepods have become ectoparasitic, mainly upon fishes. These include the families Lernaeidae and Argulidae, which are found in both fresh and salt waters; the genus *Argulus* is known from estuaries and fresh waters.

The crustacean order Cirripedia is of interest only out of our concern for the estuary. This group includes the barnacles. The young are motile, but as development proceeds they become attached to objects in the water, including other animals. Almost all barnacles are strictly marine, only a few inhabiting estuaries. *Balanus improvisus* is a widespread species typically found in mesohaline waters.

The crustacean subclass Malacostraca includes some twelve orders characterized by having abdominal appendages, eight segments in the thorax, six segments in the abdomen, and by being generally large in size. Many of the malacostracans are bottom dwellers possessing highly muscular pincers and strong, biting mouthparts; these include crayfish, crabs, lobsters, and the like. Others are small, but have similarly developed, diminutive appendages; many of these animals, such as shrimps, are swimming forms, and a few species are conspicuous components of marine plankton. Only about one third of the dozen orders is represented in fresh waters; the remainder are marine, with estuarine species.

The order Mysidacea includes about 300 species of small, shrimp-like crustaceans usually found in cool waters. Only one wholly freshwater species is known for the United States; this is *Mysis oculata*, found in cold lakes. Another species, *Neomysis mercedis*, inhabits fresh waters and estuaries of the Pacific slopes from California northward. *Neomysis vulgaris* commonly occurs in estuaries of northern Europe. Where abundant, mysids are taken in considerable numbers by fishes, especially those that feed near or from the bottom where the mysids normally live.

Sowbugs (Isopoda) are flattened malacostracans found in damp terrestrial habitats and in practically all types of aquatic communities. Of the more than 3000 aquatic species only a few are adapted to fresh water. The freshwater species, such as those of the genus *Asellus*, are often very common in certain situations, such as near the littered shores and bottoms of small streams. *Cyathura carinata* and species of the genus *Idotea* occur in estuaries of northern Europe and the Atlantic coast of North America, *C. carinata* being found as far south as Virginia. Isopods are generally nocturnal in their activity. Jetties, piers, and shores are often overrun with the animals at night.

The order Amphipoda includes about 3000 species of laterally compressed crustaceans, commonly called "scuds" or "sandhoppers," ranging to about 20 mm in length (see Figure 12•7). Most of these are marine and estuarine. A number of amphipods, including *Gammarus locusta*, and *Corophium volutator*, are known from estuaries of northern Europe and the Atlantic coast of North America. The genus *Gammarus* is represented in freshwater, estuarine, and marine communities, several species being found in a variety of habitats. In thickly vegetated spring runs *Gammarus* may occur in densities of several thousand per square meter (see Figure 14•8, Chapter 14). A common freshwater amphipod is *Hyallela azteca;* in southern lakes and streams, the dense hair-like roots of a single water hyacinth (*Eichornia crassipes*) may harbor hundreds of *Hyallela*. Amphipods are eaten in great numbers by many fishes. Some amphipods serve as intermediate hosts in the life cycle of many parasites. *Hyallela*, for example, harbors a stage of the acanthocephalan, *Leptorhyncoides thecatus*, which has its adult stage in a vertebrate animal.

Familiar to all are the shrimps, crabs, crayfishes, and other crustaceans of the order Decapoda. Of some 8000 known species, most are marine and estuarine. The freshwater representatives include only the crayfishes and some freshwater shrimps.

Various habitat affinities are exhibited by the crayfishes. *Orconectes propinquus* and *Cambarus longulus* are typically stream inhabitants. *Cambarus diogenes* is a burrowing form found from New Jersey to Texas and the Great Lakes. *Procambrus blandingi* occurs widely in various habitats from New England to Mexico. Underground waters and caves often contain crayfishes; *Orconectes pellucidus* inhabits such situations from Indiana to Alabama. There appear to be no species particularly characteristic of lakes.

The freshwater shrimps are represented by the family Palaemonidae, which also has marine members, and by the genus *Macrobrachium*, the large river shrimps. These are characteristically associated with vegetation or debris. In certain lakes of Florida, *Palaemonetes paludosus* occurs abundantly among the roots and tangled plant fragments in the underside of floating islands. In estuaries, members of the Palaemonidae commonly live in patches of grasses growing in shallow water. The river shrimps (*Macrobrachium*), some reaching nearly 200 mm in length, are erratically distributed from Virginia to Texas, and in the Ohio-Mississippi River drainage.

Many species of crabs are found in estuaries, and in southern regions the brown shrimp (*Peneus setiferus*) is especially common. Mesohaline waters are, in fact, important "nursery grounds" for the shrimp, indicating that an early phase of the life cycle must be spent in such zones.

CLASS ARACHNOIDEA

The most familiar members of this class are the terrestrial spiders and ticks. The only truly aquatic representatives, however, are the mites of fresh and marine waters, and the king, or horseshoe crabs, and sea spiders (Pycnogonida) of the marine and estuarine environments. The water mites (Hydracarina) are common in a wide variety of freshwater habitats. Hot springs, lakes, ponds, stagnant pools, and rapid streams normally support hydracarinid populations. Some species occur in salt waters. In many instances the populations are rather narrowly restricted to their particular habitats, with the result that each type of community usually has a distinctive fauna. Many species are vigorous, active swimmers; others are retiring and crawl about in masses of debris, or algae, or simply over the substrate. Some of the mites are parasitic. Mostly, however, mites are carnivorous, feeding upon small crustaceans, worms, and other animals. The mites, in turn, are consumed by a number of larger organisms, including fishes. Many hydracarinids are brilliantly colored, often with distinctive patterns. The widespread stream mite, *Limnesia,* is red with black designs on the dorsum. *Diplodontus despiciens,* a pond form found in North America and Europe, is frequently a brilliant red. The marine mites are usually dark colored.

The king crab, or horseshoe crab (*Limulus*), is a bizarre arachnoid. The appendages and body are contained beneath a broadly domed, leathery carapace. *Limulus* is a bottom inhabitant, often burrowing and pushing along through the sediments, uncovering small animals, which it eats. Pycnogonids, sometimes placed in a separate class, are found along the coasts in vegetation, and in the deep sea. In the latter habitat the animals may have a spread of several feet.

CLASS INSECTA

The number of North American insects known to spend part, or all, of their lives in fresh waters exceed 5000 species. Included in this figure is a great array of forms exhibiting a remarkable variety of adaptations to all types of fresh waters. Although the number of strongly marine forms is inconsequential, many species, derived from fresh waters, inhabit estuaries, the nature of the fauna depending largely upon salinity. In our synopsis here we can give only cursory consideration to the orders represented in aquatic communities.

On the basis of development from egg to adult, insects can be classified into essentially two categories: one group in which metamorphosis is gradual or incomplete, and another in which metamorphosis is complete (egg→larva→pupa→adult). In the first category the nymph resembles the adult, at least superficially, and changes in form are gradual. This group includes the stoneflies, dragonflies, mayflies, and true bugs. In the second category are the forms which pass through abrupt stages, including a worm-like and a pupate form, before reaching adulthood. Insects in this group include the alderflies, dobsonflies, spongeflies, caddisflies, moths and butterflies, beetles, and the two-winged flies. These developmental and physiological adaptations are of ecological interest because they pertain to seasonal occurrence and abundance of insect forms in the food and energy relationships within a given community.

Stonefly nymphs (Plecoptera) are common inhabitants of swift, cool streams and the shores of temperate lakes. They are rather sluggish creatures, usually associated with stones or aquatic vegetation. The series of nymphal stages may persist for several years, followed by mass emergence, usually in early summer. Stonefly nymphs superficially resemble those of mayflies (to be considered subsequently), but differ from the latter in characteristically possessing only two tail appendages and lacking tracheal gills on the abdomen. Some of the stoneflies (Perlidae for example) are carnivorous; others are plant and detritus feeders. Larger organisms, including stream fishes, feed on stoneflies.

Damselflies and dragonflies of the order Odonata are represented in fresh waters by nymphal stages. These forms are recognized by the high degree of development of the labium (lower lip) into a mask-like seizing device. Odonate nymphs are quite predaceous, feeding on a variety of animals, often of considerable size. These insects are widely distributed in most types of fresh waters. A number of species, such as *Anax junius* and *Pachydiplax longipennis* (Figure 12·9), are typically inhabitants of ponds and sluggish stream waters. *Erpetogomphus designatus* is primarily a stream inhabitant. In the estuary of Weekiwachee Springs, Florida, a damselfly, *Enallagma durum*, reaches maximum population density in a zone in which chloride concentrations approach 5.5‰ In many ponds and streams odonate nymphs constitute a standard item in the diet of some fishes.

The nymphs of mayflies (Ephemeroptera) are common inhabitants of most streams and standing waters containing sufficient oxygen. At least one species inhabits brackish waters. These animals are readily recognized by the presence of tracheal gills attached to the abdomen and two or three caudal filaments (Figure 12·10). Although the series of nymphal stages may last for several years, the adults live but a few days at the most, that time being spent in reproduction. Mayfly nymphs are variously adapted for living in lakes or streams. In streams, some species make their way freely over the bottom, others inhabit debris and masses of plant material, and still others are strongly associated with submersed objects in swift water. Lakes contain burrowing types and bottom dwellers. *Hexa-*

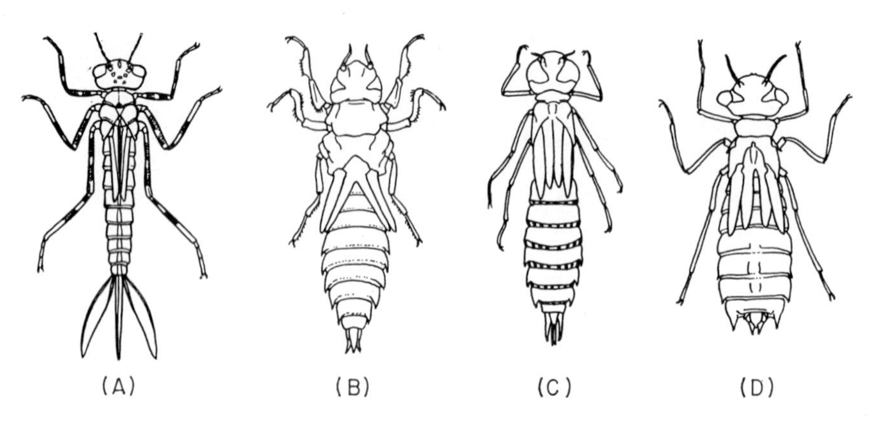

FIGURE 12·9. Some Odonata Nymphs of North American Fresh Waters. (A) *Argia emma* Kennedy, a damselfly. Dragonfly nymphs: (B) *Progomphus obscurus* Rambur; (C) *Anax junius* Drury; (D) *Pachydiplax longipennis* Burmeister. ((A) after Kennedy, C. H., 1915. "Notes on the Life History and Ecology of the Dragonflies (Odonata) of Washington and Oregon," *Proc. U. S. Nat. Museum,* Vol. 49, 259-345.)

genia (Figure 12·10) and *Ephemera* are burrowing forms widely distributed in North America. *Stenonema* is usually found wherever there is some movement of water, often along lake shores. *Baetis* is a widespread stream inhabitant. *Caenis* is a common bottom form in lakes and ponds. *Pseudocloeon* occurs widely in moving water. Mayfly nymphs are primarily herbivorous in their feeding habits, and thus form a short "chain" to many fish species which feed upon the insects.

The true bugs (Hemiptera) are characterized by the modification of the anterior wings into tough horny sheaths and the mouthparts into a rostrum specialized for sucking. Some species are wingless. The Hemiptera exhibit various modifications directed toward inhabiting the shore, the surface film, the bottom sediments, and actively swimming in the water of ponds and streams. Some forms, such as *Trichocorixa*, *Rheumatobates*, and *Halobates*, occur in (or on) marine waters. A number of species, including the "light bugs" (Belostomatidae) and the water boatmen (Corixidae) are capable of sustained flight and frequently leave the water. Some representative types are shown in Figure 12·11.

The larvae of alderflies, dobsonflies, and spongeflies (Neuroptera) are found widely, though usually not abundantly, in lakes, ponds, and

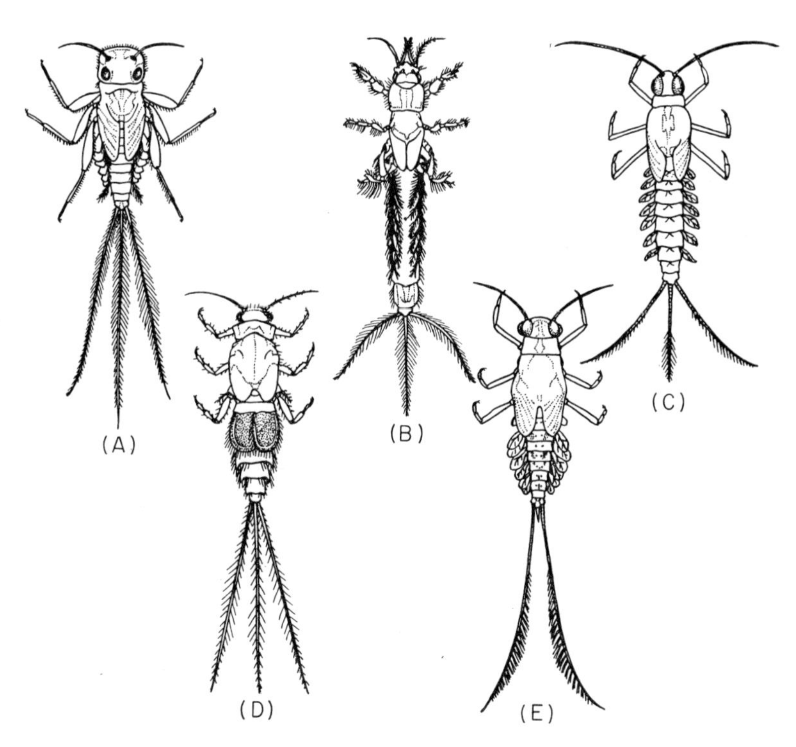

FIGURE 12·10. Nymphs of Some Examples of Mayflies Found in Fresh Waters of North America. (A) *Stenonema exiguum* Traver. (B) *Hexagenia munda marilandica* Traver. (C) *Baetis spiethi* Berner. (D) *Caenis diminuta* Walker. (E) *Pseudocloeon alachua* Berner. (Redrawn from Berner, L. M., "The Mayflies of Florida," 1950. Univ. of Florida Press, Gainesville, Fla.)

streams throughout much of North America. The most familiar, at least to fishermen, is the large, stream-dwelling larva of the dobsonfly, *Corydalis*, known as a hellgrammite; it may grow to over 80 mm in length. The larvae of the alderfly, *Sialis*, are found in a variety of habitats, including plant debris on lake or stream bottoms. Neuropteran larvae are lively predators, possessing strong mouthparts. Caddisfly (Trichoptera) larvae, or at least their cases, are well known to those who have probed into streams and standing waters, and the adults are familiar to trout fishermen. Many of the larvae build a variety of cases to protect themselves and the pupae of the developing caddisfly (Figure 12·12). These cases are usually in the form of elongate cones or cylinders constructed of cemented sand grains, sticks, leaves, and other materials, the type of case sometimes being characteristic of a particular family. In flowing waters, a number of species build nets between stones, the nets, directed upstream, serving to catch food carried in the current.

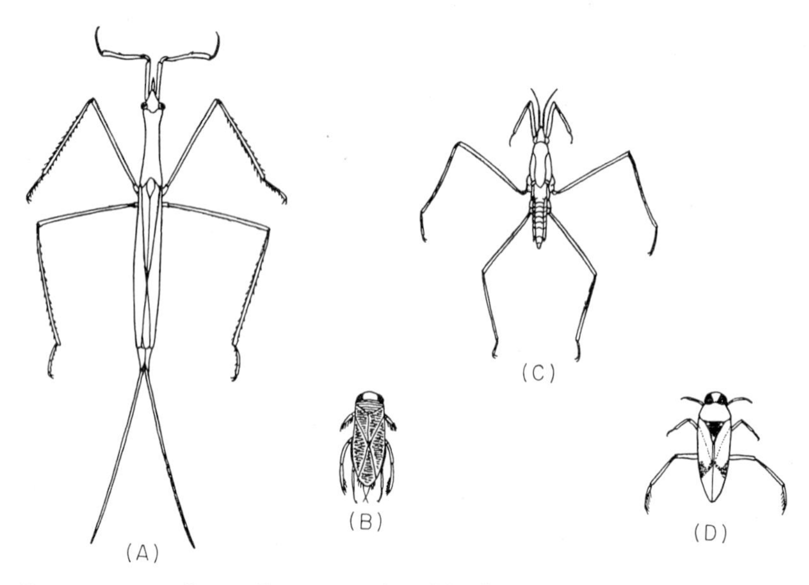

FIGURE 12·11. Some Representative Hemipterans. (A) *Ranatra.* (B) *Corixa.* (C) *Gerris.* (D) *Notonecta.*

The beetle order (Coleoptera) is represented in fresh waters by a variety of forms which have aquatic larvae, terrestrial pupae, and aquatic adult stages. Many aquatic beetles are typically carnivorous, both as larvae and as adults; others, including the Haplidae, feed on plant material. Many of the species, such as the diving beetle (*Dytiscus*) and some of the scavenger beetles (Hydrophilidae), leave the water on occasion and fly about. An impressive array of feeding, breathing, locomotion, and reproduction adaptations are found in this order. These are too numerous to describe here; however, a concise synopsis of the families is found in the appropriate section in Edmondson (1959).

The order Diptera includes a vast number of insects such as the midges, houseflies, mosquitoes, blackflies, and craneflies. All of the adults of these are aerial or terrestrial. As in the Coleoptera, the larvae of the flies display a wide assortment of morphological and ecological adaptations. Most of the species are found in fresh waters. Some, however, such as the mosquitoes of the family Culicidae, occur abundantly in salt marshes and pools. Some representative dipterans are shown in Figure 12·12.

Of particular interest to us from the general ecological point of view of food relationships are the true midges of the Tendipedidae, the biting midges (Heleidae), and the mosquitoes. Tendipedid larvae are found widely in all types of fresh waters, and many species develop dense populations. Such large populations are possible because the larvae feed

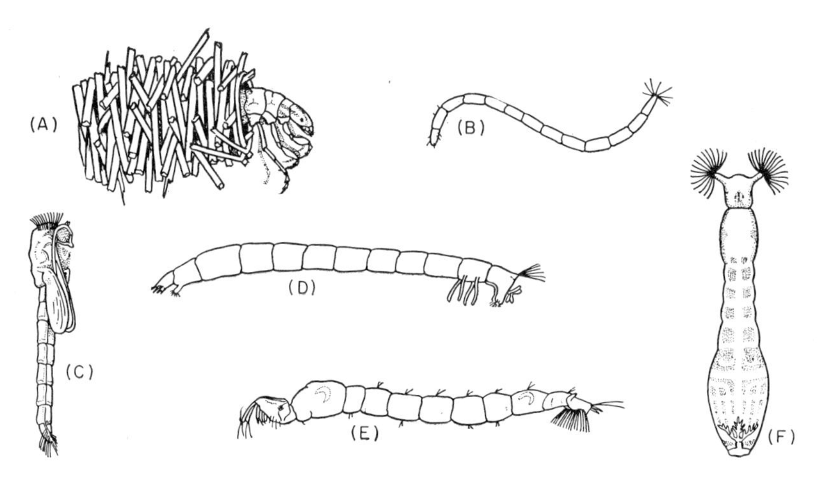

FIGURE 12·12. Some Insects of Inland Waters. (A) The caddisfly, *Oecetis*. (B) A dipteran, *Ceratopogon*. (C) The pupa of a dipteran, *Tendipes*. (D) The larva of *Tendipes*. (E) The larva of the dipteran, *Chaoborus*. (F) The larva of the blackfly, *Simulium*.

primarily on detritus and algae. Tendipedid larvae are small, ranging up to about 25 mm in length, and slender, some possessing a deep red color ("bloodworms") due to a particular blood pigment. Due probably to their size and abundance, these forms are consumed in great quantities by many fishes. *Tendipes* (=*Chironomus*) and *Calopsectra* (=*Tanytarsus*) are frequently reported as major items in fish diets. The transparent culicid larva *Chaoborus* is a predaceous form found on organically rich lake and pond bottoms. It is somewhat unique in that it can exist without surfacing for air. It does, however, migrate to the upper waters, and during this time it is captured by fishes. Although we have emphasized the importance of dipterous larvae in the food of fishes, we should point out that many pupae are also consumed. In a study of approximately 1000 specimens of the black crapple (*Pomoxis nigro-maculatus*) in Florida, it was found that over 25 per cent of the fishes had eaten diptera pupae, and over one third of the stomachs contained larvae; in the fall, an average of five pupae per fish were recorded.

PHYLUM MOLLUSCA (SNAILS, CLAMS, OYSTERS)

This phylum comprises over 70,000 species of externally diversified animals. Internally the animals exhibit common features, including a muscular foot, a mantle covering a visceral mass, and a calcium carbonate shell secreted by the mantle. Although a fairly large number of species inhabit fresh waters, the group has been most successful in the marine environment, for there the greatest variety of adaptations is found. Three of the five recognized orders are not found in fresh waters, although a few species enter estuaries. One order includes the small, slightly arched, gliding animals of the bottom, the chitons. Another order is that of the tusk shells, and the third marine order includes such animals as the octopus and squid. A small squid, *Loliguncula brevis*, commonly inhabits mesohaline waters from Chesapeake Bay southward and along the Gulf coast.

CLASS GASTROPODA

This class includes the familiar snails and other mollusks possessing a single coiled shell. Gastropods, in general, are widely distributed in fresh, estuarine, and marine waters. More specifically, however, many families, genera, and species exhibit restricted distributional patterns which may reflect the physical and chemical characteristics of the waters of the region. For example, two groups of freshwater snails are recognized on the basis of breathing mechanism: the pulmonates, which possess a lung and utilize atmospheric gases, and the prosobranchs, which have gills and make use of dissolved gases. Although the pulmonates are essentially inhabitants of fresh waters and the land, a few species are found in the marine environment; *Melamphus coffeus*, for example, occurs in estuaries of the Gulf coast. Similarly, in fresh waters of North America the large family Viviparidae is common east of the Mississippi drainage, but is absent to the west. The occurrence and abundance of gastropods are greatly influenced by pH, carbonate concentration, and dissolved oxygen. In addition to these factors, food, of course, is important. Most of the species feed on plant material, particularly algal deposits; others are omnivorous and scavengers. In the food web, snails are consumed in large numbers by certain fishes such as some of the catfishes, sunfishes, and suckers. In estuarine regions of the Florida Gulf coast, the batfish (*Ogcocephalus*) ingests considerable quantities of gastropods, including *Nassarius*, *Urosalpinx*, *Bittium*, and *Mitrella*.

CLASS PELECYPODA

The Pelecypoda are the mollusks typically enclosed between two calcareous shells. In our province this includes primarily the clams, mussels, and oysters, most of which are marine. In all situations, pelecypods appear to be most common in regions of relatively stable substrate, free from pollution, and with low silt turbidity. These factors relate, of course, to the fact that many of the animals characteristically lie buried, or partially so, with their siphons protruding into the water above. Through a ventral siphon water is taken in and food extracted; therefore shifting sediments and heavy silt load would be detrimental. In fresh waters the greatest abundance and number of species are typically found in large streams, the most numerous group being the family Unionidae. The Sphaeriidae, particularly the genus *Pisidium,* are often locally abundant. Of particular interest with respect to the estuary are the oyster reefs. The American oyster (*Crassostrea virginica*) builds extensive reefs which become inhabited by a most interesting assemblage of euryhaline plants and animals. Similarly, massive clumps of mussels, such as *Mytilus* or *Brachidontes,* come to harbor more or less characteristic associations, the nature of the associations being largely dependent upon local ecological factors and type of regional fauna.

In fresh waters, a number of kinds of pelecypods are eaten by various fishes and other animals, including muskrats and waterfowl. Man also takes large quantities of certain species for the pearl-button industry. In the estuary, small clams and oysters are preyed upon by fishes, and in many areas the large forms are eaten by starfishes and large crustaceans.

Phylum Echinodermata (Sea Cucumbers, Starfishes, Sea Urchins)

These are radially symmetrical animals possessing a spiny skeleton composed of calcareous plates. Nearly 5000 species are known, all of which are primarily marine. Many species, however, are characteristic of estuaries, the distribution of the forms being determined mainly by salinity. The sea cucumbers are soft-bodied, creeping forms, several of which, such as *Thyone* on the Atlantic coast, and *Thyonacta* on the Gulf coast, enter estuaries. A number of species of starfishes occur in estuaries; some, such as *Asterias,* are commonly associated with oyster or mussel reefs where, as noted above, the echinoderm constitutes a major menace to the mollusks. The disk-shaped sand dollars and the fragile brittle-stars also inhabit mesohaline waters and are fed upon by bottom-dwelling fishes such as flounders. Sea urchins (*Arbacia*) are widely distributed along our coasts and are often conspicuous components of the estuarine fauna.

Phylum Chordata

This phylum includes the familiar vertebrates and the less well known "invertebrate chordates," or protochordates. The protochordates possess a nonsegmented "backbone" at some stage of their life, and are essentially marine, with many species found in estuaries but none in fresh waters. Of these primitive forms, the tunicates (Urochordata) are often abundantly represented in estuaries. Ascidians, or "sea squirts," are commonly associated with oyster reefs and other submerged or intertidal objects. Along the Atlantic coast *Molgula* and *Botryllus* occur as members of encrusting communities on buoys, pilings, stones and shells, and other such structures. Cephalochordates, commonly called "Amphioxus," are found on sandy beaches below the high-tide mark. The genus *Branchiostoma* is found on both coasts of the United States, and a southern species, *B. caribaeum*, enters estuaries, at least along the west coast of Florida.

SUBPHYLUM VERTEBRATA

This subphylum covers the vertebrates, which are distinguished, among other things, by the possession of vertebrae. The classes of this great group include the jawless vertebrates, the cartilaginous fishes, the bony fishes, the amphibians, the reptiles, the birds, and the mammals. All of the classes are represented, to various extents, in fresh and marine environments and in the intermediate communities. It is this group that usually comes to mind as that of the highly predaceous animals, representing, as it were, the culmination of the aquatic food chain (see Chapter 14). Although this is generally true, the over-all ecological picture is complicated by a wide range of size and feeding adaptations among the vertebrates.

Class Agnatha. This class is represented by the jawless vertebrates, including the hagfishes and lampreys. The hagfishes are wholly marine, usually found in offshore waters. Lampreys occur in both freshwater and marine environments, but the marine forms must enter fresh water to spawn. The sea lamprey (*Petromyzon marinus*) is parasitic as an adult, and has become permanently established in some bodies of fresh water, including the Great Lakes. Some species inhabit inland streams and are not parasitic.

Class Chrondrichthyes. The members of this class, the sharks, skates, and rays, are cartilaginous vertebrates which are almost wholly marine. A few species enter estuaries, and one shark, *Carcharhinus nicaraguensis*, is apparently adapted to fresh water in Central America. These are mainly carnivorous animals. In estuaries skates and rays devour mollusks, worms, crustaceans, and other animals.

Class Osteichthyes. The bony fishes, representing this class, are widely distributed in practically all natural waters. Many are narrowly restricted to fresh waters, others to marine waters, and some from both elements freely inhabit estuaries or move from one extreme to the other. The salmon are anadromous and move from the sea into fresh waters to spawn; eels are catadromous, migrating from inland waters to the sea for reproduction. In inland waters, many species are clearly adapted to standing waters, others to running water. Within a body of water, fishes are generally associated with particular zones such as open water, bottom, or shallow vegetated regions. In the food web, fish species are variously adapted, some obtaining nutrients from organic detritus, others from plankton, and many from predation upon a variety of larger invertebrate animals and, indeed, other fishes.

Class Amphibia. None of the frogs, toads, and salamanders, members of this class, is truly marine, although the larvae of a few have been found in brackish pools, and adult toads (*Bufo*) and frogs (*Rana*) have been reported in estuaries. Adult frogs and seasonally abundant tadpoles are common to nearly all fresh waters. The tailed amphibians, or salamanders, such as the newts (*Triturus*), are less conspicuous due to their secretive habits. They are, nevertheless, often common in ponds and streams among litter or stones. The larvae of amphibians are essentially scavengers; the adults feed upon insects, worms, and other small organisms.

Class Reptilia. As a class, the reptiles are represented in marine and fresh waters and the intermediate estuary. The Galapagos iguana is apparently the only marine lizard. Snakes, particularly *Natrix*, are common in estuaries within the geographical distribution of the genus. A few snakes are fully marine, being found in the open seas. Crocodilians inhabit fresh waters and estuaries. Turtles, of course, occur in both marine and freshwater environments, but there is apparently little mixing of the two faunas in the estuary. Along the Atlantic and Gulf coasts of North America, the most typical estuarine species is probably the diamond-back terrapin (*Malaclemys terrapin*). The aquatic reptiles are primarily carnivorous, although some freshwater turtles feed upon plants to a certain extent.

Classes Aves and Mammalia. The birds and mammals are associated with inland waters and estuaries in a host of ways. Various waterfowl use the plants and animals of lakes, streams, and estuaries for food, and the marshes often associated with these bodies of water provide nesting sites. Wading birds and shore birds are characteristic of shallow zones and shores of bodies of water, and these animals also take food produced in or near the water. In inland waters beavers, muskrats, otters, and other fur-bearing mammals enter into the ecology of various communities, the dam-building trait of beavers often changing the nature of streams. Some of these mammals may be found in and around estuaries. Of the large "sea-going" mammals, some, such as the manatee and the dolphin (porpoise) enter estuaries. A considerable portion of the Amazon River of South America is inhabited by freshwater dolphins (*Inia geoffrensis* and *Sotalia*).

General References Which Include Aquatic Animals of the United States

Blair, W. F., Blair, N. P., Brodkorb, P., Cagle, F. R., and Moore, G. A., 1957. "Vertebrates of the United States," McGraw-Hill Book Co., New York, N.Y.

Edmondson, W. T. (Ed.), 1959. Ward and Whipple, "Fresh-Water Biology," John Wiley & Sons, Inc., New York, N.Y.

Hyman, L. H., 1940-1959. "The Invertebrates," McGraw-Hill Book Co., New York, N.Y.

MacGinitie, G. E. and MacGinitie, N., 1949, "Natural History of Marine Animals," McGraw-Hill Book Co., New York, N.Y.

Pennak, R. W., 1953. "Fresh-Water Invertebrates of the United States," Ronald Press Co., New York, N.Y.

PART V

Relationships of Organisms and Environment

Chapter 13

POPULATIONS IN AQUATIC ENVIRONMENTS

A GROUP OF organisms occupying a given space at a particular moment in time is called a *population*. A population may consist of a single species and thus constitute a *species population*, or it may be composed of a number of species exhibiting similar ecological traits (food requirements, breeding sites, etc.) and regarded as a *mixed population*. In addition to structure and time-space attributes, a population also exhibits a number of measurable characteristics such as birth rate (natality), death rate (mortality), density, and capacity for increase. The extent to which these traits are manifested in an environment is strongly influenced by properties of that environment.

THE PRINCIPLE OF LIMITING FACTORS

The very fact that we can recognize an environment-organism complex such as a lake community or an estuarine community attests to the premise that all organisms do not react uniformly to all combinations of physical, chemical, and biological features of the environment. Through evolutionary processes, organisms have become variously adapted to certain sets of conditions. It follows, therefore, that successful development and maintenance of a population depend upon harmonious ecological balance between environmental conditions and tolerance of the organisms to variations in one or more of these conditions. This idea suggests that one or more factors can serve to limit the areal distribution, density, and other attributes of a population. A factor that so checks or exerts some restraining influence upon a population through incompatibility with species requirements or tolerance is said to be a *limiting factor*. You will note that the limiting factor principle rests essentially upon two basic concepts. One of these relates organism to environmental supply of materials needed for metabolism and growth. The second concept pertains to the tolerance which an organism exhibits toward environmental factors and conditions.

The aspect of environmental supply in the limiting factor principle includes, among other things, the provision of nutrients and other substances (respiratory gases, for example) necessary for growth of the organisms. If, out of a broad array of materials required for growth and development, one is absent, a given species will not be able to survive. On the other hand, if the nutrient is present but only in small amounts, the population of a particular species will be limited proportionately. These conditions were described in 1840 by Liebig in what has come to be called the *law of the minimum.* This principle recognizes that the development of a population is essentially regulated by the substance occurring in minimal quantity relative to the requirement of the population.

It is frequently difficult to determine precisely which of the factors of a complex is exerting a limiting effect. This is due largely to the fact that the utilization of a certain substance may be regulated by other materials, or by physical factors in the environment, or by physiological adjustments of the organism itself. It has been shown, for example, that the uptake of phosphorus by the alga *Nitzschia closterium* is influenced by the quantity of nitrate and phosphate in the environment; nitrate utilization, on the other hand, appears to be unaffected by the phosphate. Potassium, calcium, and magnesium generally regulate interactions in plant metabolism. The assimilation of nutrients by some algae is related to environmental temperature. Temperature and carbon dioxide tension may affect the rate of oxygen utilization in certain fishes. We have previously learned that the absence of free carbon dioxide in waters is not necessarily critical for certain plants which can assimilate carbon dioxide from bicarbonates.

Dissolved oxygen is necessary in respiration of most aquatic organisms. The minimal requirements of species vary and often limit the spatial distribution of certain forms in the community. Figure 13·1 shows the law of the minimum operating under natural conditions to limit the upstream penetration of several species of fish. The water issuing from the "boil" is oxygenless, but gains the gas with downstream flow. A small amount of hydrogen sulfide is present in the boil and decreases downstream; this may also be involved in the observed distribution.

The law of the minimum pertains primarily to the reaction of organisms to factors necessary for growth and metabolism. In the case of plants, these factors include the inorganic nutrients and solar energy for photosynthesis. In addition to satisfying their metabolic needs, organisms are also confronted with physical and chemical factors which essentially regulate the distribution and numbers and the extent to which the organisms are able to utilize growth factors. Such environmental features as salinity, temperature, and currents govern many populations. The extent to which these factors limit a population depends primarily upon the *tolerance* of the organisms to a single factor, or, in some instances, to a complex of interacting factors. These ideas are the essence of the *law of tolerance* stated in 1913 by V. E. Shelford of the University of Illinois.

Tolerance to features of the environment varies widely among aquatic organisms. A population may exhibit a wide range of tolerance toward

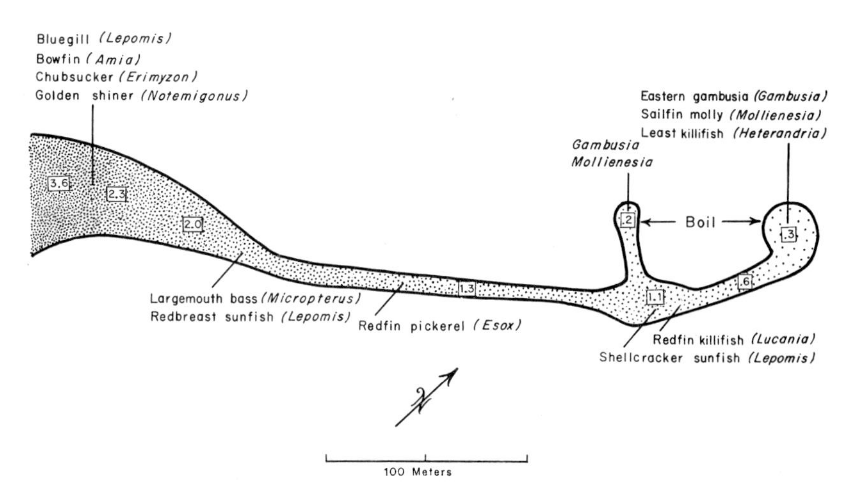

FIGURE 13·1. The Distribution of Twelve Species of Fishes in Beecher Springs, Near Welaka, Florida. The locations shown are the nearest that the fishes were observed to approach the head of the spring near noon of a sunny day in June, 1953. Dissolved oxygen values are expressed in parts per million. (After Odum, H. T. and Caldwell, D. K., 1955. "Fish Respiration in the Natural Oxygen Gradient of an Anaerobic Spring in Florida," *Copeia*, Vol. 1955, 104-106.)

one condition, and a narrow range toward another. All stages in the life history of an organism do not necessarily show similar ranges of tolerance. The range of tolerance toward a given factor may be modified by another factor. A wide range of distribution of a species is usually the result of the wide tolerance.

In describing the tolerance of an organism, the prefix *eury-*, meaning wide, or *steno-*, meaning close or narrow, is added to a term for the particular feature. We have already become familiar with the terms *euryhaline* (wide salt tolerance) and *stenohaline* (narrow salt tolerance). These conditions are shown in Figure 13·2. To these, we might add *eurythermal* and *stenothermal*, pertaining to temperature. Let us now consider some examples of tolerance relationships, for such are of utmost importance in regulating the abundance and distribution of organisms.

Within our provinces of fresh waters and the estuary, salinity constitutes a striking example of an environmental factor which limits organisms, primarily by interactions with physiological processes other than the uptake and utilization of nutrients. Although we have defined salinity in terms of total dissolved solids, some of which are plant nutrients, our reference here is directed toward chloride concentration in relation to density of the environment. It is this factor, and the various biological adaptations to it, that essentially distinguish marine and freshwater organisms.

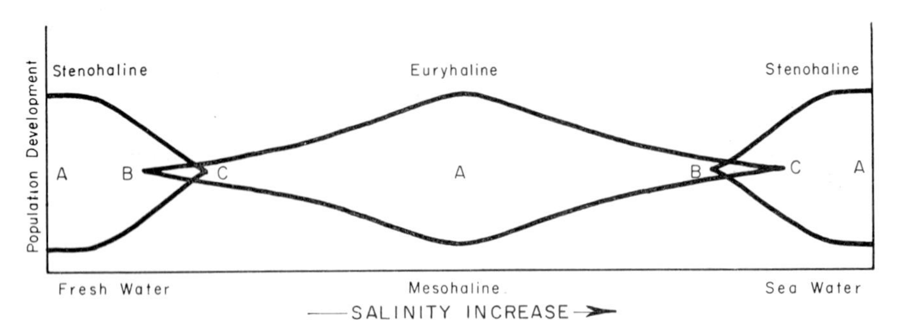

FIGURE 13·2. Schematic Comparison of Population Development (Growth, Reproduction, Migration) of Stenohaline and Euryhaline Organisms. Tolerance to a salinity gradient is indicated by the width of the figures. Optimum conditions are shown at (A), minimum limits at (B), and maximum limits at (C). Stenohaline organisms may be tolerant to either a high salinity or low salinity, and population development is often affected by slight changes in salinity.

The extent to which inhabitants of fresh or marine waters can invade the contrasting environments, and, indeed, the ability of certain euryhaline populations to occupy estuaries permanently, is related to the limit of tolerance of each organism to variation in salinity. Among animals, generally, tolerance to salinity changes varies greatly, as do the mechanisms associated with adjustment. The salt concentration in the body fluids of most marine invertebrates is the same (isotonic) as that of the environment. These forms are essentially stenohaline and are restricted to regions of relatively stable, near-seawater salinities. Freshwater organisms, on the other hand, inhabit a medium in which the concentration of salts is usually less (hypotonic) than that of the body juices; this is also true of regular inhabitants of brackish waters. According to diffusion principles, these organisms continually take up water across exposed, water-permeable membranes. Thus, freshwater animals must possess some structure or mechanism for maintaining proper internal balance of salts and water as well as for ridding the tissues of excess water.

Adaptation to estuarine existence demands various modifications of these osmoregulatory organs, as a few examples will indicate. Contractile vacuoles are characteristic of freshwater protozoans, but not of their marine relatives. In certain flatworms (*Gyratrix*, for example) the flame-cell mechanism serving in osmoregulation and excretion is more complexly developed in freshwater forms than in estuarine ones; in marine flatworms, major parts of the system are absent. Freshwater crayfish possess long, well developed nephridial canals; in the estuarine and marine lobsters these structures are reduced and poorly developed.

The fluids of marine bony fishes are hypotonic to the environment, and therefore the animals tend to lose water to the medium. In contrast to freshwater inhabitants, marine fishes actively drink water, excrete a concentrated urine, and excrete excess salts (obtained by drinking sea water) against an environmental gradient through specialized cells in the gill region. In the estuarine environment, these marine forms must function in the manner of freshwater species. The young of certain species of marine fishes inhabit estuaries and possess glomerular kidneys; with migration to higher salinity as adults, the glomeruli degenerate. Highly euryhaline fishes, particularly those adapted to extreme migrations from fresh to salt water (the eel (*Anguilla*), for example), are usually capable of performing as either marine or freshwater fish.

We have dwelt on this subject of osmoregulation because it demonstrates some of the forms of adjustment often necessary in order for organisms to tolerate certain controlling factors in the environment. In many instances, however, the limit of tolerance of a given organism may be modified by other factors. It has been shown, for example, that the shrimp (*Crangon crangon*) and estuarine crab (*Carcinus maenas*) are able to tolerate lower salinities during summer seasons of warm water. In Japanese waters, the oyster (*Ostrea gigas*) exhibits a wider range of salinity tolerance at winter temperatures than during the warm season. The tendency for marine species to tolerate lower salinities at higher temperatures accounts, in part at least, for the generally greater number of species in estuaries of the tropical zones than in the cooler climes.

As in the case of salinity, temperature also acts as a controlling factor related to range of tolerance of species. In this sense, temperature serves to regulate growth and metabolic rates of various organisms, and often determines the time of reproduction. It is immediately apparent, therefore, that temperature is highly important in delimiting the rate of utilization of nutrients and light by plants, and the tempo of food intake by animals, to satisfy metabolic demands. These relationships are contained in the "Q_{10} Rule," or Van't Hoff's principle, a chemical law applied to some physiological processes. The principle holds that the rate at which biological processes proceed is increased nearly two-fold with each 10° rise in temperature (the rate is a linear function of temperature). Obviously, this principle operates only within the range of tolerance of a given species, and is further restricted by an optimal point within the over-all range. For example, if sufficient nutrients are available, an algal population may grow rapidly and "bloom" at the optimal temperature for that species; beyond this point, and before the maximum tolerance limit is reached, the population declines.

Throughout the living world, organisms have become adapted to temperature ranges and fluctuations in a spectacular variety of ways. In the first place, we recognize two broad divisions: (1) the "warm-blooded" (*homoiothermic*) forms, the birds and mammals, whose body temperatures are maintained at a uniform level independent of environment; only a few of this group are wholly aquatic, and (2) the "cold-blooded" (*poikilothermic*) organisms whose body temperatures approximate that of the environment and vary accordingly. Among the latter group are such extremes as the algae that live in ice water of the polar regions and bacteria that inhabit hot springs at temperatures near 90°C. Between these extremes are found many species of plants and animals variously adapted to thermal conditions.

Great numbers of species are unable to exist as physiologically active stages throughout the range of temperatures experienced in their environment; these typically form spores, or other "resting stages," which exhibit a wider range of tolerance than do the active forms. Stenothermal and eurythermal species are found in all major taxonomic groups. The European turbellarian, *Crenobia alpina*, is apparently restricted to a temperature regime below 12°C; sessile rotifers are usually found at temperatures above 15°C; tardigrades are active in a wide range from about 0° to 40°C; the optimum temperature for the bryozoan, *Pectinatella*, is between 20° and 23°C, the lower limit of tolerance being near 15°C; the oyster (*Crassostrea virginica*) tolerates temperatures from 4° to 34°C. We quickly call to mind typical "cold-water" and "warm-water" fishes, many of these being restricted to such environments by the temperature tolerance of the developing eggs; the eggs of the eastern brook trout (*Salvelinus fontinalis*) generally fail to hatch at temperatures above 12°C, the optimum being near 8°C. The optimum temperature for growth of the adult trout is about 15°C. Thus, several optima may be operative in the life history of a given species.

Light appears to be a controlling factor in the distribution of certain animals in time and space. Diurnal vertical migrations of planktonic crustaceans have been noted and studied widely. These movements characteristically take the organisms to near-surface regions during the middle of the night, and to the depths during midday. Numerous explanations have been suggested, and, indeed, the causes may be numerous. However, the correlation between the depth of penetration of the blue segment of the light spectrum and the position of greatest abundance of a copepod species, *Cyclops strenuus*, as shown in Figure 13·3, would indicate a strong influence of the light. Note particularly the conspicuous change in space occupancy with increasing day length from April to June.

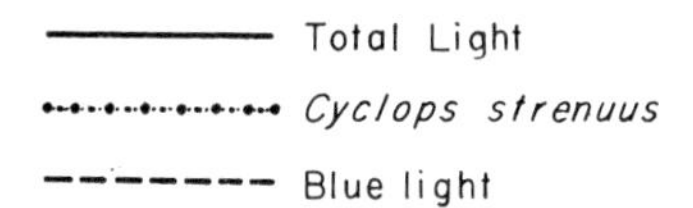

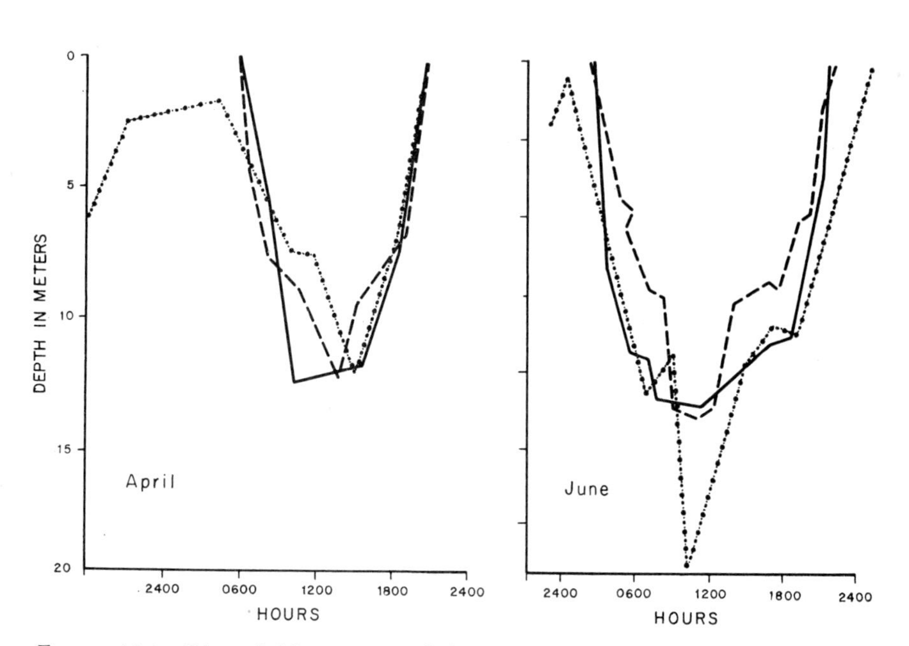

FIGURE 13·3. Diurnal Movements of the Copepod *Cyclops strenuus* Relative to Penetration of Light in Lake Windermere, England. The graph of total light shows at hourly intervals the depth to which light of an intensity of 32,800 erg/cm²/sec in April, and 108,000 erg/cm²/sec in June penetrated; the dashed line shows blue light treated in the same fashion, the intensity recorded being 9600 erg/cm²/sec in April and 305 erg/cm²/sec in June. The dotted curve shows at hourly intervals the depth at which most individuals were caught. (From Macan, T. T. and Worthington, E. B., 1951. "Life in Lakes and Rivers," *The New Naturalist Series*, Wm. Collins Sons & Co. Ltd., London, after Ullyot.)

From consideration of the "law of the minimum" as it pertains to nutrients, and the "law of tolerance" as it relates to environmental "controls" we can now appreciate the broader *principle of limiting factors*. This principle embodies the concept that population growth and success, in terms of abundance and distribution, are determined by a set of environmental factors, any one of which may, through scarcity *or overabun-*

dance, be limiting. We have seen a number of examples in which short supply of a requirement could hinder the success of a population; this is not difficult to understand. What is sometimes more difficult to comprehend is that an excess of normally important factors such as nutrient salts or sunlight can exert lethal effects upon organisms. For example, high concentrations of nitrate and phosphate have been shown to be toxic to both plants and animals.

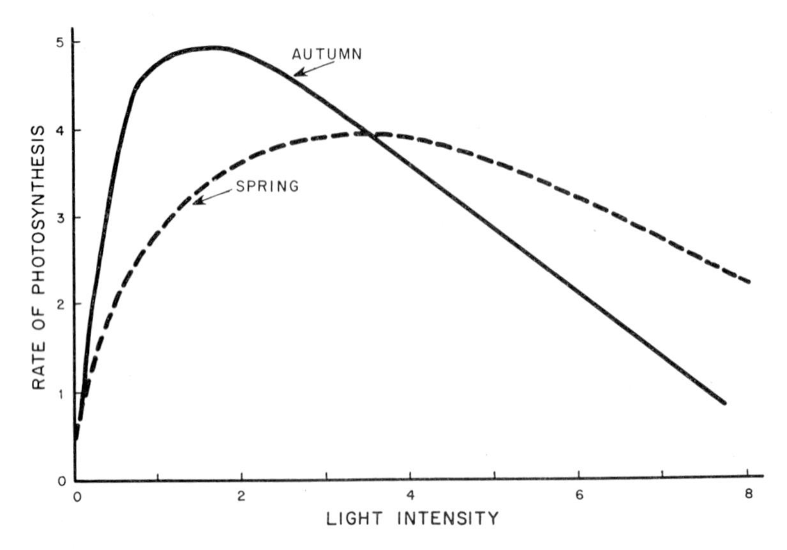

FIGURE 13·4. Photosynthesis (μmoles CO_2 absorbed per 10 μliters of plant matter per hour) Against Light (milliamperes, Weston photronic cell) in Natural Communities in Two Seasons in Western Lake Erie. (From Verduin, J., 1956. "Energy Fixation and Utilization by Natural Communities in Western Lake Erie," *Ecology*, Vol. 37, 40-50.)

Several exciting aspects of the limiting factors principle can be shown in the reactions between sunlight and phytoplankton in photosynthesis. Consider, for example, the curves of photosynthesis rates of spring and autumn phytoplankton assemblages in western Lake Erie shown in Figure 13·4. Note first that in both cases the rate of photosynthesis (measured by rate of uptake of carbon dioxide by plants) is inhibited by "too much" as well as by "too little" light. Light conditions optimum for photosynthesis are indicated at the peak of each curve, but notice that these conditions differ for the two seasonal assemblages. It appears also that the spring phytoplankters exhibit a wider range of tolerance to the light intensities encountered than do those of autumn. Of particular interest is the higher rate of photosynthesis in the dim autumn light than in the relatively bright light of spring. This would seemingly indicate that the autumn community utilizes light more efficiently than do the spring populations. In all probability, however, the species comprising the communities are different.

The principle of limiting factors is not built upon physical and chemical factors alone. Biological features of the environment are often quite influential in determining the abundance and distribution, or over-all success, of a specific population. Once we introduce other organisms and their demands and activities into a set of factors, we immediately perceive greatly increased complexity of ecological relationships. For in addition to *reactions* between a population and the physicochemical factors, we must also recognize *coactions* between, or among, the several species living together. We must also recognize coactions among the individuals of any given population.

Interspecific relationships involve a broad spectrum of coactions, including several forms of symbiosis, tolerance, competition for one or many of a great variety of necessities, antibiosis, parasitism, and predation. Within this list of activities are found those that pertain to shelter, food supply and procurement, and breeding activities. These often relate plant to plant, or plant to animal, or animal to animal in a broadly graded series of coactions ranging from those in which one member of the relationship suffers extreme harm at the expense of another, to those in which both of the participants are benefited. Singly, or in combination, interspecific relations may, and often do, act as a limiting factor. They are also basic in the structure and dynamics of communities, and we shall consider certain of these coactions in the following chapter.

The ability of a species to successfully occupy a place in the economy of a community is also determined by intraspecific relations of the individuals and certain inherent biological qualities of the population. Intraspecific competition for such requirements as food, territory, and breeding partners may serve to limit a given population. Similarly, overcrowding, which may in some cases lead to stress symptoms, can be limiting in nature. Inherent biological qualities such as frequency of breeding activity, number of offspring produced, survival rate, and mortality rate are often correlated with physiological tolerance to environmental factors. Upon these adaptations depend, at least in part, the abundance and distribution of a population. Some of these population attributes will be examined in the following paragraphs.

Although we have somewhat dissected the set of limiting factors for the sake of illustration, it is important to bear in mind that a single factor can become limiting only when it is near the maximum or minimum of the range of tolerance of a population. In nature, under usual conditions, a given population reacts to a complex of factors. The composition of this complex is not normally fixed; some of the components change seasonally, and a number of interdependencies exist, as we have seen. A specific organism or population, on the other hand, presents a different story; its tolerance to each factor is more or less firmly established by inheritance and evolution. Thus the position and place of a species in the economy of a community is fairly well defined and delimited by relations between organisms and environmental factors. This thought brings us to another important concept in population ecology, that of the *niche*.

POPULATIONS AND THE ECOLOGICAL NICHE

We have just witnessed how an imposing series of environmental conditions and processes combine to circumscribe the total functions of a given animal or plant population. As a result of evolution of both the complex of environmental factors and the population, the functional reactions and coactions of a particular species have become essentially unique to that species. Thus, each population (characterized by certain rather definite morphological and physiological parameters) and its environmental complex constitute a sort of system, the attributes of which distinguish it from other such systems. Once two systems overlap significantly, the population of each is thrown into competition for nutrient, or other, requirements, and one system breaks down. This "systems" example is a homely analogy to the functional concept of the *niche*.*

Can more than one population occupy the same or overlapping niches in natural communities? This question has intrigued ecologists for many years. As yet, the problem has not been solved to the satisfaction of all. The idea is highly important in ecological thinking, however, because, among other things, it offers a point of entry into the study of the factors limiting a single population, and it also relates to events commonly observed in animal communities maintained in the laboratory. Figure 13·5 depicts the events following the introduction of two species of cladocerans, *Daphnia magna* and *D. pulicaria*, into a single medium in the laboratory. In nature, these forms inhabit small ponds rich in organic matter, and the distributional ranges of the two species overlap. In laboratory cultures, *D. magna* consistently lost in the competition with *D. pulicaria*. Note that in both algae food and the yeast food culture, the population of *D. magna* reached its peak abundance at about 21 days, after which there was a rapid extinction due to interspecific competition. Observe also

* In much ecological literature, "niche" is used in the sense of *location* within a habitat, a *microhabitat*, as it were. For example, within a stream (habitat) the turbellarian, *Dugesia*, often inhabits the undersurfaces of stones on the stream bottom, this locality sometimes being designated as the niche. However, we need a term to indicate the functions of populations. Because this seems to be the sense of the word as used by Joseph Grinnell of California in 1917 (see Grinnell, J., 1917), we should employ the term "niche" in the functional meaning.

the effects under the different nutrient conditions. In algae, the initial growth rate of *D. pulicaria* was quite rapid; in yeast, the rate was less rapid, being exceeded, as shown in the figure, by that of *D. magna*. Two limiting factors, food and oxygen, were operative in this experiment. The results clearly demonstrate that two species do not both persist in a simple environment exhibiting minimal niche diversification. Under natural conditions, of course, food is but one of many factors for which competition by similar species may exist. Intense competition between closely related or ecologically similar species is often avoided by adapta-

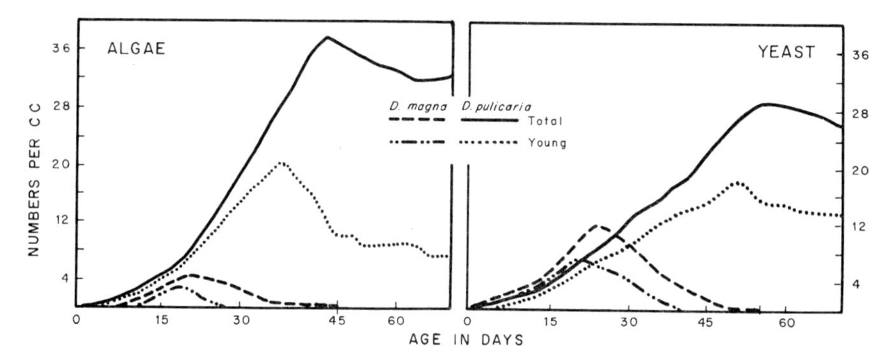

FIGURE 13·5. Competition Between Two Species of Cladocerans in Laboratory Cultures. *Daphnia pulicaria* exhibits dominance over *Daphnia magna* in both algae and yeast food media. The larger population of *D. magna* in the yeast food culture is due to the less favorable nature of this food for *D. pulicaria*. (After Frank, P. W., 1957. "Coactions in Laboratory Populations of Two Species of *Daphnia*," *Ecology*, Vol. 38, 510-519.)

tions such as differing feeding times or breeding seasons, or by geographical distribution.

Another important aspect of the niche concept describes the "role" of a species population in the food and energy relationships in a community, i.e., as a "producer" or a "consumer." Although the dynamics and levels involved in these relations will be discussed more thoroughly in the next chapter, it seems worthwhile to mention briefly how these relate to our present topic. The role which a population performs in the economy of a community involves, again, a given set of reactions and coactions. The role of a diatom, for example, as a producer of the "original" energy-containing substance, is based upon the ability of the plant to carry on photosynthesis. The position, or niche, of the diatom, with respect to solar radiation, raw materials, and herbivorous organisms, is essentially the same throughout time. Consider, on the other hand, that the role of a great number of animals differs greatly with various stages of life cycles. The important point here is that organisms may be rather narrowly and firmly established in a given niche, or, depending upon adaptations, may function in a number of niches during a life span.

ATTRIBUTES AND DEVELOPMENT OF POPULATIONS

In the foregoing discussions our interest has centered mainly on some of the major physicochemical and biological features of the environment and their effects on populations. How, in the face of such an imposing array of limiting factors, are abundance and integrity of a species population maintained; and, beyond this, how do numbers increase? The answer is found partly in environmental characteristics and partly in the unique attributes of the population. The essential properties of the population which govern its growth, structure, and spatial distribution are *natality*, *mortality*, and *dispersal*. These qualities of the species are, in a sense, the "counterweights" set against limiting factors, the "degree of balance" militating, as it were, for or against species survival.

Natality and Mortality

Natality is a population parameter which describes the rate at which new individuals are produced; it is, in essence, the birth rate. It is often important in certain population problems to consider two aspects of this attribute; these are: (1) *potential natality*, or the maximum rate of population increase if no limiting factors are operative, and (2) *realized natality*, or the rate experienced under natural conditions of reactions and coactions. In estuaries, a population of the American oyster exhibits a very high potential natality, for a single individual is capable of spawning over 100 million eggs. The realized natality, however, amounts to only a small fraction of the potential. Estuarine populations of the sea catfish (*Galeichthys felis*) have a low potential natality; a single female seldom produces more than about 25 or 30 eggs at a time. Parental care, in which the male carries the eggs in his mouth until hatched (oral incubation), contributes to an apparent high realized natality for this species.

The value of the two aspects of natality is found in their usefulness in studying population growth and development in relation to environmental factors and the niche. Potential natality, determined from the rate and number of eggs produced per individual, serves as a standard with which a census of an existing population can be compared. This measure alone, would not, of course, account for the momentary structure of a population. Natality is always a positive quality leading toward continual increase in population size. This effect is essentially offset by population traits such as mortality and dispersal.

Mortality, as a population attribute, pertains to the death rate of individuals of the population. In contrast to natality, mortality has a negative effect on population development, tending to bring about its decline. As in the case of natality, two aspects of mortality may be recognized; these are: (1) *physiological longevity*, or that experienced under the most favorable set of ecological conditions, and (2) the *realized mortality*, or that due to environmental effects prior to fulfillment of the inherited potential.

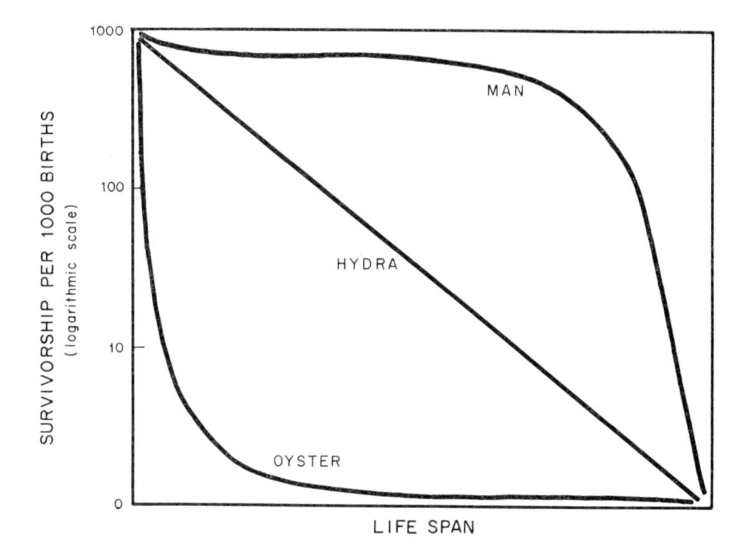

FIGURE 13·6. Mortality (Survivorship) Curves of Three Organisms. The curves are based on survivors per thousand births (log scales) and age in relative units of mean life span. (After Deevey, L. S., Jr., 1950. "The Probability of Death," *Sci. American*, Vol. 182, 58-60.)

The realized mortality of a given population can be illustrated clearly by means of a "survivorship" curve derived from plotting the number, or percentage, of individuals still living at successive time intervals against the known life span. Three such curves are shown in Figure 13·6. If a great majority of individuals of a particular population survived to the limit of the life span of the species, i.e., if physiological longevity were to be experienced, the curve would extend horizontally to physiological longevity and then drop steeply; such a condition is not found in nature. The curve for man in Figure 13·6 approaches such a pattern. Note, however, that this curve shows a decided decrease at the low end, indicating high mortality in early years. As medical science keeps proportionately more people alive, man's curve will more closely approximate that of physiological longevity. If a high percentage of the original population dies during early life stages, the curve resembles that of the oyster in

Figure 13·6. The oyster survivorship curve is so shaped because of high mortality among the motile larvae; with attachment of the larvae to the oyster reef and subsequent development as a sessile form, the slope of the survivorship curve decreases. The survivorship curve for hydra indicates a relatively constant percentage of mortality with age of the population.

Dispersal

Population dispersal refers to shifting or rearranging of individuals, small groups of individuals, or the entire population with respect to space and to time. Most species possess some adaptation for dispersal at some stage of a life cycle. In some forms, dispersal is accomplished during the seed or spore stage; in others the larval stage may be moved by active swimming or through passive transport by currents or other agents. Very few natural populations exist in which there is not some immigration and emigration of individuals. The degree of either of these may be quite subtle with respect to numbers and to the rate at which the movement occurs. In this case, dispersal would probably exert little effect on the over-all structure of the aggregation. Conversely, rapid emigration or immigration by large numbers may result in depletion or near-depletion, or in overcrowding of the original stock.

The effects of loss or gain of individuals are important considerations in management of natural populations, such as fishes, for human exploitation. For example, the removal of individuals by fishing of moderate intensity is usually followed by replenishment of the stock. The rate at which the population is rebuilt is, of course, greatly dependent upon the natality-mortality relationship. Extreme exploitation may reduce the numbers below a minimum level necessary for population regrowth. This condition can result in extinction of the population. Rapid immigration of large numbers of individuals often has deleterious effects on the original population, primarily through increased competition for breeding territory, cover, and food. For many years this idea was not appreciated in fishery management circles, and many bodies of water already at their maximum "carrying capacity" (see below) received truckloads of fish fry and/or fingerlings. One result was overpopulation and stunting of all individuals of that particular species; another result was simply that the introduced fishes failed to survive or were eaten by carnivorous inhabitants of the lake or pond.

Under certain conditions natality alone may not be sufficient to maintain observed population numbers. In estuaries, for example, there is a net seaward flow of mixed water sufficient to compensate for the contribution of fresh water by the stream. Thus, estuarine plankton populations are subject to seaward flushing as determined by the rate of water circulation in the estuary. In order to maintain a given population density, natality, mortality, and dispersal must be attuned to seaward transport. In Great Pond, an Atlantic estuary in Massachusetts, the summer population of copepods (*Acartia*) has been studied. It was found that the reproduction rates in the uppermost parts of the pond were not adequate to account for the numbers of animals there, that immigration from deeper layers contributed to the population density of the upper region. In the middle portion of the estuary, natality was apparently sufficient to replace the loss by seaward transport. At the lower end of the estuary, great mortality was experienced, the small population present there being maintained only by immigration of copepods from the upper regions.

As indicated earlier, dispersal, natality, and mortality together constitute a set of interacting factors which essentially regulate growth and structural dimensions of a population. We have seen that there is a tendency for populations to grow, due to the fact that the potential natality normally exceeds the potential mortality. However, the extent to which such growth proceeds would be tempered by dispersal traits. Should the realized natality exceed the realized mortality under conditions of minimum dispersal activities, the population would increase. Under the same natality-mortality conditions, and with excessive emigration, the population could be expected to decrease. With equal birth and death rates, the population would tend toward stability, but again, this is influenced by movement of individuals. If the realized mortality exceeds the realized natality, extinction of that population would be expected; recruitment from nearby aggregations of the same species could serve to maintain the "unbalanced" population. Having given attention to these "forces" which mold population growth and density, let us now turn to consideration of these two population parameters.

Growth of Populations

All natural species populations possess an *innate capacity for increase* at a maximum rate—the *biotic potential*. For a given species, this maximum rate is set by inherited potential natality-potential mortality values. These values differ greatly with species. Compare, for example, the figure given previously for a single oyster, with that of man. In nature, the biotic potential is not normally fulfilled, due to the limiting features of many environmental factors. Thus, opposing the innate tendency of a population to increase is a set of conditions which, in effect, offer resistance to population growth. We recognize, therefore, that the growth and success of a species population depend upon the degree of "harmony" struck between biotic potential and *environmental resistance*.

The actual, or realized, rate of growth of a population is determined by the initial number of individuals, plus additions, minus the number lost per unit time. If, during a period of one year, two individuals produce 24 offspring from which four are lost, the population density would stand at 22 at the beginning of the following year. If similar conditions of natality and mortality prevail during the second year, the census should be 242 individuals. In other words, the rate of population growth is geometric, and is represented by the relationship:

$$\frac{dN}{dt} = rN$$

in which N is number of individuals and t is time. The symbol r represents the rate of natural increase of a population, in other words, the dif-

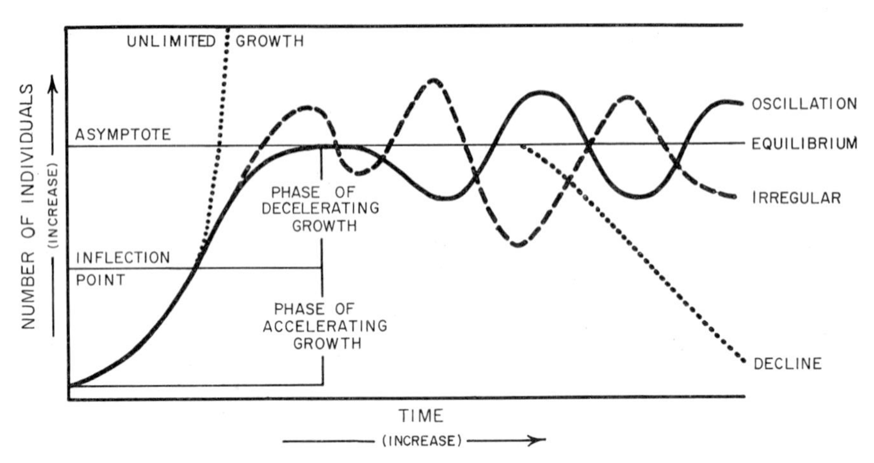

FIGURE 13·7. Stylized Representation of Various Stages of Population Growth and the Changes in Population Form. (After Allee, Emerson, Park, Park, and Schmidt, "Principles of Animal Ecology," 1949. W. B. Saunders Co., Philadelphia, Pa.)

ference between natality and mortality at a given moment; $\frac{dN}{dt}$ gives the average rate of change of N per unit time. If such growth continues unlimited, it takes the form indicated by the curve in Figure 13·7.

Since a particular community can support only a given number in any one species population, it is apparent that the maximum biotic potential can not be maintained in nature. Consequently, the form of the growth curve of populations subject to limiting factors must differ from the "unlimited" curve in Figure 13·7, and so it does. As a population begins to increase under favorable environmental conditions, growth is typically rapid and passes through a *phase of accelerating growth* (Figure 13·7). This is especially characteristic of populations exhibiting high natality and under conditions of little intraspecific competition. In time, the rapidly increasing numbers do create problems of maintenance, and the growth rate begins to slacken, the population passing through a *phase of decelerating growth* as shown in Figure 13·7. The point on the curve at which the rate of increase slackens is called the *inflection point*. Eventually the population size becomes somewhat stabilized through essential equilibrium of reactions and coactions. This size level may be at, or near, the *carrying capacity* of the community. On the curve, the carrying capacity (number of individuals) is designated the *asymptote*. Note that the curve representing population growth under natural conditions is typically in the form of a shallow "S," or the *logistic curve*. Given self-regulating growth processes, the curve may be derived from the differential equation:

$$\frac{dN}{dt} = rN\frac{(K - N)}{K} \quad \text{or} \quad \frac{dN}{dT} = rN\left(1 - \frac{N}{K}\right)$$

where r is the biotic potential for each individual in the population; N is the total momentary population size; t is time; and K is the maximum number of individuals possible in the population under prevailing conditions. The expression $\frac{(K - N)}{K}$ describes the extent of resource utilization by increasing population; in other words, the growth of the population itself causes the environment to become more limiting. As a result of this condition, the potential rate of reproduction is decreased as the population size grows toward the asymptote. This equation tells us that the population growth rate is equal to the maximum possible rate of increase times the degree of realization of that maximum potential rate.

The equilibrium which a population maintains about the asymptote is primarily a function of natality (n) and mortality (m), and, under certain conditions, dispersal. At equilibrium, $n - m = 0$, meaning that losses due to mortality are offset by additions derived from births. Deviations above the asymptote occur when $n - m$ is positive (population increase); deviations below the asymptote follow when $n - m$ is negative (population decrease).

In Figure 13·8, data on the growth of a laboratory population of a small fish, the guppy (*Lebistes reticulatus*), are fitted to the logistic curve. Note the characteristic phases of growth until the asymptote is attained. Observe also that the population increase initially "overshoots" the asymptote, but returns to the carrying capacity; this phenomenon is often encountered in population studies.

Upon reaching its asymptote, a population does not normally remain at uniform density, but rather fluctuates in numbers in response to one or more ecological or inherited factors. Irregular fluctuations (Figure

13·7), or relatively asymmetrical departures from equilibrium, may result from variations in any critically limiting environmental factor. The periodic fluctuations of about 65 weeks' duration in Figure 13·8 are partly ascribable to artificially produced temperature fluctuations. A great number of observations on populations of plants and animals have shown symmetrical fluctuations, or oscillations, of population numbers in response to factors regulated by daily rhythms, tidal cycles, seasonal and annual cycles, and to predator-prey and host-parasite relationships. Some further aspects of the cyclic phenomena will be considered in the following chapter.

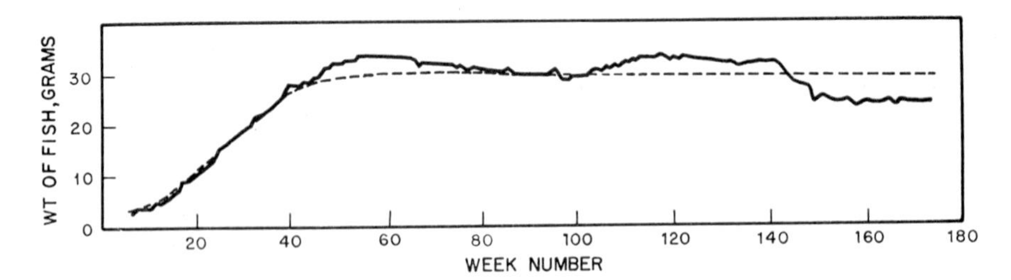

FIGURE 13·8. Growth in Weight of Laboratory Population of the Guppy (*Lebistes reticulatus*) Fitted to Logistic Curve. (From Silliman, R. P. and Gutsell, J. S., 1958. "Experimental Exploitation of Fish Populations," *U. S. Fish and Wildlife Serv. Bull.*, Vol. 133, 215-252.)

The third departure from equilibrium illustrated in Figure 13·7 is that of population decline and extinction resulting from negative growth. We have seen how "too much" and "too little" of a physical factor (Figure 13·4) and interspecific competition(Figure 13·5) contribute to decline in populations. These will suffice as examples of many possible causes of the action.

Within a given population, the relative number of individuals in each age level is important in determining the growth form, as well as serving as a basis of diagnosis of the condition of the population. Figure 13·9A illustrates a relatively stable population with respect to growth; in this structure one might expect $n - m$ to be near zero. Figure 13·9B represents a declining population in which there is a disproportion of young and old, or $n - m$ is negative. Figure 13·9C is representative of a population in which $n - m$ is positive; this, therefore, is a population increasing in size. These general diagnoses assume that the distribution of age groups in the population is stable. The importance of age distribution in a population lies in the cumulative effect of the natality-mortality values for each of the age classes on the general population. As indicated above, stability of population structure depends upon nearly equal rates of natality and mortality in each of these classes, deviation from this resulting in increase or decrease of population size.

In our consideration of population growth form, we have endeavored to show that a somewhat common pattern of increase is manifested by most organisms. This pattern, governed to a great extent by natality and mortality characteristics, includes rather clearly defined phases. These phases vary from species to species depending upon reactions between the population and the physicochemical nature of the environment, upon co-

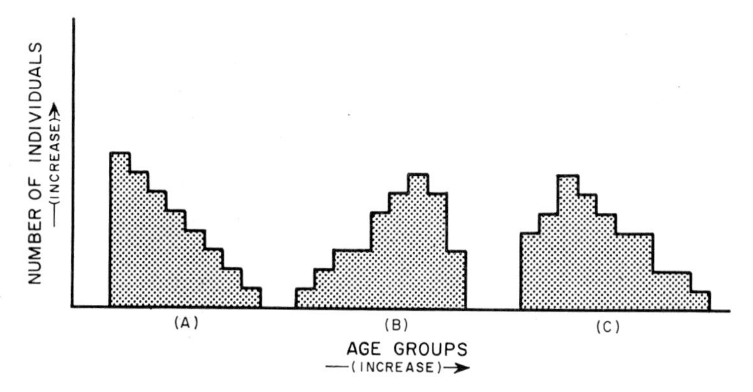

FIGURE 13·9. Distribution of Age Classes in Three Hypothetical Populations. (A) represents a relatively stable population (birth and death rates equal) in which there is a moderate proportion of age groups from young to adult. (B) depicts a declining population suggested by the high frequency of adults and few young. (C) represents an expanding population in which there is a large number of young compared to the number of older age groups.

actions between the species and other inhabitants of the community, and upon inherent qualities of the life history of the species. This last-mentioned qualification pertains to certain mammals, and also to other animals such as many crustaceans and most insects which exhibit complex life cycles. The growth form of populations of these animals usually deviates from the logistic curve. Thus, the logistic formula is a useful tool in describing growth of some thoroughly studied populations, but it should not be considered universally applicable to all forms of growth.

POPULATION DENSITY **

Population density refers to the number or mass of individuals of a given population occupying a unit of space. Several different space-relative units of density measurements can be used. We have, for example,

** The author is grateful to Paul G. Pearson, Rutgers University, for writing this section.

made reference to an oligochaete population with a density of 19,000 individuals per square meter of bottom surface in the saline region of an estuary; Figure 13·3 shows the number of copepods of the genus *Cyclops* per cubic centimeter; Figure 13·8 relates to the grams of fish mass per volume of the container; in fish-management programs, the yield is often expressed in pounds (or kilograms) per acre (or hectare) of pond or lake surface. Population density is also time-relative in that the number, or mass, of individuals may, and often does, vary temporally. Daily vertical movements of certain zooplankton, for example, result in changes in population density at a given level in a lake; seasonal migrations of some fishes and other animals cause fluctuations in population size. Certainly, marked changes in population density are observed during breeding times.

An observed population size represents the product of a number of environmental features and relationships working through biological processes. Thus, analysis and study of density yield much enlightening information on growth form and the effects of specific environmental factors on population development. It is often difficult, however, to recognize and measure subtle effects operative under natural conditions. Ecologists frequently surmount such difficulties by laboratory studies under controlled conditions, and by the judicious use of statistics to test the meaning and validity of laboratory and field observations.

Although influenced to a considerable extent by biological attributes of the species and the biogeochemical features of the environment, population density, per se, exerts effects on the members of the population and upon the environment. The density of a given population in a prescribed space may affect the population rate of increase, the proportion of males and females, the growth rate of individuals, the rate of utilization of nutrient substances and respiratory gases, and other processes. Thus, there is an optimal density level for the growth or success of a population, and this implies that there must also be proportional limitation of growth or success under overcrowded or undercrowded conditions. It has long been known that there are deleterious effects on growth, reproduction, survival, and other biological processes which accompany overcrowded situations. It remained for W. C. Allee and his students to demonstrate that at all levels of the animal kingdom there is added safety in numbers up to the optimal population level; there are adverse effects accruing to undercrowded populations. This is known as the *Allee Effect*. One example of this is obtained by noting, in Figure 13·10, the relative rate of population increase. If data from a logistic population curve are replotted to show the change in rate of increase (dN/dt) against density, one can see that at the low and high densities there is a low rate of increase while the highest rate of increase occurs at some optimal population level.

There may be several mechanisms whereby aggregations of individuals can grow and survive better than isolated individuals in the same medium. One such mechanism involves decrease of exposed surface. In the group, there is less surface area per individual exposed to the environment than would be if the individuals were dispersed. For example, the individuals in a group of flatworms have less total surface exposed to adverse effects

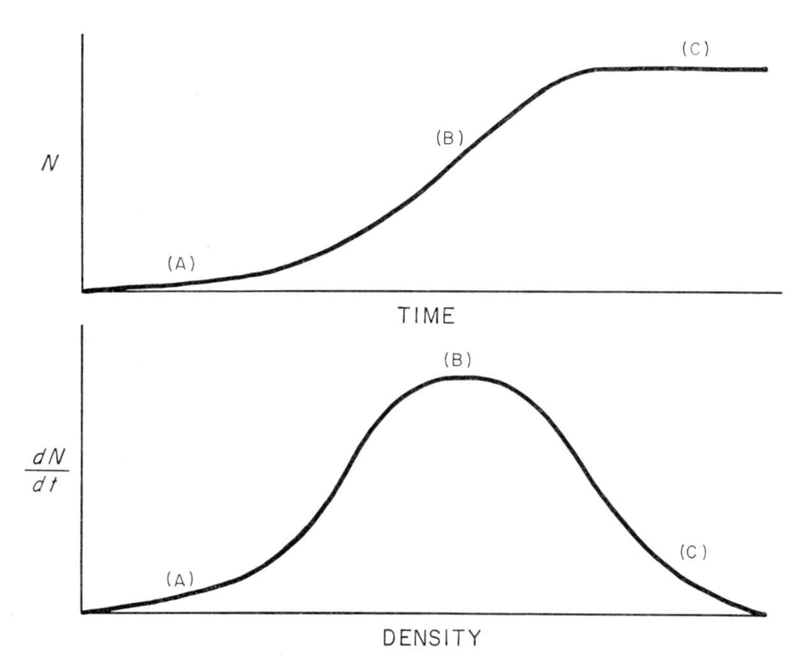

FIGURE 13·10. Schematic Representation of the Allee Effect, in which data from the logistic curve (above) are plotted to depict rate of population change (*dN/dt*) as a function of density (below). According to the lower figure, both low density and high density may inhibit rate of population growth.

of ultraviolet light or other extreme environmental factors than would be the case if these individuals were dispersed and exposed. Similarly, masses of sea-urchin sperm survive longer than dispersed cells since there is less environmental exposure to the sperm inside the mass. It might be noted, also, that there are undercrowding effects on the embryonic development of sea-urchin eggs.

Another aspect in the success of groups versus individuals is that the group can more efficiently condition the medium by removing poisons, changing the salinity of the water, or by adding growth-promoting factors. It has been demonstrated that groups of goldfish can survive longer than isolated individuals when placed in media poisoned by silver nitrate; the group more quickly removes and detoxifies the poison, whereas individuals usually die. The groups of goldfish also grow faster as a result of sharing particulate food that spews from their mouths, and as a result of a higher concentration of a growth factor associated with the slime produced by the fish. Other studies have demonstrated that groups of protozoa have a higher per-individual growth rate than do isolated individuals due to the greater production of some growth-promoting factor, and also because the group can better control the population of bacteria.

There may also be benefits to the group resulting from more efficient coordination of activities that produce greater synchrony in reproduction. This heightened psychological state that leads to greater synchronization of activities and consequently to less mortality, has been shown for shore-bird colonies.

Distribution of Individuals of a Population

We have previously considered dispersal of population individuals as a process and condition contributing to population growth form. Of utmost importance to population dynamics, generally, is the manner, or pattern, in which members of a population are distributed within the community. This quality of populations is also of primary concern to the study of intrapopulation relationships.

Within the space occupied by a given population the individuals may become dispersed in a *random* pattern, in an *evenly-spaced* pattern, or as *aggregations*. Obviously, the pattern exhibited by a given population is subject to considerable variation depending upon the census of the population, the number of sites available for occupancy, food accessibility, and a host of other factors. Equally important in setting the dispersal pattern is the inherent sociological nature of the species at various seasons. Important as the problem of animal distribution is, it has received insufficient attention, especially as pertains to aquatic organisms. Future researches may well reveal additional patterns of intrapopulation distribution.

Completely random distribution of individuals is apparently uncommon in nature. The degree of "randomness" in the dispersal of individuals is determined statistically by several methods. One of these involves the fitting of observed distributional data to the "Poisson series," a statistical measure of the frequency with which aggregations of 0, 1, 2, 3, 4, 5, . . . n can be expected in plots or quadrats. Deviation from the Poisson indicates various patterns of nonrandom distribution. It has been suggested that the lack of random dispersals in nature results from a generally inherited tendency of most animals to aggregate to some degree.

Evenly spaced distribution patterns are probably best explained on the basis of tendencies of many species toward *territoriality*. As an animal trait, territoriality refers to defense of a given area by overt aggressive action. The rather euryhaline cyprinodont fish, *Cyprinodon variegatus*, exhibits marked territoriality during breeding season. At this time, the male fish come to occupy vigorously defended areas in the shallow zone of marsh streams. The territories are small, usually being about 30 to 50 cm in diameter, and the aggressive behavior lasts for several days; any male intruder is immediately attacked by a territory holder. The females are usually somewhat concentrated a short distance from the area of territories, and show little excitement. With considerable frequency females singly enter the territory of a male, where a brief breeding encounter takes place. Territoriality is often associated with dominance-subordinance levels developed in many animals. Green sunfish (*Lepomis cyanellus*) in a laboratory colony develop hierarchies and defend their respective positions. Depending upon numbers of individuals and available territories, the most subordinate fish may not be able to establish a place in a "peck-order" such as do members of chicken flocks.

The *home range* of an animal is the region in which the individual normally travels. The important distinction between territoriality and home range lies in the aggressive response to invasion of the area shown in territoriality; home range carries no connotation of aggressive behavior. The concept of home range has been investigated less in aquatic animals than in terrestrial forms. However, a fair volume of literature on the movement of tagged fishes in lakes and streams strongly points to homing and home range in these animals. For example, of 4557 fishes marked and released at Grapevine Point, Douglas Lake, Michigan, from 1937 to 1939, not a single one was caught from any other point. Investigations in an Indiana stream revealed little movement of fishes between pools. Most recently, results from a study of homing in the newt (*Taricha* (= *Triturus*) *rivularis*) in a California stream convincingly attest to observance of home range in this semiaquatic species (Figure 13·11). Even blinded individuals returned to their original locality after having been displaced some 0.8 km downstream.

The tendency to form aggregations is widespread in both plants and animals, and results in the most frequently encountered pattern of distribution. These aggregations, however, may be somewhat randomly scattered over a given space. Aggregation has been described as "a uniform response to a nonuniform environment"; it may also be a response to inherited social traits. One can call to mind a wide array of aggregation patterns, ranging from pronounced "schools" of certain fishes to tightly formed clumps of flatworms on the undersurface of a stone on a stream floor. Time and space do not permit us to give due consideration to the many aggregation patterns. As to causes of this general type of distribution, we can list: annual, seasonal, or nocturnal-diurnal variations in

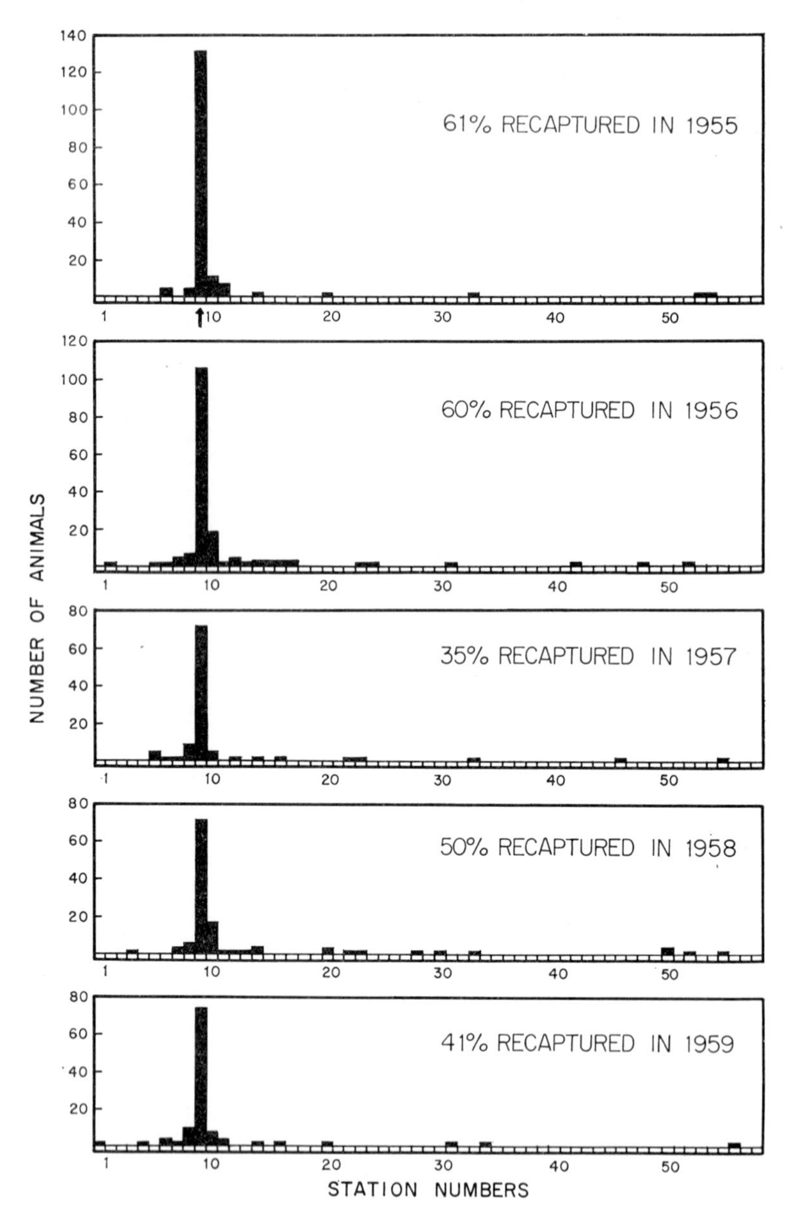

FIGURE 13·11. Home Range as Exhibited by the Newt (*Taricha rivularis*) in California. In 1953, 262 adult newts were captured at Station 9, marked, and released in the same stream station. The graphs show the locations at which the animals were recaptured in subsequent years beginning in 1955. The stations shown on the horizontal axes are 50 yd apart; the arrow at top indicates the point of release. (From Twitty, V. C., 1959. "Migration and Speciation in Newts," *Science*, Vol. 130, 1735-1743. Reproduced by permission.)

physicochemical qualities of the environment; coactions among biological components of the community; physiological processes such as reproduction; and the above-mentioned sociological attributes of the species. The effects of aggregative tendencies are varied and manifold; some of these have been presented in the preceding discussion of population density.

Chapter 14

AQUATIC COMMUNITIES

> Every oyster-bed is thus, to a certain degree, a community of living beings, a collection of species and a massing of individuals, which find here everything necessary for their growth and continuance, such as suitable soil, sufficient food, the requisite percentage of salt, and a temperature favorable to their development—*Möbius*

THUS, in an 1883 translation of an earlier paper, Karl Möbius set forth the concept of an ecological *community*. To such a collection of species inhabiting a given set of environmental factors Möbius applied the term "biocoenosis." This concept stresses the pronounced interrelationships of the organisms to one another and to physicochemical features of the medium in which the plants and animals live. These interrelationships include harmful and beneficial interactions, as well as those involving only tolerance.

Under given environmental conditions any oyster reef community is composed of characteristic species populations; but so also is a riffle community of an upland stream, or the community of the marginal zone of a lake, or that of a wharf piling in an estuary, or a host of other relatively distinct aggregations of organisms. Each of these generally distinct assemblages has gained its distinctiveness through evolutionary adaptation to conditions determined by the fitness of the environment. In the process, mutual adjustment and harmonious interactions of the organisms have reached a level such that, given nutrients and living space, the distinctiveness of the aggregation is maintained. From these ideas we see that a community typically possesses a certain degree of structural unity insofar as the species populations composing it are concerned. This does not mean, however, that the composition of a community is static and fixed in time and space. In the preceding chapter we learned of the temporal and spatial variations in density of populations; it follows, therefore, that community composition similarly varies. We shall deal presently with some aspects of intracommunity dynamics.

Composition, however, is only one of two important aspects of the community concept. The second aspect relates to organization of communities. A given community is more than simply an assemblage of species living together without pattern. There is organization within the community with respect to food and energy relationships, or *metabolism*. The arrangement, or location, of populations of a community is generally organized and manifested in *stratification* of the inhabitants. The activities of the components of a communtiy are strongly time-related, lending to the community temporal organization as seen in *periodism*. Change within a community is, under normal conditions, orderly, and often with predictable direction; this aspect of organization is called *succession*.

On the basis of these ideas, general though they be at the moment, we may now broadly define a community as a local assemblage of species populations maintained in an area delimited by environmental features. Note that, in addition to "community," two other key ideas, "maintenance" and "environment," are incipient in the definition. *Community* refers to the living components; *environment* embodies the total of the physical and chemical factors which exert an effect upon the living assemblage; the dynamic interaction and coaction of environment and community, involving nutrient cycles and energy flow (*maintenance*), combine in the concept of *ecosystem*.*

On the basis of our definition, a community may consist of only a few ecologically related individuals, or it may be composed of a vast assemblage. A small puddle of water with its plants and animals, temporary though it may be, constitutes a community; at the other extreme the total biota of the ocean, with its shore-zone inhabitants, its bottom assemblages, and its plankton of the open waters, is also a community. We may, therefore, recognize major communities, such as those of the sea, the estuary, a lake, or a pond. As we have just seen, major communities are essentially self-sufficient and generally independent of other major communities. Within a major community are found varying numbers of minor communities. The undersurface of a lily pad, for example, forms the substrate for a somewhat unique community, the *aufwuchs*. The bottom muds from which the lily pad arises contain a bottom community, the *benthos*. The water phase between the lily pad and the bottom harbors the *nekton* and *plankton*. In some instances the species composition between two communities is marked and conspicuous. For example, the gradation from stream type to pond type, encountered where a fast brook enters a

* In much of the literature, "community" and "community metabolism" embody the ecosystem concept. Although the functional interrelationships of organisms and environment have been recognized for some time (see, for example, the classical "The lake as a microcosm," by S. A. Forbes, 1887), the term "ecosystem" is of more recent origin. (See Tansley, 1935.)

pond, is often abrupt. Similarly, the intertidal community of a wharf piling in an estuary is usually decidedly different from that below the low-tide mark, the latter usually containing a greater number of species even though some forms are common to both zones on the piling. On the other hand, some communities grade insensibly, and recognition of boundaries and components is often difficult. This condition is frequently encountered in the transition from freshwater to marine com-

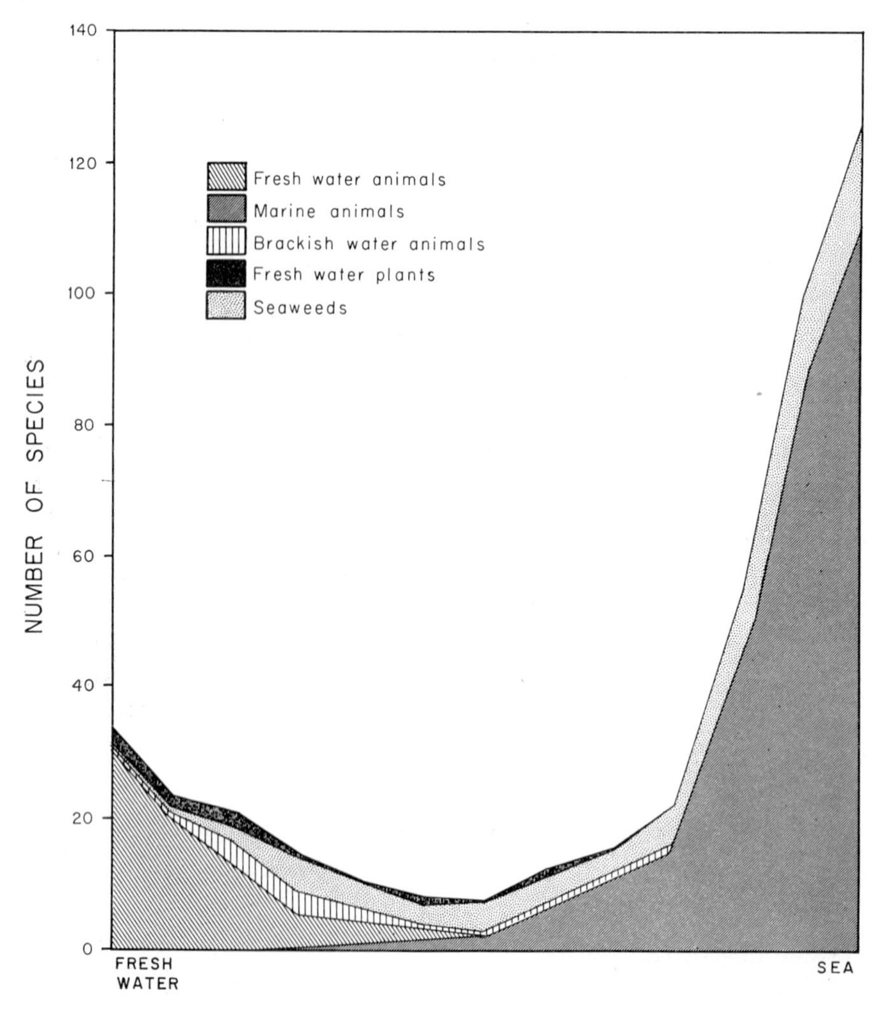

FIGURE 14·1. Comparative Composition of Freshwater, Mesohaline, and Marine Plants and Animals Along the Tees Estuary, England. (After Macan, T. T. and Worthington, E. B., 1951. "Life in Lakes and Rivers," *The New Naturalist Series*, Wm. Collins Sons & Co. Ltd., London, from Alexander, Southgate, and Bassindale.)

munities in relatively stable estuaries of considerable length. In estuaries, the community composition grades quantitatively, with respect to number of species (Figure 14·1), as well as qualitatively in terms of kinds of organisms. There is also evidence that the number of individuals is lowest in the estuarine transition between freshwater and marine communities.

Unlike species populations, which are delimited by inherited biological traits, communities are bounded, generally, by local environmental factors or geomorphological features of the land or water. The *type* of community developed in a certain locality depends, to a great extent, upon the relationships between environmental factors, species characteristics, and species functions. For example, an oyster reef community and a stream riffle community harbor predictable populations of ecological kinds of organisms performing particular roles, and therefore may be thought of as types of communities; many other cases exist in nature. Observe that we refer to *ecological kinds* of species. The species composition of a given community may vary taxonomically from one geographic locality to another. The types, or ecologically adapted kinds, however, are quite characteristic wherever similar environmental conditions prevail. These kinds may, indeed, belong to the same genera. The macrofauna of a stream riffle in California is ecologically similar to that in a riffle community in a New Jersey stream, in that insects are conspicuous and dominant organisms. Many genera of mayflies, beetles, and dipterans are shared by the two widely separated communities.

Regardless of the type of community, however, certain general principles pertaining to community composition, to the activities of the component populations, to the ecological roles of groups of populations, to energy relationships, and to evolutionary processes are common to all. In this chapter we wish to deal with selected principles pertaining to some of these aspects of the community concept. For more comprehensive treatment of these and other community principles, the student is invited to consult Allee, *et al.* (1949) and Dice (1952).

COMPOSITION OF SOME REPRESENTATIVE COMMUNITIES

In our preceding chapters we have given a great deal of attention to geological, physical, chemical, and biological features of natural waters. We have repeatedly emphasized that one of the major points of view in the consideration of such features derives from our interest in the "fitness of the environment" with respect to habitation by plant and animal populations. In the opening pages of this chapter we suggested that communities are predictable assemblages of organisms inhabiting a given environment or segment thereof. Let us now view some of the characteristics of a few selected aquatic communities.

STANDING WATERS

The conspicuous regions of a lake as shown in Figure 14·2 are normally inhabited by typical, but by no means rigidly delimited, associations of

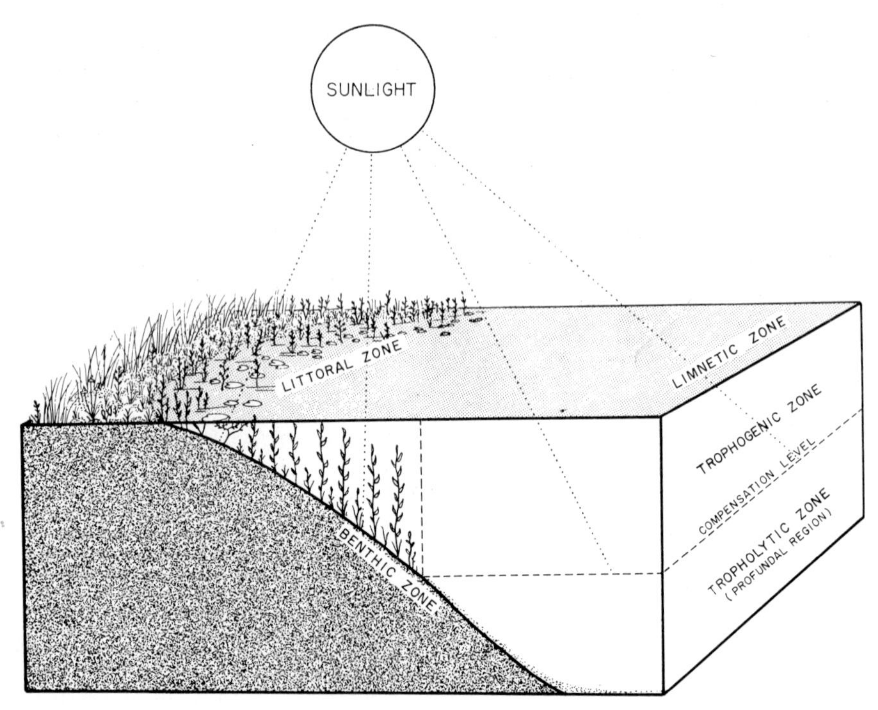

FIGURE 14·2. The Major Horizontal and Vertical Life Zones of a Lake.

species populations. Some of these are recognizable as minor communities. Within the minor communities are to be found even more restricted assemblages. Among the regions which support more or less characteristic communities are:

(1) The lake surface
(2) The limnetic zone
  (a) The trophogenic stratum
  (b) The tropholytic stratum
(3) The littoral zone
(4) The profundal zone
(5) Submersed structures

THE LAKE SURFACE

The lake surface, as the interface between air and water, serves as the substrate upon which an interesting and diverse community has developed. The organisms associated with the surface film are collectively referred to as *neuston*. Some snails and flatworms and, seasonally, the larvae and pupae of dipterous insects occupy the undersurface of the water film; these are the *infraneuston*. On the surface, freely floating plants such as *Lemna* and *Azolla*, and the floating leaves of rooted species, combine in a distinctive plant community. Add to this a number of species of Hemiptera (*Ranatra, Hydrometra*, and others) which tread upon the surface film, swimming "whirligig" beetles (Gyrinidae), and a number of other insects lighting upon the plants, and a colorful community, the *supraneuston*, results. As living, mature individuals, at least some of these organisms might conceivably be considered as belonging to something other than the aquatic realm. The hemipterans named above, for example, contribute little, while living, to the food web of the water phase of a lake or stream.

THE LIMNETIC ZONE

The limnetic zone of a lake is that region of open water, the horizontal extent of which is bounded peripherally by the zone of emergent vegetation. Shallow bodies of water in which vegetation extends across the basin, of course, lack a limnetic zone. Vertically, the limnetic zone extends from the surface to the bottom and includes both the trophogenic and tropholytic zones, if the latter is present (see below). Recall from earlier considerations that these zones are usually marked by quite different regimes of physical and chemical conditions. We should expect, therefore, that the animal and plant communities would also differ.

The upper, or trophogenic, zone normally corresponds to the lighted zone and serves as environment for a community of microscopic photosynthetic plants and a plant-based animal community. The lower, or tropholytic, zone is characterized by an abundance of bacteria and a paucity of organisms generally. The animals which inhabit the tropholytic zone are typically those adapted to low-oxygen conditions, many of the species obtaining food from the rain of organic particles from the trophogenic zone above. In shallow lakes and ponds, or in relatively clear waters, the tropholytic zone may be poorly defined or missing. On the basis of locomotion, the inhabitants of the limnetic zone are classified as *nekton* and *plankton*. The former includes the large free-swimming animals capable of sustained, directed mobility. The nekton (Greek: *nektos*, "swimming") includes primarily the fishes and certain insects. The second grouping, the plankton (Greek: *planktos*, "wandering"), comprises an abundant and varied assemblage of essentially microscopic to submicroscopic plants and animals.

Although numerous references have been made previously to reactions between physical and chemical factors and plankton, and to coactions between plankton and other plants and animals, it seems appropriate to consider, briefly, some general features of the composition of a limnetic plankton community. In subsequent pages we will give attention to food and energy relationships and to community dynamics. Both zooplankton and phytoplankton range in size from the smallest protozoans and bacteria to algae and crustacea easily seen without magnification. On the basis of size, *nannoplankton* includes the organisms that pass through a collecting net of bolting silk. The usual material (No. 25 silk) has a mesh opening of 30 to 40μ. Nannoplankton is collected by centrifuging water samples (see Welch (1948) for details). *Net plankton* is that held in the collecting net.

The phytoplankton community of lakes and ponds is composed mainly of diatoms, blue-green algae, green algae, and photosynthetic flagellates. It must be recognized that the quantitative and qualitative composition of phytoplankton communities vary greatly from one body of water to another, and that no truly "typical" community exists.

In a small Minnesota pond, *Fragilaria* and *Navicula* were found to be the most prominent diatoms, the population density increasing during colder months, and reaching a maximum of nearly 4000 cells per liter in December. In a Colorado lake (Gaynor Lake) some 12 genera of diatoms were reported, and on two occasions, in April and in May, the number of cells in a mixed population reached nearly 12 million and 20 million per liter, respectively. Seasonal "pulses" in diatom populations have been reported for many lakes. These pulses may occur as a single vernal increase in numbers, or as two conspicuous blooms, one in early spring and another in the autumn. Sometimes three or more pulses per year are common. The precise mechanisms involved in such population fluctuations are not fully understood. They appear, however, to involve complex interactions between environmental factors, such as temperature and nutrient supply, and the physiology and reproductive potential of the plants. It does seems that diatoms are characteristically the most abundant algae of larger lakes.

The blue-green algae are often abundant and represented by a notable variety of kinds in ponds and lakes. In the Minnesota pond, this group was predominant. *Microcystis*, *Anabaena*, and *Oscillatoria* were common, and a bloom of *Aphanizomenon flos-aquae* in the early summer produced over 1.5 million cells per liter. In large lakes, blue-green algae typically exhibit a late summer pulse, and sometimes an additional spring pulse. In Boulder Lake, Colorado, a spring pulse, mainly of *Coelosphaerium*, produced over 13 million cells per liter. During an autumn pulse, chiefly of *Schizothrix*, the population density reached nearly 41 million cells per liter. In another lake of the same region, *Chroococcus* was the predominant blue-green in a pulse that reached a peak of over 62 million cells per liter in summer.

Although in ponds and lakes a considerable variety of kinds of green algae may occur, their densities are usually lower than those of diatoms

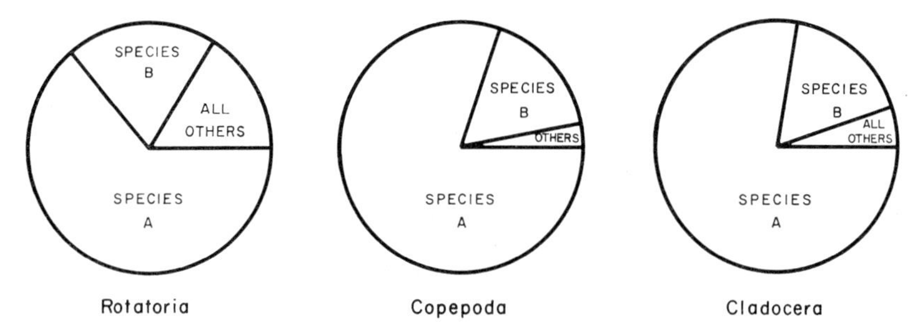

FIGURE 14·3. Momentary Species Composition of a Typical Limnetic Zooplankton Community. The relative proportions of the sectors indicate percentages of both dense and scanty zooplankton populations. (After Pennak, R. W., 1957. "Species Composition of Limnetic Zooplankton Communities," *Limnol. Oceanog.*, Vol. 2, 222-232.)

and blue-green algae. *Pediastrum*, *Cosmarium*, *Scenedesmus*, and *Staurastrum* are commonly found in both small and large bodies of water. Small-scale blooms of green algae have been reported, some occurring simultaneously with pulses of blue-green algae and diatoms.

Green flagellates constitute a conspicuous segment of pond and lake plankton, at least seasonally. Among this group, *Synura* often becomes abundant in small lakes and ponds. *Trachelomonas* and the common lake plankter, *Ceratium hirundinella*, are characteristic of most large and small lakes of the temperate zone.

The zooplankton of ponds and lakes is composed predominantly of rotifers and microcrustaceans (Figure 14·3). In the Minnesota pond, 29 species of rotifers were collected. These reached maximum abundance in late summer; during one year a population density of over 4000 individuals per liter was recorded. Two common, cosmopolitan species of rotifers, *Keratella cochlearis* and *K. quadrata*, were predominant during winter, the maximum density being attained during summer. The rotifer genus *Brachionus* was represented in the summer community by six species. Several other forms occurred commonly. Fifteen species of rotifers were found in the Colorado lakes, the most common forms being *Keratella cochlearis* and *Polyarthra trigla*. In neither the pond nor the lakes did rotifer populations follow any predictable pattern with respect to seasonal population density and occurrence of species. Although a considerable literature suggests cyclical trends in seasonal abundance of particular species, intensive studies such as those of the above-mentioned pond and lakes indicate that irregular phenomena of rotifer activity are probably the more usual.

Among the plankton crustaceans of the limnetic regions of ponds and lakes, Cladocera and Copepoda constitute the conspicuous elements. Although cladocerans are characteristically more abundant in ponds than in lakes, certain species, such as *Bosmina longirostris*, *Daphnia longispina*, and *D. pulex* are commonly found in both types of waters. In the Minnesota pond, *B. longirostris* was the predominant cladoceran throughout the year, the greatest density occurring in late summer. Eight other species inhabited the pond, some being primarily summer inhabitants. The Colorado lakes were inhabited by five limnetic species, only three of which also occurred in the Minnesota pond. In the open waters of large lakes, copepods are frequently more abundant than cladocerans, although the number of species of the former may be less. *Cyclops bicuspidatus* is a common inhabitant of both ponds and lakes in temperate regions of North America. In one of the Colorado lakes, this species reached a population density of over 2000 individuals per liter. Seasonal maxima in both ponds and lakes may be erratic with respect to the number of pulses; in some lakes one, two, or no pulses may occur. The limnetic plankton of the Colorado lakes contained only one other species of copepod; this was *Diaptomus siciloides*. In addition to *C. bicuspidatus*, the Minnesota pond supported *Diaptomus eiseni* and *Canthocamptus staphylinoides*, a bottom littoral form, the latter population arising rapidly in spring and existing for only one month. Although the densities of the various species populations in the limnetic zooplankton of a given lake often exhibit unpredictable and considerable variation, the species composition of the open-water zone remains relatively constant in time. A considerable amount of research has revealed that the number of dominant microcrustacean species in the limnetic plankton is typically small (frequently less than a half dozen). Furthermore, the component populations of any lake, under normal circumstances, apparently persist year after year. Studies of the plankton of Lake Michigan, for example, have shown that little change in the animal and plant composition occurred over a 40-yr period. Of the species abundant in 1887-1888, only one, the copepod *Epischura lacustris*, was absent from the 1926 investigations. Associated with ecological succession (see below), the gradual change in community structure, however, is a natural order of replacement of species. Depending upon a number of biogeochemical factors, this process may be exceedingly slow or quite rapid.

The preceding descriptive consideration of the limnetic plankton of lakes and ponds serves to illustrate some of the general principles pertaining to plankton dynamics. We have seen, for example, that the composition of plankton varies both quantitatively and qualitatively from lake to lake, and seasonally within a given body of water (Figure 14·4). These variations are related to a number of environmental factors, including

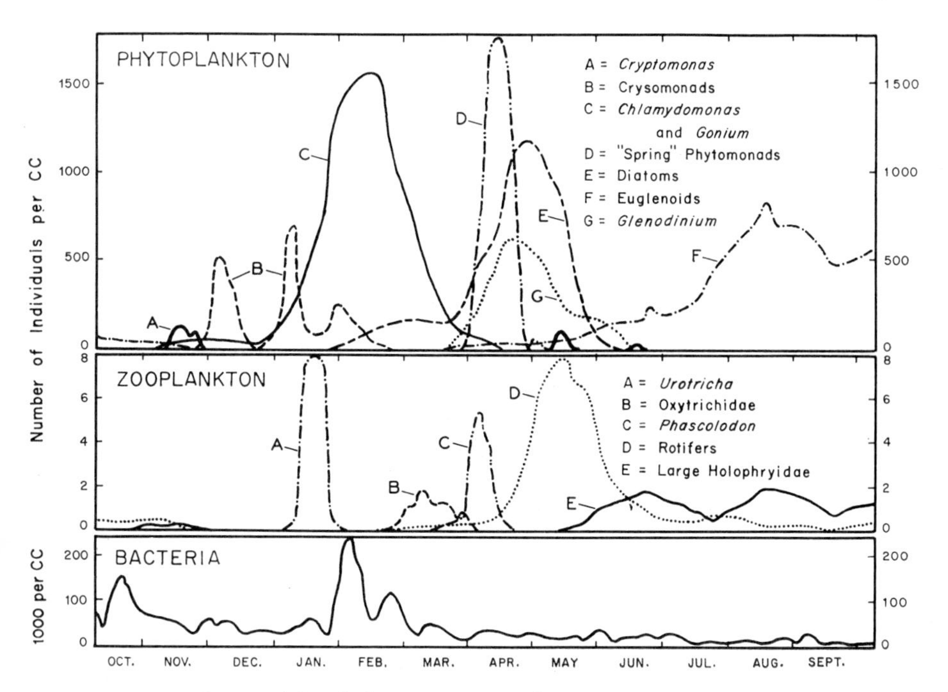

FIGURE 14·4. Seasonal Variation and Succession of Dominant Organisms in a Small Artificial Pond. The curves are smoothed by fifteen-point moving averages. (After Bamforth, S. S., 1958. "Ecological Studies on the Planktonic Protozoa of a Small Artificial Pond," *Limnol. Oceanog.*, Vol. 3, 398-412.)

light, temperature, and dissolved substances (particularly nutrients necessary for phytoplankton production), and to biological features such as competition and population growth attributes. In general, fertile ponds and small lakes contain a richer plankton than do large lakes. The volume of nannoplankton is greater than that of net plankton, in some bodies of water the ratio reaching three, or more, to one. In ponds and lakes of temperate North America, phytoplankton volume usually exceeds that of zooplankton by two to six times. The relationships between phytoplankton and zooplankton pulses are not clear. Although it has been suggested often that sudden, conspicuous increases in population density of zooplankton follow phytoplankton blooms, a number of studies have shown a lack of significant correlations between the two events. Furthermore, in some lakes, at least, food (phytoplankton) does not appear to constitute an important limiting factor with respect to the over-all zooplankton population.

The composition of phytoplankton communities can often be correlated with trophic stages of lakes. Oligotrophic lakes are generally characterized by a relatively low quantity of total plankton, and population pulses are uncommon. Eutrophic lakes, on the other hand, typically support a large quantity of phytoplankton composed of few species; pulses are common and frequent. Table 14·1 lists algal species which

TABLE 14·1. APPROXIMATE TROPHIC DISTRIBUTION OF DOMINANT LIMNETIC ALGAE IN LAKES OF WESTERN CANADA

(From Rawson, D. S., 1956. "Algal Indicators of Lake Types," *Limnol. Oceanog.*, Vol. 1, 18-25.)

OLIGOTROPHIC	MESOTROPHIC	EUTROPHIC
Asterionella formosa	*Fragilaria crotonensis*	*Microcystis flos-aquae*
Melosira islandica	*Ceratium hirundinella*	
Tabellaria fenestrata	*Pediastrum boryanum*	
Tabellaria flocculosa	*Pediastrum duplex*	
Dinobryon divergens	*Coelosphaerium naegelianum*	
Fragilaria capucina	*Anabaena* spp.	
Stephanodiscus niagarae	*Aphanizomenon flos-aquae*	
Staurastrum spp.	*Microcystis aeruginosa*	
Melosira granulata		

contribute a large percentage of the phytoplankton community in lakes of Western Canada during summer.

In the temperate regions of the world two more or less characteristic assemblages of phytoplankton may be recognized. One community, in which diatoms predominate, appears to be prevalent in cold-water lakes of the more northern latitudes. The other community, characterized by dominance of blue-green algae, is more commonly found in warmer climes in which the summer surface temperature of lake waters exceeds about 19°C. Temperature also relates to annual cycles of production of plankton. As a rule, plankton abundance is greater in summer than in winter. Some species, however, exhibit winter pulses, often producing large numbers underneath ice cover. In some North American lakes the population density of plankton may increase four-fold or more during a short time in spring.

The very integrity of the plankton community and, indeed, much of the organic maintenance of standing waters are ultimately dependent upon flotation of the organisms in the medium. In order to carry on anabolic activities, autotrophic species must remain within the lighted zone. It follows, therefore, that a great proportion of the herbivorous organisms must be suspended in somewhat the same region. Since the specific gravity of protoplasm and the various body coverings secreted by plants and animals is greater than that of water, most organisms tend to sink. The rate at which a body sinks is determined by the relationship between weight beyond the specific gravity of water and frictional resistance between the surface of the organism and the environment (see Stokes' Law, Chapter 6). Increased surface area increases friction. Thus the small size of plankton organisms is of distinct advantage in flotation, for the ratio of area to volume is thereby reduced. In addition to inherent small size, various morphological and physiological adaptations contribute to suspension of plankton in water. The development of spinous projections, such as those of *Ceratium* and *Scenedesmus* (Figure 11·1), increase surface area and decelerate sinking rate. Needle-shaped (*Synedra*, Figure 11·3) and curved (*Closterium*, Figure 11·1) body forms tend to reduce sinking. The formation of radial and chain-like colonies of cells such as *Asterionella* and *Tabellaria* (Figure 11·3) also contributes positively toward flotation. Organelles such as flagella (see *Synura* and *Eudorina* in Figure 11·1) are aids in maintaining position. Many flagellates, diatoms, rotifers, and plankton crustaceans produce oil globules which serve to decrease the specific gravity of the organisms. The zooplankton typically possesses some swimming ability. This, coupled with produced filaments and spines, oil droplets, and air bladders (in the larva of the dipteran *Chaoborus*, for example) found variously among the animal plankton enable these forms to inhabit the limnetic zone. Of further importance in maintaining position in open water is the movement of water itself. We have seen previously that turbulence, development of eddy viscosity, and other water motions are characteristically a part of the dynamics of standing waters. Depending upon the intensity of these factors in particular lakes and ponds, the plankton components may be benefited with respect to maintaining position in the medium.

The plankton of standing waters is rarely distributed uniformly throughout the vertical extent of the water. We have already considered the fact that lakes typically contain two regions with respect to biological activity. The upper, lighted stratum, in which photosynthesis exceeds total plankton respiration, we termed the trophogenic zone; the lower, unlighted region, where decomposition predominates, was called the tropholytic zone. Between these two lies a level at which there is a balance between algal photosynthesis and total plankton respiration; this is the *compensation level* (Figure 14·2). In highly turbid or strongly colored lakes this level may lie within a meter or so of the surface, or at considerable depths in clear lakes. With regard to vertical distribution of plankton generally, the compensation level marks the lower limit of occurrence of populations. Phytoplankton is restricted by light requirements to the zone above the compensation level. Since the zooplankton is directly or indirectly dependent upon the plants, the animals are, likewise, found in greatest abundance in the trophogenic zone. Because of diurnal migration patterns, some zooplankton is often more abundant in the trophogenic zones during the night and in the tropholytic during the day.

It is now apparent that the entire water mass of a lake is not uniformly inhabited by plankton. Vertical stratification, determined by light, temperature, water movement, and nutrient supply, is characteristic of the limnetic zone. Even within the euphotic zone, the vertical distribution of plankton is typically marked by stratification of various populations of the community, this zonation being determined by a number of environmental factors. Many plankton species populations undergo vertical movements, or migrations, thereby changing the community structure at various levels. We have previously seen that excessive sunlight can limit photosynthesis; thus, the zone of maximum density of phytoplankton is normally at some depth below the lake surface rather than in the uppermost waters.** The level at which the greatest abundance of phytoplankton is found is usually determined by transparency. In highly turbid lakes, for example, phytoplankton may be narrowly restricted to near-surface regions; in deep, clear lakes the photosynthetic zone may extend to considerable depths. In summer, thermal stratification may serve to limit the vertical distribution of both animal and plant components of the plankton, an important factor in this instance being the lack of oxygen in the hypolimnion. In winter, with more uniform conditions throughout the water column, various planktonic forms may be more uniformly distributed to greater depths. Nocturnal-diurnal migrations of many plankton species change the nature of the limnetic community regularly. Figure 13·3 illustrates vertical movements of one such migratory species.

THE LITTORAL ZONE

The littoral zone is that portion of a body of water extending from the shoreline lakeward to the limit of occupancy by rooted plants. In many shallow lakes and ponds this zone extends completely across the basin, particularly during the growing season. In the deeper bodies of water the development of a littoral zone is influenced by factors which

---

** An exception to this distribution is seen in some blue-green algae. Recall that these plants typically possess gas-filled vacuoles and tend to float at the surface. Some of these such as *Anabaena* and *Microcystis* often develop dense floating masses at the surface of lakes and ponds.

are limiting to plant growth. These may include depth of water, vertical extent of effective light transmission, movement of water (particularly wave action), and fluctuations in water level. To these may be added such obvious features as nutrient supply and texture of the substrate. Of the limiting factors, depth of the water is important to rooted emergent species and those with floating leaves. Given sufficient light, submersed plants may exist at any depth. The amount of wave action is, of course, important, for we are all aware of the lack of plant growth in the shallow, wave-washed regions of lakes, as compared with the great profusion often found in quiet, protected areas. Fluctuating water levels are usually very damaging, and reservoirs and lakes subject to such fluctuations generally have little rooted aquatic vegetation.

It is now apparent that the littoral zone is typically dominated by "higher" plants. These plants normally occur in relatively distinct associations and form conspicuous physical and ecological "subcommunities" within the littoral zone. Without reference to any specific example, let us view the general features of plant zonation and the animal associations in the shallow regions of a lake. It is in the littoral that we find the greatest variety of species of any of the lake regions, for here occurs abundant cover for protection, substrate for travel and attachment, food, and dissolved substances.

From the shoreline lakeward, to a depth determined basically by physiological limits of wetting of the plant inhabitants, exists the *zone of emergent vegetation*. Throughout most of the world, this border is typically dominated by grasses, rushes, and sedges possessing long leaves and stems. In the temperate zone, bur-reed (*Sparganium*), spike rush (*Eleocharis*), bulrush (*Scirpus*), and cat-tail (*Typha*) combine in a fairly typical and commonly found assemblage. These may be joined by herbaceous types such as arrowhead (*Sagittaria*), water plantain (*Alisma*), and, in eastern North America, arrow arum (*Peltandra*). The plants of the emergent zone comprise a somewhat "in between" stage in energy relationships. The photosynthetic organs are typically above water, and there utilize atmospheric oxygen and carbon dioxide; gases given off by photosynthesis and respiration are contributed to the atmosphere. Mineral nutrients and water, however, are taken from the lake substrate. Upon death, the emergent structures fall into the water where, as litter, they form a habitat for numerous plants and animals. Decay of the fallen plants contributes to nutrient supply and cycles of the lake. Some of these emergents are used as food by waterfowl and mammals, and as nesting sites and cover by certain birds.

Lakeward from the zone of emergent vegetation, an association of plants bearing long stems or petioles and floating leaves typically parallels it. This second community is termed the *zone of floating-leaf plants*,

Sometimes the zones are abruptly and clearly marked; in other instances a subtle transition from one to the other is seen. The assemblage in this zone characteristically includes rooted plants with floating leaves, such as the familiar (at least in central and eastern North America) white water lily (*Nymphaea*), the more widely distributed yellow water lily (*Nuphar*), water shield (*Brasenia*), and several species of pondweed (*Potomogeton*). Frequently this zone includes a number of free-floating plants; for example: water fern (*Azolla*), duckweed (*Lemna* and *Spirodela*), and in some of the southern states, water lettuce (*Pistia*) and water hyacinth (*Piaropus*). The free-floating hydrophytes and rooted plants with floating leaves influence energy relationships and community composition in various ways. Both groups may contribute to a surface massing of such extent as to shield the underlying water from sunlight, thereby inhibiting organic production. In warm regions the water underneath such mats has frequently been found to be nearly oxygenless. The roots of free-floating plants extract dissolved nutrients from the water, and, together with the undersurfaces of floating leaves, often harbor rich communities of organisms (aufwuchs) and serve as the substrate for egg deposition of a number of animals. Certain snails (*Physa* and *Planorbis*, for example), beetles (*Donacia* and *Gyrinus*), hemipterans, and at least one mayfly deposit their eggs on the undersurfaces of lilypads. This zone is commonly used as a breeding and nesting area by sunfishes and various other fishes.

The *zone of submersed vegetation* typically forms the innermost belt in the lake border of plants. The plants in this zone are characterized by long, sinuous leaves, or by a bushy growth-form with fine, highly branched leaves. These plants may be considered as wholly aquatic, for they derive gases and nutrients for photosynthesis and respiration from the water and, in turn, most of their substance returns to the water through decomposition. The genus *Potamogeton* includes over 40 species of typically aquatic plants, many of which occur widely throughout North America. Hornwort (*Ceratophyllum*), naid (*Najas*), milfoil (*Myriophyllum*), waterweed (*Anacharis*), and certain species of arrowhead (*Sagittaria*) are commonly found in the zone under consideration. In certain lakes, stonewort (*Chara*) may be prominent in the submersed vegetation zone; we have previously considered some of the conditions conducive to stonewort development (see Chapter 9 on chemical factors).

The plankton of the littoral zone is typically rich in number of species, and sometimes in number of individuals. A considerable amount of this plankton may, however, consist of organisms which have been displaced from an aufwuchs community. These elements, termed *tychoplankton*, include many filamentous algae as well as larger animals normally found on some substrate. Diatom species are abundant in the littoral plankton, often more so than in the limnetic. These may be supplemented, particularly in stagnant waters of embayments, by filamentous blue-greens, such as *Oscillatoria* and *Anabaena,* and form dense blooms. Green algae, *Spirogyra, Ulothrix,* and *Oedogonium,* for example, frequently form dense mats in these waters. Desmids such as *Closterium, Micrasterias,* and *Cosmarium* occur commonly in the still waters of the littoral of lakes and ponds. The zooplankton of the littoral zone includes many forms which do not occur commonly in the limnetic region. Mites (Hydracarina) and ostracods, not typically found in the open waters of lakes, are often common in the littoral zone. Cyclopoid and harpacticoid copepods are more characteristically littoral than limnetic, and frequently abound in the shore zone. Certain cladocerans are more abundant in the littoral plankton. The nekton of the littoral zone includes many fishes, beetles, hemipterans, and dipterans not normally found in the limnetic region. Larger animals such as water snakes, turtles, newts, and frogs, are conspicuous members of the littoral community.

BOTTOM COMMUNITIES OF LAKES

The association of species populations of plants and animals that live in or on the bottom of a body of water is called the *benthos.* Probably no other community within a lake exhibits greater variety of kinds and numbers than does the benthos. Grading, as it does, from shore to considerable depths and often consisting of a number of different substrates, the bottom frequently presents an impressive variety of environmental complexes. These complexes, in turn, are normally inhabited by more restricted communities which are quantitatively and qualitatively rather different. The composition of benthic communities is controlled by a number of factors, some of which appear to operate directly to limit certain species. In other instances, the influences are more subtle, suggesting interrelationships of several factors. Among the more conspicuous and better understood of those influencing the kinds of organisms and their distribution in lake bottoms are: (1) physicochemical features of the water such as temperature, transparency, dissolved oxygen content, and water currents (these are typically related to lake-basin morphology and the nature of the drainage basin and local climate), and (2) biological factors such as food, protection, and competition.

The lake bottom is usually considered as including three zonal subdivisions: (1) the littoral bottom, which extends from shore to the lakeward limits of the zone inhabited by rooted hydrophytes; (2) the sublittoral bottom, or the zone between the littoral and deep (profundal) region; and (3) the profundal bottom, or the area of the lake bottom contiguous with the hypolimnion of the limnetic region. Depending upon the area and slope of shoal, these zones and their communities exhibit extreme diversity, not only within the limits of a single lake, but from lake to lake. In shallow, clear bodies of water, the sublittoral or profundal zones may be quite reduced or even entirely absent. As an alternative, the benthos is often described and delimited by depth alone; in such instances the three zones given above are designated, somewhat arbitrarily, by depth, or are disregarded. In any case, the biotic differentiation of littoral and profundal is usually more distinct than is that of the sublittoral, the last constituting an *ecotone*, or "buffer" zone, between two communities. Of major importance is the fact that each of these zones, if present in a lake, typically supports a relatively distinct community. Furthermore, each of these communities usually contains a number of assemblages based primarily on depth and the nature of the bottom sediments.

The profundal bottom of eutrophic lakes is characterized by absence of light, low oxygen content, and high carbon dioxide concentration. The sediments are typically soft, flocculent, or ooze-like materials rich in organic substance. Fungi, bacteria, and certain protozoans are usually abundant near the mud surface. Green plants are normally absent. The macrofauna consists predominantly of tubificid oligochates, bivalved mollusks of the family Sphaeriidae, and dipteran insects of the genera *Chaoborus* and *Tendipes* (= *Chironomus*). Some species of *Chaoborus* occasionally take the form of plankton in that they inhabit the bottom during the day and migrate to the lake surface at night; the migration is often rapid, being about 20 m per hr. As indicated, the number of kinds of animals adapted to bottom dwelling in the deeper profundal region is relatively small. The population densities of these species may, however, be high. Table 14·2 gives the average number per square meter of the major groups of animals at various depths in Lake Simcoe, Ontario. Note, especially, that at depths greater than 20 m the benthos is dominated by dipterans (*Chaoborus* and several species of the family Tendipedidae) and by oligochaetes (*Tubifex*). Although both gastropod and pelecypod mollusks inhabit the profundal bottom, the latter (represented by Sphaeriidae) is much the more abundant. The miscellaneous animals include nematodes, leeches, hydrachnids, and odonates and other insects. These data illustrate clearly a characteristic feature of deep eutrophic lakes of the temperate regions, that is, a typically rich community (in terms of numbers or mass) maintained by a small number of species. As shown in Table 14·2, there does not appear to be a great difference between the average mass of organisms produced in the littoral and that in the profundal. In some lakes, however, this differential is more pronounced. In Lake Texoma, on the Oklahoma-Texas border, the annual mean mass of bottom fauna in the profundal was found to be 102.9 kg per hectare, while in the littoral the annual production amounted to 21.75 kg/ha.

Table 14·2. Average Number of Bottom Organisms (of the Major Groups) per Square Meter, at Different Depths in Lake Simcoe, Ontario, 1926-1928

(Data from Rawson, D. S., 1930. "The Bottom Fauna of Lake Simcoe and Its Role in the Ecology of the Lake," *Univ. Toronto Stud., Publ. Ontario Fish Research Lab.*, No. 40.)

	Tendipedid	Ephemerid Nymphs	Gastropoda *	Pelecypoda *	Amphipoda	Oligochaeta	Chaoborus	Trichoptera	Miscellaneous	Avg No. of All Organisms per sq m	Avg Dry Wt of All Organisms * in mgm per sq m
Shore Zone, 0 to 1 m	152	48	54	22	28	16	0	1	84	405	1028
0 to 5 m	300	90	124	82	112	36	0	14	30	788	1280
5 to 10 m	450	54	130	78	138	30	2	10	34	926	1480
10 to 15 m	240	52	54	88	94	24	9	4	9	574	654
15 to 20 m	540	28	82	118	10	26	15	8	17	844	1170
20 to 25 m	620	0	9	114	2	80	30	0	92	947	1420
25 to 30 m	780	0	8	102	0	106	62	0	10	1068	1340
30 to 35 m	860	0	7	84	0	98	74	0	11	1134	1220
35 to 40 m	760	0	5	52	0	118	70	0	7	1012	950
40 to 45 m	740	0	6	34	0	120	72	0	6	978	852

* Weight of the mollusk shells deducted.

Analysis and description of the benthos of the littoral zone in most lakes are complicated by kinds and abundance of both plants and animals present, and by variations in the nature of the bottom sediments. The presence of plant growths offers opportunities for some animals to

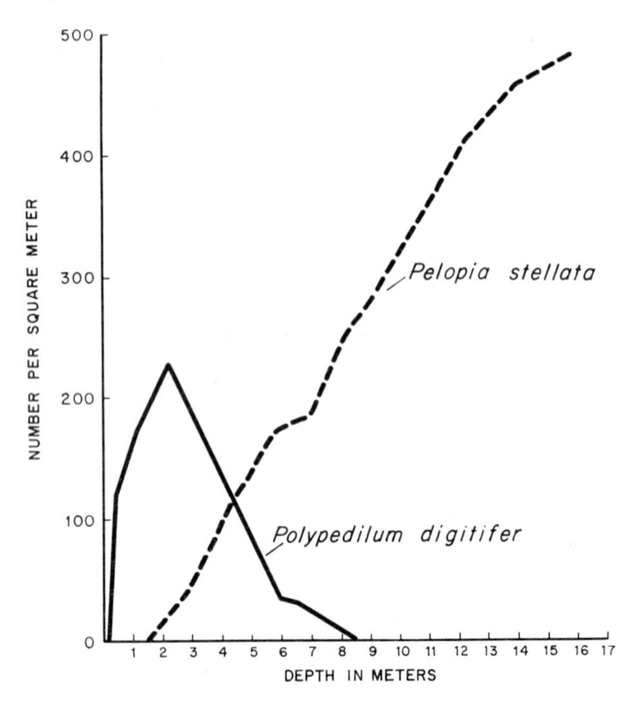

FIGURE 14·5. The Depth Distribution of Two Species of Tendipedids in Lake Texoma, Oklahoma and Texas, Showing Zones of Overlap of the Two Populations. This overlap is construed to be an ecotone. Summer populations (1949) are represented in the smoothed curves. (After Sublette, J. E., 1957. "The Ecology of the Macroscopic Bottom Fauna in Lake Texoma (Denison Reservoir), Oklahoma and Texas," *Am. Midland Naturalist*, Vol. 57, 371-402.)

desert the benthic environment and to climb, thus adding an essential vertical stratification to the over-all littoral pattern. Variations in vegetation density in the shallow zone also make for irregularities in animal distribution and abundance. Water movement, especially wave action, is a major factor influencing the nature of the sediments and, in turn, the composition of communities.

Within a given lake there usually exists a high correlation between the nature of the substrate and the number of species and population density.

Furthermore, each substrate type typically supports a relatively distinct community of organisms. In Lake Texoma, for example, four minor communities based upon bottom sediments in the littoral zone have been recognized. A gravel shoal, or bar, kept clean of fine sediments by wave action, contained a community of nine species of macroscopic animals, including a sponge, a gastropod, and insects. Three of the species populations were not found in any other situation in the lake; these were a mayfly, a damselfly, and the sponge. A pure sand substrate was inhabited by only one species, a caddisfly (*Oecetis inconspicua*) that builds sand cases. Where the substrate was composed of a mixture of sand, clay, silt, and organic detritus the community contained nearly 60 taxonomic groups; of these, 25 were restricted to the particular substrate. A clay substrate was inhabited by only one species, a large blood-red dipteran larva (*Xenochironomus festivus*). Beds of the spermatophyte, *Potamogeton,* were inhabited by 28 species of animals, of which only six were restricted to the habitat. The depth distribution of some species was found to overlap substrates, thereby demonstrating the position of an ecotone between the habitats (Figure 14·5).

COMMUNITIES ON SUBMERSED STRUCTURES: THE AUFWUCHS

The communities of the aufwuchs include all the organisms that are attached to, or move upon, a submersed substrate, but which do not penetrate into it.*** The benthos is, by definition, excluded from this community. The physical (and probably the ecological) base of the aufwuchs is composed, typically, of an assortment of unicellular and filamentous algae such as that shown in Figure 14·6. Located among the plants may be various attached protozoans, bryozoans, and rotifers. The free-living members of the aufwuchs may include representatives of most of the animal phyla found in water. Free-living protozoans, roundworms, rotifers, annelid worms, crustaceans, and insects are variously found in the aufwuchs, the kinds and numbers of these animals differing with substrate, water movement, depth, and chemical composition of the water.

Although we are considering the aufwuchs under the discussion of lakes, the concept and principles are applicable to stream and estuarine conditions also. No fixed object, living or nonliving, submersed in natural waters is devoid of aufwuchs. The nature of the object (substrate) fre-

*** In much of the English literature the term "periphyton" has been used. This term, however, has usually been applied to the total assemblage of *sessile* or *attached* organisms on any substrate. Under this definition, the creeping and crawling forms commonly found in association with the attached organisms are not included. The connotation of the German term *Aufwuchs* is broader and includes both attached and free-living plants and animals. It is adopted herein and somewhat Anglicized by dropping the capitalization of the first letter.

quently determines the composition of the community, however. Bryozoans seldom inhabit the alga *Chara*, due probably to the sulfur content of the plant. The bryozoan *Plumatella* typically colonizes on broad, flat plant surfaces to a greater extent than on other surfaces; *Fredericella*, on the other hand, occurs more commonly on finely branched leaves. The hydra, *Pelmatohydra*, is seldom found in a dense aufwuchs dominated

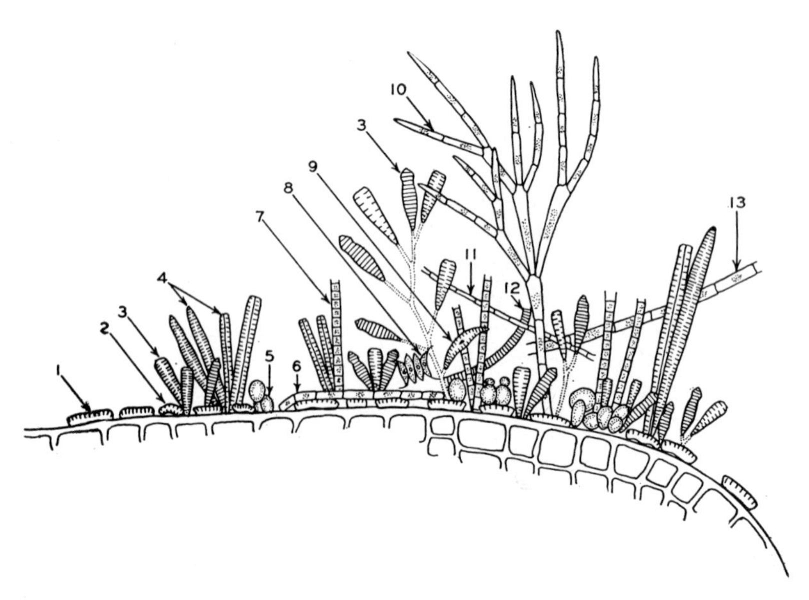

FIGURE 14·6. Schematic Presentation of the *Cocconeis-Stigeoclonium* Aufwuchs Community of Hard, Freshwater Springs. The algae, represented on the upper surface of a leaf of *Sagittaria*, are shown in typical position and abundance. Succession follows from left to right. (1) *Cocconeis*, (2) *Achnanthes*, (3) *Gomphonema*, (4) *Synedra*, (5) *Xenococcus*, (6) *Pseudoulvella*, (7) *Lyngbya*, (8) *Scenedesmus*, (9) *Cymbella*, (10) *Stigeoclonium*, (11) *Melosira*, (12) *Fragilaria*, (13) *Microspora*. (After Whitford, L. A., 1956. "Communities of Algae in Springs and Spring Streams of Florida," *Ecology*, Vol. 37, 433-442.)

by algae. In the marine and estuarine environment, the composition of the aufwuchs on wood substrate differs greatly from that on concrete. Clumps of algae often serve as habitat for free-swimming protozoans, the community relationships becoming complex. Continued population increase of the protozoans leads to destruction of the algae, ending in breakdown of the community. Various species of rotifers exhibit different affinities for certain substrates. Generally, plants possessing finely divided leaves support a greater number of rotifer species than do broad-leaved plants. Of particular interest with respect to substrate occupancy is the aufwuchs developed on the carapaces of a number of species of turtles. Some half-dozen species of filamentous algae are known to occur on such a substrate. Two species, comprising the genus *Basicladida*, are found only on turtle carapaces; these algal growths may be sufficiently dense to support a number of other organisms. Certain crustaceans (copepods and crayfish) sometimes serve as substrate for a community of sessile protozoa and protists.

The great importance of the nature of the substrate in the establishment and maintenance of aufwuchs is seen by comparison of a stone and a leaf, for example, in an upland brook of moderate current in summer. A portion of the stone is apt to be covered by a dense growth of filamentous algae and diatoms. Mayfly nymphs and blackfly larvae are common on the exposed surfaces of the stone. Other animals, such as caddisflies, flatworms, and snails may also be present. The aufwuchs of the leaf, on the other hand, may consist of a sparse association of algae, stalked protozoa, and, perhaps, hydra. The aufwuchs of the stone gives, in general, an appearance of "permanence" and complex interrelationships born of the more lasting substrate. The community on the leaf is typically less dense, giving to the observer an impression of temporariness, reflecting, actually, the impermanence of the substrate.

The development, maintenance, and composition of aufwuchs are also influenced to a great extent by movement of water and by fluctuation in water level. In streams, maximum growth and density of the community are usually found in zones of clear water and moderate current velocity. In pools where rate of flow is reduced, silt deposition inhibits community development. In zones of rapid water, the scouring effect serves to prevent extensive colonization of bottom structures. Indeed, freshets in streams may essentially remove the established aufwuchs. In the littoral zones of lakes, objects in shallow regions of considerable wave action normally support a meager aufwuchs, the more productive zones being those of extensive growths of larger plants. The aufwuchs of submerged plants, or plant parts, in lakes and ponds is especially rich and varied.

The chemical characteristics of the environment, operating under the principle of limiting factors, act upon the biotic structure of aufwuchs, especially as the various members of the community exhibit different reactions to dissolved substances in the water. This effect is particularly noticeable in estuaries where the aufwuchs changes considerably from the high-salinity lower reaches to the freshwater upper regions. Although salinity may not be the sole factor in all instances, it is certainly a very influential one. The relationships between aufwuchs animals and the chemistry of the medium are not completely known. This is due, in part, to the often complex interactions of dissolved substances, such as we have studied in earlier chapters, and the interrelationships among organism, substrate, and chemistry. There exists a considerable literature on the tolerance of a great number of aufwuchs organisms to various chemical factors, especially with respect to pH. It has been shown that certain species exhibit great affinities for acid waters, others for waters of high pH. Since, as we have seen, pH involves a number of processes and forms of carbon dioxide, it is difficult to recognize particular factors involved. There is good evidence, however, that certain sessile rotifers, *Ptygura melicerta* and *Collotheca algicola* for example, are absent from aufwuchs of certain lakes due to the high bicarbonate content of the water. Dissolved oxygen and chemical pollution limit the distribution of a great number of aufwuchs animals.

Studies of the aufwuchs algae of four types of Florida springs have revealed that each spring type supports a relatively characteristic aufwuchs community. The composition of these communities appears to be determined to a great extent by the chemical nature of the water. As is a frequent practice, a community may be named for its dominant components; with respect to algae kinds and water chemistry, the following relationships have been shown for the Florida springs: (1) a *Cocconeis-Stigeoclonium* aufwuchs of hard, freshwater springs of constant temperature and flow (this community is illustrated in Figure 14·6); (2) a *Cladophora-Cocconeis-Enteromorpha* community of "oligohaline" springs, these springs being similar to the hard, freshwater kinds except for greater content of chlorides (100 to 1000 ppm); (3) an *Enteromorpha-Lyngbya-Licomphora* dominated aufwuchs of "mesohaline" springs (chlorides: 100 to 10,000 ppm); (4) A *Phormidium-Lyngbya* community found in sulfide springs, these springs having constant temperature, an absence of oxygen, and high concentrations of sulfates and hydrogen sulfide.

Stream Communities

The communities of streams are under the influence of several major environmental features not encountered by lake inhabitants. One of the most important of these is current, including, as it were, the numerous hydrological processes related to stream flow. Moving water poses problems in maintenance of position for most organisms, for animals caught up in the current are apt to be carried to unsuitable environments. The corrosive and erosive action of flowing water modifies the chemical and physical characteristics of given habitats. Variable discharge may deposit quantities of silt at one moment and scour a zone at another time. On the other hand, the considerable exchange between land and stream often serves to enrich the nutrient supply. Current-created turbulence tends to maintain a relatively uniform set of physicochemical conditions (at least within a given segment) normally free from stratification such as found in lakes. Sorting of bottom materials resulting from variable stream velocity produces a great variety of substrates for colonization and the development of communities. In this last-mentioned aspect is found the keynote of stream communities: namely, astounding variety. From the slow-moving, often lake-like lower stream course to the rapidly flowing upper reaches, an impressive array of communities exists throughout the length of a stream. These include pool, run, riffle, and various shore and bottom variations.

STREAM PLANKTON

There seems to be general agreement among aquatic biologists that no "true," or distinctive, plankton community exists in streams. This does not mean, however, that streams do not possess plankton, for many of them do. The plankton may be derived from headwater lakes or ponds, or from quiet backwaters of the stream. Plankton developed in quiet waters and introduced into a stream is often lost rather rapidly. As we have already learned, streams frequently contain an abundant tychoplankton consisting of organisms dislodged from the bottom or from submersed objects. Among the more common plants of stream plankton (potamoplankton) of the temperate regions are: *Asterionella formosa, Fragilaria capucina, Synedra ulna, Tabellaria fenestrata, Melosira granulata,* and species of *Pediastrum, Scenedesmus, Ulothrix,* and *Navicula.* Zooplankton in upper stream courses is quite sparse and, indeed, may be nonexistent in rushing brooks. Rotifera and Cladocera, common planktonic animals of standing waters, seldom occur in swift streams. In pools and slower waters, Copepoda and Cladocera may become prominent.

THE SHORE ZONE OF STREAMS

In the middle and lower courses of many streams shallow-water communities may develop. Often these communities do not differ greatly from those of lakes and ponds in the same drainage area. The growth and maintenance of the near-shore communities are acutely dependent upon favorable substrate, its slope and sedimentary character, and upon stability of discharge. Where present under optimum conditions, the community contains a zone of emergent vegetation near shore, a zone of floating-leaved plants lying parallel and streamward to the emergent zone, and an inner zone of submersed vegetation (Figure 14·7). Where the shore slopes steeply the emergent zone may be absent. Swift current, depth, and silt content limit the extent of submersed vegetation.

In an extensive shallow-zone community such as described above, a "true" plankton may be meager or absent, but a rich tychoplankton, derived from normally abundant benthos and aufwuchs, is usually present. The nekton among the plants contains many insects and a number of fishes; turtles and water snakes are often common. Current-delivered nutrients and continual flow of water across the community contribute to an often luxurious aufwuchs. This latter factor is an important one, and

FIGURE 14·7. Zonation of the Floating and Emergent Plants Along a Small Tidal Tributary of the St. Johns River, Florida. In the background, at the shore line, a zone of sawgrass (*Mariscus jamaicensis*) predominates. A zone of dark-leaved pickerel weed (*Pontederia lanceolata*) lies in front of the sawgrass. Next, and mostly left of center, is a zone of smartweed (*Persicaria portoricensis*). Water hyacinth (*Eichornia crassipes*) and parrot's feather (*Myriophyllum proserpinacoides*) comprise the zone bordering open water. The pronounced zonation shown here is probably due to the relatively stable water level. (Photograph by A. M. Laessle.)

deserves greater study with respect to microclimate. In standing waters, for example, the rate of delivery of nutrients and respiratory and photosynthetic gases, and the rate of removal of metabolically produced substances are relatively slow; maintenance of sessile members of the aufwuchs is probably limited thereby. In streams with current sufficient to overcome the "drag" at the surface of each organism, the environment in contact with each individual is constantly moved, thus favoring maximum community growth.

THE BENTHOS OF STREAMS

We have previously emphasized the overwhelming variety in numbers and kinds of organisms inhabiting stream floors from the upper reaches to the lower stream course. This over-all change (or *longitudinal succession*) results, for the most part, from manifestations of the dynamics of flowing water described in Chapter 3. Downstream differences in velocity, discharge, and load contribute to the formation of numerous habitats differing with respect to water movement, composition of substrate, and chemical characteristics. The occurrence of plants and animals along a stream course can be limited by any one of a number of factors encompassed by the foregoing categories, or by interactions of several factors. We have seen, for example, in Chapter 13 that oxygen in a spring run can limit distribution of fishes (Figure 13·1). A small-scale example of the eff[illegible] of velocity and bottom composition on the distribution of animals in a spring run is shown in Figure 14·8. Lander Springbrook, New Mexico, is a short (77-m), relatively chemostatic and thermostatic stream inhabited by a generally depauperate fauna. It serves well to illustrate relationships between community structure and certain environmental features.

Along the course of a large stream we find environment-community interrelationships magnified and multiplied many times over those of Lander Springbrook. The nature of bottom sediments along the course of a stream has been described more fully in Chapter 3. We can, however, review this as we consider, briefly, some general aspects of benthic communities in running water.

The substrate of lower stream courses is characteristically of fine sand, silt, mud, or mixtures of these. The water is usually turbid. Animals of the benthos are not unlike those of the same community in lakes. Tendipedid larvae, inhabiting self-made tubes, burrowing mayflies, odonate nymphs, and the larvae of the alderfly are common insects of this region. The annelid, *Tubifex*, may be abundant. Mollusks, such as the small clams, *Sphaerium* and *Pisidium*, and the larger forms, *Unio* and *Anodonta*, occur commonly in or on the substrate. Under favorable conditions, the population density of *Sphaerium* becomes exceedingly high, on the order of 5000 or more per square meter. In regions of the Mississippi River over 30 species of pelecypods may be found. The number of molluscan species decreases toward the upper reaches of streams. Several species of mud-grubbing or detritus-feeding fishes commonly frequent the bottom community of lower stream courses. These include the carp, suckers, catfishes, and, in estuarine regions, mullets (*Mugil*) and young croakers (Sciaenidae). All of these forms feed primarily on detritus.

The upper stream course is characterized by pools, stretches of fast

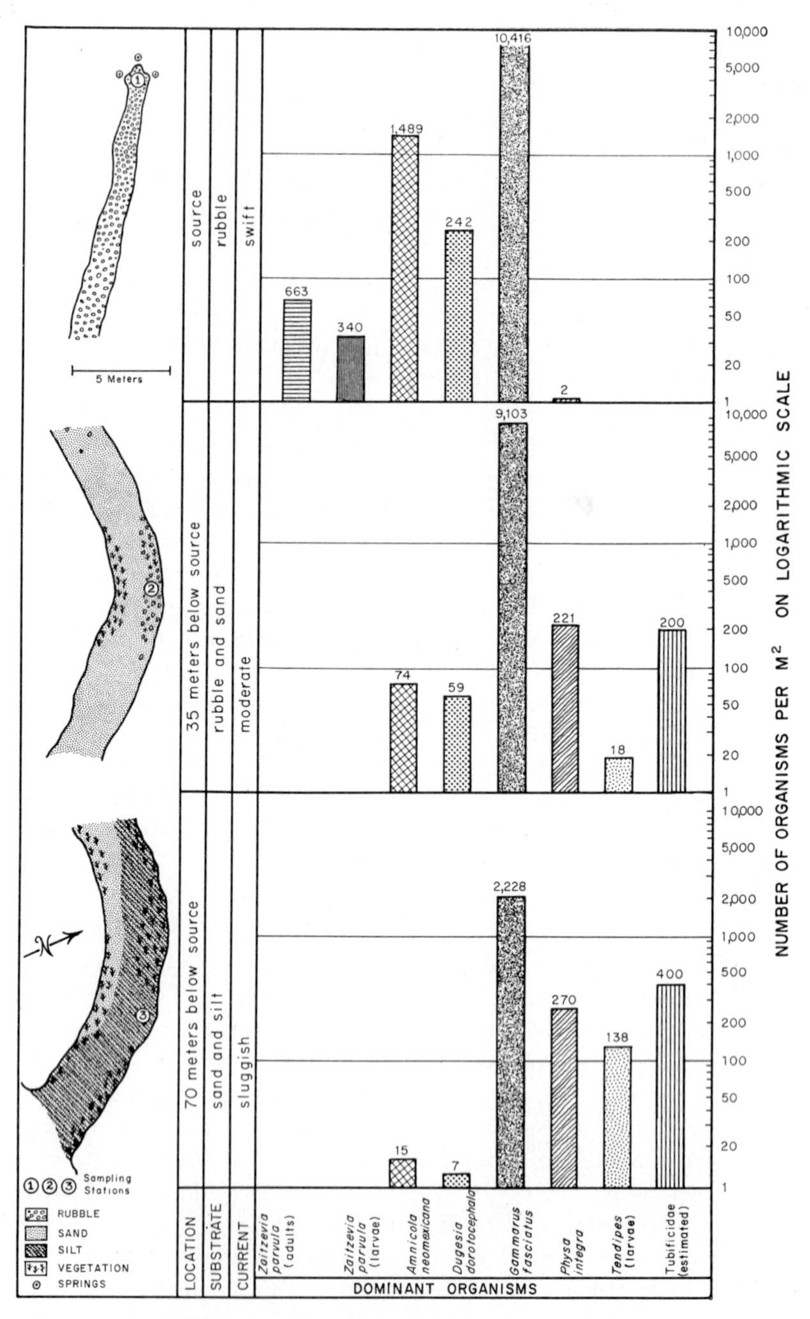

Figure 14·8. Habitat Characteristics and Mean Annual Populations of Some Animals in Lander Springbrook, New Mexico, 1950-1951. (Data from Noel, M. S., 1954. "Animal Ecology of a New Mexico Springbrook," *Hydrobiologia*, Vol. 6, 120-135.)

water, and riffles; each of these harbors rather distinct communities. In pool communities a supraneuston commonly occurs, and the nekton consists of fishes, such as the sunfishes (Centrarchidae) and minnows (Cyprinidae). The riffle sections are inhabited predominantly by insects, the community frequently containing a large number of species. A single riffle area in a California stream was found to be inhabited by nearly 40 kinds of insects.

The benthic communities of rapids and mountain brooks are composed mainly of organisms adapted for living in the swift current. The plants of the community are mostly filamentous algae which are firmly attached to the substrate, or diatoms which grow as crustose masses usually covered by slippery mucous secretions. The moss (*Fontinalis*) occurs in this region. The benthic animals exhibit a variety of adaptations which fit them to the rapid waters. Streamlined or flattened bodies reduce friction with the current or permit the animal to inhabit cracks and interstitial spaces among the rocks; rapids-dwelling mayflies, stoneflies, and dragonflies exhibit these adaptations. The larvae of the blackfly (*Simulium*), a common inhabitant of swift water, attaches to upper surfaces of stones by suckers on the posterior end of the body (Figure 12·12), or by means of strands secreted by the animal. Some caddisfly members of this community add small stones to their pupal cases, presumably for additional weight; the larvae of some species may not build cases but cling nakedly to objects in the stream. Many insects possess strong claws for holding and for locomotion against the current. Snails and flatworms are adapted by means of adherent surfaces and the secretion of mucilaginous material. Some animals, such as crayfishes and some insects, inhabit regions of fast water by more or less avoiding the major factor of current. These forms live behind stones or on the down-stream surfaces of stones, thereby gaining protection from the current. Fishes of the benthos are predominantly darters of the family Percidae. These have streamlined bodies and strong pectoral fins used for bracing against the current. Sucking mouthparts on amphibian tadpoles and the armored catfishes of South America are further adaptations to life in mountain streams.

Pollution and Communities of Pollution

In many parts of the world there remain but few primitive streams unaffected by the ways of man. Throughout much of North America, most streams have, to varying extents, been modified from their original state. In many instances, entirely too many, these man-made changes are the results of pollution, i.e., the introduction of substances which make the waters abnormal in comparison with natural, undisturbed waters. Pollution, and the subsequent change in community composition, comes about in essentially three ways: (1) by the introduction of erosional products, such as silt and clay, through improper control of soil in mining, timbering, and agricultural practices; (2) through inflow of materials from industrial operations which directly poison the water or otherwise make the environment uninhabitable; and (3) by the dumping of domestic sewage or industrial substances which enter into biological processes, resulting in a lowering of the oxygen concentration below the limits of tolerance of the original inhabitants.

Many agencies concerned with water use have been prone to disregard or minimize the roles of turbidity and sedimentation in pollution. In the sense that these factors can exert adverse effects upon the original aquatic communities, as well as upon the proper use of natural waters by humans, suspended and deposited substances should be considered as pollutants. We have already seen (Chapter 6) that turbidity is a major influence in determining development of phytoplankton and rate of photosynthesis which, in turn, influence over-all productivity in bodies of water. In earlier sections of this chapter, we learned that communities based upon submersed plants contain a great number of species and are highly productive; deposition of silt on these communities can eradicate the entire assemblage. Examples of this effect are common throughout the land; one has only to look at downstream sections of streams located near mining operations, housing developments, road-building operations, farming, or lumbering activities.

Limnological studies concerned with detection and control of sedimentation as a pollution process would involve considerable application of principles of stream hydrology and limiting factors. In practice, the establishment of rigid standards of turbidity and sedimentation may be difficult, due to the complexities of stream flow and the biology of communities encountered from one area or drainage basin to another.

Communities found in turbid and sedimented stream regions are generally those described earlier for lower stream courses; these typically include a small number of species, such as annelids, dipterans, mollusks, and fishes. The development of such communities is, however, highly dependent upon rate of sedimentation and degree of turbidity. Under extreme conditions of either of these factors, the community may be severely reduced.

Polluting effluents from industrial plants are highly varied and may affect aquatic communities in many ways. One category of substances might include those which impart disagreeable odors or taste to the receiving waters; the effects of these might affect not only the aquatic communities, but nearby human assemblages as well. A second group of materials includes chemicals such as lead, phenolic compounds, sulfite, acids, and numerous others which could be directly toxic to all organisms or to certain ones which may be important in food relationships in the community. In either case, there are apt to be deleterious effects on the animal and plant groups. Chemical effluents may also act to make the environment untenable by changing the density or chemistry of the water. Brine from oil fields causes streams to become highly saline, and, more recently, it has been found that wastes from mining of uranium ore, if unrecovered, may greatly lower the pH of streams in the vicinity. Radioactive wastes from research and industrial installations would fall into this general category; the effects of these materials are poorly known at present but are certainly worthy of considerable study. A third category of industrial substances encompasses organic compounds which through rapid decomposition utilize great quantities of oxygen, or through slow biochemical digestion form flocculent masses which increase turbidity and suffocate organisms. Included here are fats and coal-tar derivatives common to a great number of manufacturing processes, and cellulose carbohydrates of paper-making. A fourth category is concerned with heating of stream waters brought about by the use of the water in cooling processes in large industries. Water returned to streams after circulating through cooling systems may often be of considerably higher temperatures than when taken in.

The dumping of domestic sewage and organic substances often exerts great effects on stream communities, primarily through uptake of oxygen beyond the "normal balance" of natural photosynthesis and respiration processes in streams. Depending upon local conditions, the inflow of oxidizable matter may be of sufficient magnitude as to bring about complete depletion of dissolved oxygen, in which case aerobic bacteria fail to function in decomposition of the material; this latter process is continued, however, by anaerobic forms. As a result of such conditions, gases (such as hydrogen sulfide and methane) are formed, and a species-depauperate community exists in the highly polluted region.

It is well established that the effects of pollution decrease with distance downstream, assuming of course, that no additional pollution occurs. This process is reflected in a form of longitudinal succession of stream communities, the composition of each being determined by the degree of pollution throughout a stream segment. A number of schemes devoted to classification of pollution zones have been devised. One such system, applied to the River Trent in England, is shown in Figure 14·9. This figure also serves to illustrate longitudiual succession associated with certain chemicals introduced into the stream. Because the natures of pollution and of the streams themselves differ greatly, it has been difficult to establish a standard scheme for describing zones of pollution. It is generally agreed, however, that the following zones are found in heavily polluted streams. (The names of these zones have been selected from several proposed classifications.)

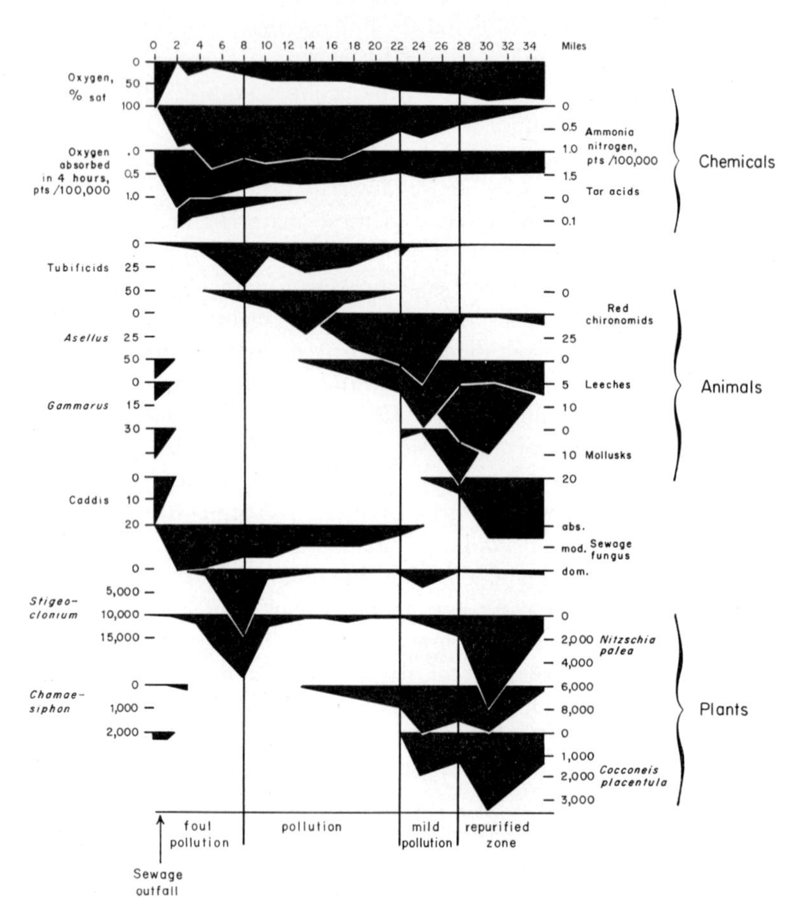

FIGURE 14·9. Longitudinal Succession Associated with Pollution in the River Trent, England. Beginning at the area of sewage outfall, four zones are recognized: (1) The zone of foul pollution, where sewage fungus flourishes: *Stigeoclonium* and tubificids also inhabit the zone. (2) The zone of pollution, where tubificids are abundant but other animal life, except for chironomids, is rare; fungus is common and a few algal species become more common. (3) The zone of mild pollution, characterized by great numbers of the isopod, *Asellus*, and algae; sewage fungus declines in quantity. (4) The zone of repurification in which, other factors being similar, the biota resembles that living above the outfall. (From Macan, T. T. and Worthington, E. B., 1951. "Life in Lakes and Rivers," *The New Naturalist Series*, Wm. Collins Sons & Co. Ltd., London, after Butcher.)

(1) *Zone of Recent Pollution:* Much organic material present. Early stages of decomposition. Dissolved oxygen content usually high. Green plants present. Fish may be abundant.
(2) *Zone of Active Decomposition–Septic Zone, Polysaprobic Zone:* Great amount of oxidation occurs here. Zone may be nearly depleted of dissolved oxygen. Much carbon dioxide and hydrogen sulfide. Bacteria abundant. Green plants mainly absent (blue-green algae may be present). Many kinds of protozoans live in this zone. The rat-tailed maggot, *Tubifera* (=*Eristalis*), abounds. If a small amount of oxygen is present, tubificid worms may be found, often in great abundance. Through decomposition of organic matter, this zone grades into the
(3) *Strongly Polluted Zone–Alpha-Mesosaprobic Zone:* Green algae present, although in reduced numbers. Bacteria continue abundant. Dissolved oxygen content low, particularly at night. Fauna more varied and a greater number of species present than in the preceding zone.
(4) *Mildly Polluted Zone–Beta-Mesosaprobic Zone:* Green algae and "higher" plants common in this zone. Dissolved oxygen usually above about 5 ppm. Diminishing oxidation of organic matter. Conditions within range of tolerance of many animals. Fishes such as eels, carp, and minnows may inhabit this zone.
(5) *Zone of Cleaner Water–Oligosaprobic Zone:* This section is essentially free of pollution, although decomposition occurs. Dissolved oxygen content generally above 5 ppm even at night. Green plants are abundant. Animal populations are generally those of typical "healthy" streams of the region.

Determination and measurement of pollution may be made in the following ways:

(1) By direct chemical analyses of the water.
(2) By bioassay, in which test organisms are placed in water samples and their reaction compared with controls.
(3) By the use of "indicator" organisms, i.e., the presence of certain plants and animals which experience has shown to be significantly characteristic of kinds and degrees of pollution. It has also been suggested that the absence of organisms known to be highly intolerant of pollution might serve as an indication of unnatural conditions. At present, the concept of indicator organisms is receiving much reconsideration and appraisal.

Another closely related ecological approach to recognition of pollution involves the effect of the condition upon community composition. This approach relates the limiting factors principle to species number and population density of a community. In other words, when a factor, oxygen for example, becomes limiting, a number of narrowly tolerant species are lost, thus decreasing community-wide competition for the factor. The remaining, more tolerant, species are then able to undergo population growth. Comparison of community structure, when well known, provides, therefore, an index to pollution.

(4) By measurement of the amount of oxygen required to stabilize the demands from aerobic biochemical action in the decomposition of organic matter. This is not the amount required to completely oxidize all organic matter, but rather the volume necessary to restore balance between oxidation and bacterial activity. This measure, known as the *biochemical oxygen demand* (BOD) is widely used in sanitary engineering work. For details of this and other methods of water analysis see A.P.H.A. (1955); for further information on various aspects of pollution consult publications from the U. S. Department of Health, Education, and Welfare, especially those of the Robert A. Taft Sanitary Engineering Center, Cincinnati, Ohio.

COMMUNITY METABOLISM; ENERGY RELATIONS WITHIN THE ECOSYSTEM

The maintenance of communities such as those described in the preceding section is dependent to a great extent upon food relationships and energy flow. These dynamics involve interactions between both the community and the environment (the ecosystem). Some major communities, such as ponds and lakes, are essentially self-sustaining. That is, given radiant energy and organisms acting the roles of producers, consumers, and decomposers (see below), the community is maintained more or less independently of other communities. Other assemblages, such as the stream-riffle community and the oyster-reef community, are largely dependent upon nutrients delivered by flow of water. In this section, we wish to consider some of the ecosystem interrelationships and the organization involved in community metabolism.

Trophic Levels, Food Chains, and Pyramids

The fundamental operation in community metabolism rests upon the roles which organisms perform at different nutritional (trophic) levels in maintaining transfer of energy (in food) through a series of individuals. These roles may be classified and defined as follows:

Producers: the organisms capable of synthesis of energy-containing organic substance through utilization of solar radiation and inorganic materials. These organisms include the chlorophyll-bearing phytoplankton, larger green plants, and photosynthetic bacteria.

Consumers: the organisms, mainly animals, that are incapable of synthesis of matter from the sun's energy, and which depend directly, or indirectly, upon the producers. Within this group, we recognize *herbivores,* which feed upon the green plants, and the *carnivores,* which eat the herbivores or other carnivores.

Decomposers: the heterotrophic bacteria and fungi which break down organic substance to the elemental state, thereby returning nutrients into the cycle for use by producers.

We see, therefore, that the "links" in the food chain represent levels of feeding and being fed upon; these are termed *trophic levels.* Correlating trophic level with the above classification of functions, we see that green plants occupy Level 1; grazing herbivores, Level 2; small-sized carnivores, Level 3; and large carnivores, Level 4. In a lake, for example, diatoms would be considered in the first level, copepods in the second, small sunfishes in the third, and bass (secondary carnivores) in the fourth; man, eating the bass, would conceivably represent the fifth level. Generally speaking, the more distant a trophic level from Level 1, the less the dependency upon the next lower level. Under natural conditions, trophic relationships are often much more complicated than indicated by the short "chain" analogy. In most instances, feeding habits of various species differ with age, vary with seasons, and often cut across more than one trophic level. The result is that food and feeding relationships among the members of a particular community include many branchings, "short circuits," and interconnections. Thus the condition comes to be a complex *food web* such as that illustrated in Figure 14·10.

Although many complicating factors are to be found in studying trophic relationships and energy flow in ecosystems, it is often possible to represent conditions in a relatively simple fashion. The Eltonian *pyramid of numbers* (named for the ecologist Charles Elton) provides a simple illustration of population density relationships within and between trophic levels. It was early noted that the numbers of individuals decrease at each level due to differences in population growth rates and to predation upon small organisms by large ones. Thus, in the example given above, the diatoms of Level 1 would be represented by the greatest number of individuals, copepods of Level 2 would be less abundant, and small sunfishes and bass of Levels 3 and 4, respectively, would exist in successively fewer numbers. It is of interest to note that a pyramid of numbers for parasitic organisms is the reverse of that for free-living forms; the levels operating from a host increase in numbers with decreasing size.

In addition to the pyramid of numbers, we also recognize in nature a *pyramid of energy*. The energy pyramid indicates that in each trophic level there is less available energy than in the next lower level. This condition results from the loss of both energy and substance with each transfer through the food chain. The data in Table 14·3 illustrate the principle that there is a greater amount of energy in the producer level than in the herbivore (primary consumer) level, and that the primary

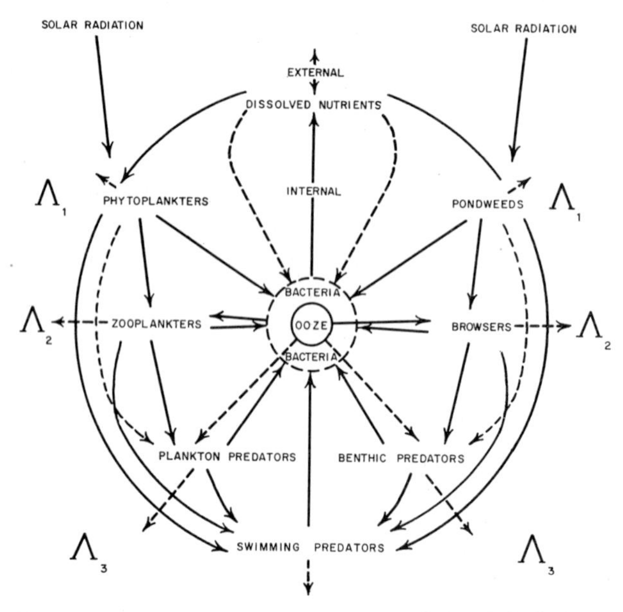

FIGURE 14·10. Generalized Scheme of Food Relationships in a Lake. Efficiency in the utilization of food increases progressively in the higher consumer levels. (After Lindeman, R. L., 1942. "The Trophic-Dynamic Aspect of Ecology," *Ecology*, Vol. 23, 399-418.)

consumer level contains more energy than the carnivore (secondary consumer) level. In other words, the percentage loss of energy by respiration increases progressively from lower to higher trophic levels. Note also that the *efficiency* of conversion of the energy in food to animal protoplasm increases among the higher trophic levels. Of especial interest in Table 14·3, is the per cent utilization and transfer of solar energy. Although the producer community in the pond received approximately 118,872 g-cal/cm²/yr, the plants produced only 49.9 g-cal/cm²/yr. This is to say, the producers utilized only 0.04 per cent of the incoming energy. Furthermore, the top trophic level ultilized only 0.0028 per cent of the original solar energy.

If the momentary total *standing crop*, or *biomass*, of the trophic

TABLE 14·3. ANNUAL RATE OF PRODUCTION AND EFFICIENCY OF TROPHIC LEVELS IN A MINNESOTA POND, CEDAR BOG LAKE, AND LAKE MENDOTA (g-cal/cm^2/yr)

(From Dineen, C. F., 1953. "An Ecological Study of a Minnesota Pond," *Am. Midl. Nat.*, Vol. 50, 349-376.)

	Pond		Cedar Bog Lake		Lake Mendota	
	Rate of Production	Efficiency (%)	Rate of Production	Efficiency (%)	Rate of Production	Efficiency (%)
* Radiation	118,872	...	118,872	...	118,872	...
Producers	49.9	0.04	111.3	0.10	480	0.40
Primary Consumers	9.2	18.4	14.8	13.3	41.6	8.7
Secondary Consumers	3.4	36.9	3.1	22.3	2.3	5.5
Tertiary Consumers (Fishes)	...	...	...	...	0.3	13.0

* Taken from Juday, C., 1940. "The Annual Energy Budget of an Inland Lake," *Ecology*, Vol. 21, 438-450.

levels in a given community of successively dependent links is measured and plotted, a *pyramid of standing crop* becomes evident. This suggests that the total mass of living substance tends to decrease from one level to the next in the same fashion as numbers and energy. Figure 14·11 illustrates the standing crop pyramid of organisms in the community of Weber Lake, Wisconsin. Several factors operate to produce such a

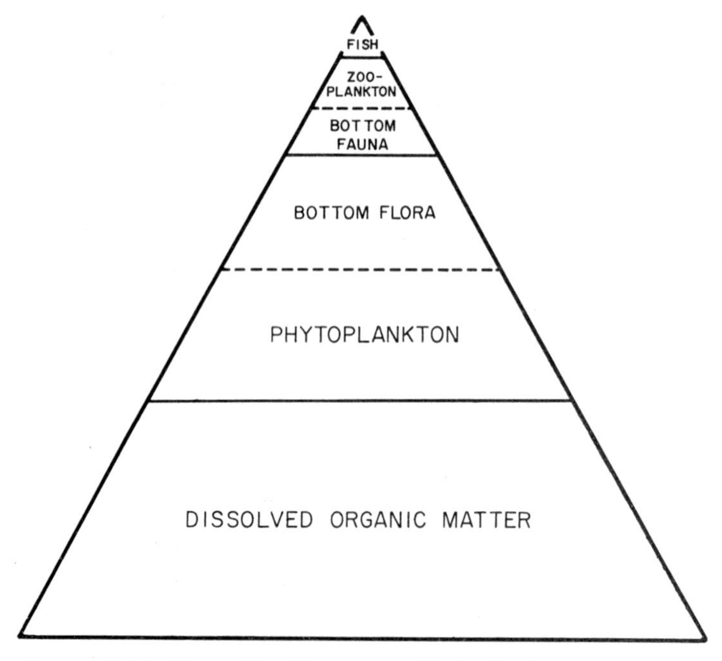

FIGURE 14·11. Pyramid of Biomass and Dissolved Organic Matter in Weber Lake, Wisconsin. Proportional areas of the triangle are based on the weight relationships of the various components. (After Juday, C., 1942. "The Summer Standing Crop of Plants and Animals in Four Wisconsin Lakes," *Trans. Wisconsin Acad. Sci.*, Vol. 29, 1-82.)

pyramid; these include growth rate and density of the various populations, and the likelihood of death of individuals of lower trophic levels before reaching the top consumers.

Thus, with respect to certain aspects of community structure and function, we have observed the organization of plants and animals in functional roles giving rise to food chains and webs. In these roles, the photosynthetic forms constitute the base of community anabolic activity which may be illustrated by pyramids of numbers, energy, and standing crop. Aquatic food webs are often complicated, however, by levels in which "feed-backs" occur. Some species that develop considerable standing crop utilize basic plant substance and organic detritus in their diet, thus "short-cutting" the food chain. These include certain herbivorous turtles in fresh water, and the mud-grubbing mullet (*Mugil*) of estuaries and some streams and lakes.

Up to this point we have been considering what might be called a *predator food chain,* or one in which energy contained in food substance, synthesized originally by plant activity, is transferred through a series of steps, or trophic levels, to successively remote levels. With consumption at each level, energy is utilized in animal activity and growth, and much potential energy is dissipated through respiration to the environment. In addition to the energy channeled through the predator food chain and that passed to the environment, some energy is utilized by heterotrophic bacteria and fungi. The source of this energy is nonliving organic matter resulting from the death of any of the organisms in the predator food chain. In the course of their nutritional activity these heterotrophs decompose, or transform, organic substances to the elemental state. Still another energy pathway is found in the chemo-autotrophic bacteria, or those forms which derive energy from inorganic chemical sources.

Concepts of Productivity

The productivity of an ecosystem connotes a quality whereby living substance is manufactured through interactions of community and environment. Three measures of this quality are commonly used and, as concepts, make up the core of productivity as a community attribute; these are: *standing crop, rate of removal of material,* and *rate of production.*

STANDING CROP

As defined earlier, standing crop (biomass) represents an instantaneous quantity of organisms. In Chapter 13 we considered a number of factors and processes relating population growth to standing crop and recognized that both rate of growth and maximum density are dependent upon a number of variables. Although standing crop is a measure of productivity, it is a poor one because it does not include the time (rate) factor concerned with the development of the crop. A high standing crop does not necessarily mean that the ecosystem has a high rate of production. For example, an abundant momentary phytoplankton crop may have been a long time in the making, indicating a low rate of turnover of nutrients. Thus, what is important, in terms of production, is the knowledge of the rate at which the densities of populations of the community are produced. In communities such as the benthos and aufwuchs, where the effects of limiting factors are minimal and where seasonal patterns are pronounced, measurement of changes in standing crop over a period of time are satisfactory indices of production.

RATE OF REMOVAL

Rate of removal of material pertains to the yield of an ecosystem per unit of time. The importance of this aspect lies in the recognition of the rate at which substance can be removed ("harvested") from an ecosystem without interfering with natural biogeochemical cycles. We have already learned that nutrient flow in a lake or pond is essentially cyclic. Thus, the loss of material must be recompensed by replacement of material if the system is to be maintained. Because of loss of material (and energy) at each trophic level, the income of nutrients and organic substance must normally be greater than their removal. These principles find application in human economic endeavors such as commercial fishing. Commercial harvesting is not the only way in which materials are removed from ecosystems, however. In many cases, such activity constitutes only a small part of the over-all removal of material from a lake or pond, and is often much less than the income of nutrients. Migration of organisms out of the area, predation by terrestrial animals, outflow of organisms through effluent streams, and loss of nutrients to sediments, along with other processes, serve to remove various proportions of the yield.

Of great interest, both economically and ecologically, is the magnitude with which yield of an ecosystem per unit time can be increased. In the United States, the yield of stocked fish in unfertilized farm ponds ranges from about 4.5 to 16.8 gm/m^2/yr (40 to 150 lb/acre/yr); in fertilized waters the yield is from 22.4 to 56.0 gm/m^2/yr. Such a conspicuous increase in yield and the rate with which fish substance may be removed are the result of the addition of basic nutrients into the ecosystem. Obviously, in order for the increased yield to be sustained there must be continued input of the nutrients. The ratio of yield to input is a valuable measure determining the removal rate in commercial and sports fisheries. It would indicate whether a given community is being overexploited or underexploited, not only by man but also with respect to predator-prey relationships and other ecological aspects.

RATE OF PRODUCTION

The third aspect of productivity, rate of production, pertains to the amount of organic substance synthesized in a prescribed space per unit of time. This concept involves the cycle of organismal growth, death, and decomposition, or the *turnover* of nutrients and energy food in an ecosystem. The rate of turnover of organisms in the community varies greatly. Phytoplankton may turn over within a period of several days; spermatophytes, such as waterlilies, exhibit only one turnover per year.

We have already touched on one phase of this activity in our reference to yield of fish substance. The fish production, however, represents synthesis of energy-containing substance at a higher trophic level. What is often of great importance to the total metabolism of an ecosystem is the rate of production at the plant, or producer, level; for the entire trophic structure of an ecosystem is ultimately dependent upon this level.

Of the total solar energy that enters a body of water, a pond for example, a small portion enters into the photosynthetic process of green plants and is stored in organic substance (chiefly carbohydrates) in plant bodies. The rate at which energy-containing material is formed by plants represents the *rate of primary production.* Since plants utilize some of the stored energy for their own metabolism and growth, the amount of energy available to consumers is less than the original stored quantity. In measuring rates of production it is necessary, therefore, to recognize the possible "pathways" and to designate two rates of primary production. The total production of the producer organisms in an ecosystem is termed *gross primary production,* and refers to the total photosynthesis. The amount of plant substance produced per unit time and space is defined as *rate of gross primary production.* As would be expected, considerable variation in rate of gross primary production is found in natural waters. In an oligotrophic lake of Wisconsin, the average rate (in mass of dry organic material per unit space and time) was found to be 0.7 gm/m^2/day; in Lake Erie in winter: 1.0 gm/m^2/day; in Lake Erie in summer: 9.0 gm/m^2/day; in Silver Springs, Florida: 17.5 gm/m^2/day; and in a polluted Indiana stream in summer: 57.0 gm/m^2/day. In general, the rate of gross primary production in eutrophic lakes ranges from about 0.5 to 5.0 gm/m^2/day during the most favorable growing seasons.

In actuality, plants utilize some of the gross production in their own respiration. Thus, *net primary production* represents the amount of organic substance remaining in plant bodies after withdrawal of some of the substance for respiration, and *rate of net primary production* represents the rate at which the organic matter is stored. This latter measure is essentially a statement of growth. Net production (or rate) is of great value to limnologists, for it represents the actual addition of energy-containing organic substance to the ecosystem for use in community metabolism.

Comparison of rates of gross and net production in a single community provides an interesting datum on utilization of plant food by the plants themselves. In Silver Springs, Florida, the rate of gross production has been measured and found to be 17.5 gm dry organic matter/m^2/day; the rate of net production is 7.4 gm/m^2/day. Thus, considering the difference between the two as the amount of production lost to plant utilization, it is evident that some 42 per cent of the gross production is available to the herbivores of the community. In most instances only a small part of the net primary production is consumed by organisms in the second trophic level or by decomposers during a given period of time; the unused portion, therefore, contributes to the increase in the amount of standing crop. On the other hand, overgrazing of the net primary production leads to decrease in the total standing crop.

The energy food actually assimilated by heterotrophic organisms, such as primary, secondary, and tertiary consumers and the decomposers, represents *secondary production* (or *assimilation*, as it is termed by some ecologists); the rate of energy storage in these levels is called *rate of secondary production*. As indicated in Table 14·3, the rates of production decrease at each higher level from herbivore to top carnivore. We have here considered secondary production as it relates to primary production. We see, however, that secondary production is the same as yield, with respect to both quantity and rate.

Within the past 20 years, interest in community metabolism and productivity of ecosystems has increased greatly. Concurrent with this growth of interest has been the development of methods for measuring rates of primary and secondary production. At the level of primary production, the development of techniques has, and continues to be, directed toward determining the rate at which raw materials are used, carbohydrate is synthesized, and by-products released, in the familiar equation for photosynthesis simplified as:

$$CO_2 + H_2O \longrightarrow CH_2O + O_2$$

It is not within our province to describe here in detail the methods used in measuring production rates. It would seem appropriate, however, to mention briefly some which are presently popular among ecologists. For a comprehensive critique, and references to methodology, see *Limnology and Oceanography*, Vol. 1, No. 2, 1956. Basically, the methods make use of certain principles and processes described in earlier chapters of our introduction to inland waters and estuaries.

(1) *Rate of Change of Standing Crop:* This method involves periodic measurement of the growth of macroscopic populations which develop in a fairly regular seasonal fashion. It is not useful in measuring production of populations (phytoplankton for example) which are subject to predation and removal by other processes.

(2) *Production of Oxygen:* From the equation for photosynthesis we see that there is a direct, quantitative relationship between the amount of carbon dioxide used in the synthesis of carbohydrate and the amount of oxygen produced as a by-product. If environmental consumption of oxygen is taken into consideration, a measure of dissolved oxygen changes

should reflect gross primary production. The "light and dark bottle method" is widely used and involves measuring the rate of change of dissolved oxygen in bottles of water suspended at depths from which the original water samples were taken. Since both organismal respiration and photosynthesis occur in the suspended bottles, a covered ("dark") bottle is used to correct for respiration. Another method, the "diurnal oxygen curve," entails periodic determinations of dissolved oxygen content over a 24-hr period. This approach is especially useful *in situ* in lakes and takes into consideration the difference between oxygen produced and used during the day, and that used (but not produced) at night. A third technique, the "hypolimnetic oxygen deficit," is based on the idea that the rate of oxygen utilization in the hypolimnion of deep lakes is proportional (per unit area) to the amount of organic matter raining into the hypolimnion from the trophogenic zone above. In employing this method, incomplete oxidation of the organic matter, bacterial respiration, loss of oxygen to effluents, and other factors must be considered.

(3) *Uptake of Carbon Dioxide:* From the expression for photosynthesis we see that there is equivalence between uptake of carbon dioxide and the synthesis of organic carbon. Therefore, a measure of carbon dioxide uptake should represent production of organic matter. When relationships in the carbon dioxide buffer system are considered, pH may be used as a measure of carbon dioxide in the system. In deep lakes, hypolimnetic carbon dioxide production, rather than oxygen consumption, has been used to measure productivity.

(4) *Uptake of Nutrients and Radioactive Elements:* Lumped under this one heading are methods of productivity measurements which involve the absorption of minerals necessary in photosynthesis. Although it must be kept in mind that elements are released through decomposition and may be combined with environmental materials as well as taken up by primary producers, measurements of such nutrients as phosphorus and nitrogen are useful tools. Thus, use of radioactive tracers (for example, carbon-14 in carbonate, and phosphorus as P^{32}) have great value in productivity measurements. The use of P^{32} has been discussed previously in Chapter 10.

(5) *Concentration of Chlorophyll:* It would appear that the chlorophyll content of plants would bear a relation to the total organic material produced by the plants, and thus the chlorophyll concentration in water would serve as an index of productivity. A number of investigations have been made along this line of reasoning with various results. It has been shown, for example, that chlorophyll production and content in plants varies considerably with light, nutrients, temperature, and other factors. On the other hand, it has been found that a constant relationship between photosynthesis and chlorophyll *a* appears to exist at constant light intensity; with consideration of certain variables and further experimenta-

tion, this relationship may be a useful measure of production and rate of production.

(6) *Transformation of Energy by Consumers* (*Secondary Production*): Of major importance in understanding the over-all energy relationships in an ecosystem is efficiency of energy transformation within the consumer level. When a preadult herbivore feeds, the energy input is directed toward growth and respiration (daily activity), and some energy is passed off in egestion. The difference between the energy of ingested substance and that of egested material is assimilated energy, and this contributes to growth and respiration. By determining the calorific content of the food algae and the protoplasm of preadult cladocerans (*Daphnia pulex*), data on efficiency of energy transformation by individuals of that species have been determined. It was found that at four levels of food concentration, the energy of growth and the energy of respiration were nearly constant. Further, as the food consumption increased, the energy of egestion also increased and assimilated energy decreased. The efficiency of growth as percentage of energy *consumed*, that is, turned into new protoplasm, decreased (13.22 to 3.87 per cent) as food supply increased. The efficiency of growth as percentage of energy *assimilated* and turned into new protoplasm remained essentially constant (55.36 to 58.64 per cent) at various levels of concentrations of food. The major portion of stored energy of adults went into the production of young. From this information we gain some idea of the proportions of energy directed toward several pathways in the primary consumer level of an ecosystem.

One final consideration will serve to summarize and to point up several important concepts pertaining to community metabolism. Root Spring is a small spring about 2 m in diameter and 10 to 20 cm deep, located in Massachusetts. A study of this ecosystem showed that the producer level consisted of benthic algae and the duckweed (*Lemna*). Approximately 50 species of animals inhabited the spring, most of these being herbivores or detritus-feeding forms. Community respiration and net production were measured monthly by "light and dark" glass cylinders pushed into the bottom. Energy transformation and the composition and size of the community standing crop were also determined. Inflow and outflow of organic matter (including the emergence of insects) were computed. From these data and the calculation of respiration rates of the various levels and the total community, the energy flow could be diagrammed as shown in Figure 14·12. The energy contained in each level is shown as kilocalories per square meter per year in the numbers in the compartments of the diagram. It is interesting that primary production (algae) accounted for little of the energy of the higher trophic levels; 76 per cent of the energy transformed by the animals entered the system as allochthonous detritus, while only 23 per cent of the energy was derived from photosynthesis within the spring. Of the total annual energy inflow, 71 per cent was transformed to heat, 28 per cent was deposited in the community, and 1 per cent was lost through insect emergence.

Among the important aspects of community metabolism illustrated in Root Spring data, the distinction between biomass and energy flow is especially evident; the most abundant herbivore species (in terms of biomass) was third in amount of energy assimilated and transformed.

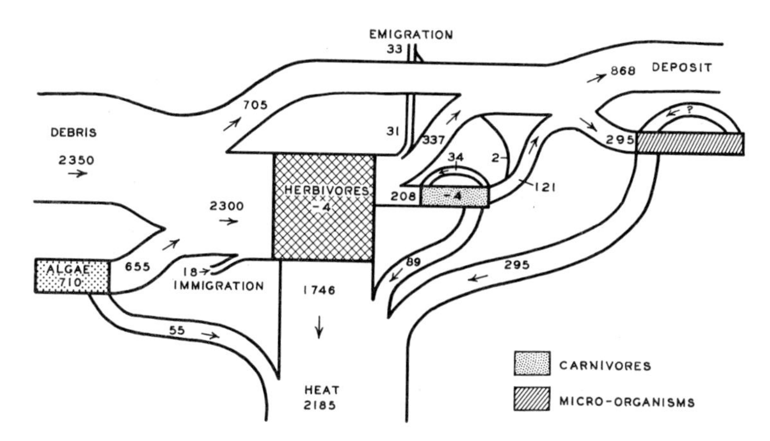

FIGURE 14·12. Diagram of Energy Flow in Root Spring, Massachusetts, 1953-1954. Trophic levels are represented by boxes, the numbers indicating changes in standing crop. Energy is expressed in kilocalories per square meter per year, and the direction of flow is shown by arrows. (After Teal, J. M., 1957. "Community Metabolism in a Temperate Cold Spring," *Ecol. Monographs*, Vol. 27, 283-302.)

Unlike the cyclic flow of basic nutrients in an ecosystem, energy flow is unidirectional. Energy flows through a community and is used only one time by the components. Because of the incompleteness of our present state of knowledge, a number of assumptions must go into studies of community metabolism. Nevertheless, we do gain considerable insight into the subject through functional analyses of various types of aquatic ecosystems.

European limnologists have used productivity and associated physicochemical features as major criteria in recognizing three basic types of lakes. Although universal application may prove difficult, the classification is valuable in making certain general distinctions. (We have considered some aspects of oxygen in two of these lake types in Chapter 9.) *Oligotrophic lakes* are usually deep, lack an extensive littoral zone, and are poor in dissolved nutrients such as phosphorus, nitrogen, and calcium.

These lakes have a low content of electrolytes and organic matter. Plankton is scarce, and production of organic carbon amounts to about 250 metric tons/km^2/yr (Weber Lake, Wisconsin). Oxygen is usually abundant throughout the depth of the water. In time, and through succession (see below), these lakes may become eutrophic. *Eutrophic lakes* are generally shallow, possess an extensive littoral zone with plant growth, and are rich in basic nutrients. The content of electrolytes varies and the organic composition is typically high. Primary production of organic carbon is on the order of 480 metric tons/km^2/yr (Lake Mendota, Wisconsin). Oxygen may become depleted in the hypolimnion during summer, and under ice in winter. *Dystrophic lakes* † are found mainly in regions of bogs and old mountains. The water in these lakes is characteristically brown in color, and high in concentration of nutrients, organic matter, and humic materials. The pH of the water of dystrophic lakes is usually low, and oxygen may be nearly or entirely absent in the deeper areas. These lakes often develop into peat bogs.

STRATIFICATION IN AQUATIC COMMUNITIES

In the first part of this chapter we considered some selected communities of lakes and streams with emphasis on their plant and animal components. We arrived at the acknowledgement that communities are organized to the extent that they are composed of generally predictable assemblages of organisms. We then turned attention to trophic aspects of the community and recognized within them organization with respect to food webs and energy relationships. A third concept embodying organization within the structure of communities is that of *community stratification.*

Community stratification pertains to the horizontal or vertical arrangement of populations, their activities (responses to environmental processes), or their effects on the environment. We have previously considered many cases of both vertical and horizontal stratification in aquatic communities. Thus we need now only to recall some of them in illustration of this very outstanding ecological principle. Thermal stratification, so characteristic of most lakes of the temperate zone in summer, serves to bring about vertical zonation of the lake community. Associated with this phenomenon we have seen the stratification of a lake with respect to heat content, dissolved substances, and, indeed, the animal and plant inhabitants. Community stratification resulting from thermal stratification

† Although "dystrophic" waters generally present certain common and distinctive features, specific bodies of water are often difficult to designate. In recent years many limnologists have come to disregard "dystrophic" lakes as a distinct type.

breaks down during vernal and autumnal overturn, but is re-established the following summer, attesting to the pliant nature of organic communities. Associated with water transparency and light transmission is the layering of a body of water with respect to production and decomposition of organic matter; we term the layers the tropholytic and tropho-

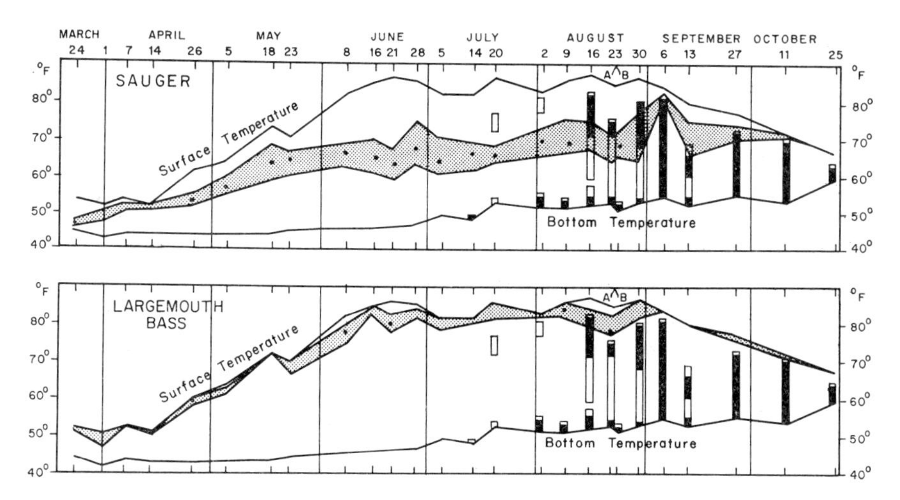

FIGURE 14·13. Vertical Stratification of Two Species of Fishes in Relation to Temperature in Norris Reservoir, Tennessee, 1943. The shaded area shows the temperature range in which the middle 50 per cent of the individuals in the sample were caught. Vertical bars indicate the dissolved oxygen concentration; open areas represent a range from 1.6 to 3.0 ppm, and the solid areas 0.0 to 1.5 ppm. Note that in early September the sauger responded to low oxygen concentration (accompanying a density current) by moving into warmer, upper waters. (After Dendy, J. S., 1945. "Fish Distribution, Norris Reservoir, Tennessee, 1943. II. Depth Distribution of Fish in Relation to Environmental Factors, Norris Reservoir," *J. Tennessee Acad. Sci.*, Vol. 20, 114-135.)

genic zones. In these layers we see stratification of organismal activity and the effects of the organisms on the environment.

Within a given community, certain species populations may occur in strata, usually as responses to tolerance of the effects of limiting factors. In the plankton community of Wisconsin lakes during summer stratification, each layer contains a rather distinct population of cladocerans, but modified according to given time of day or night, depending upon light intensity. This arrangement lends to each stratum a certain amount of temporary internal integrity. Some populations of the nekton community exhibit sharp vertical stratification as illustrated by two species of fishes in Norris Reservoir, Tennessee, shown in Figure 14·13. The sauger

(*Stizostedion canadense*) is inherently a "cool-water" species and less tolerant of warm temperatures. The large-mouth bass (*Micropterus salmoides*) thrives best in waters of warmer climates. During the winter months of uniformly low vertical temperatures, both species exist together. During summer stratification, however, the species essentially segregate according to the most favorable temperature conditions. Since, as we already know, thermal stratification is correlated with depth, the fishes also exhibit, during summer stratification, a marked layering of population density. Note, however, that in early September the sauger population was "forced" into warmer waters by an oxygen minimum which developed with a density current passing through the impoundment. Parenthetically, we might point out that here is an example of the operation of the "principle of limiting factors"; the sauger population appears to be "preferring" to move into the less optimal temperature regime in order to avoid the low oxygen. These data have practical application in that fishermen may adjust their line length and depth of bait according to vertical temperature and the desired species of fish. Newspapers in the vicinity of Norris Reservoir have published these data in the interest of better fishing.

Horizontal stratification of substrates and their communities is generally characteristic of natural bodies of water. We have considered the general tendency of higher plants to occur in rather distinct zones about the shallow waters of lakes and stream. We have learned, furthermore, of the relative distinctiveness of the plankton of the vegetated zone and that of the adjacent limnetic zone. Similarly, we have noted how benthic communities grade horizontally from the littoral to the nonvegetated bottom of the deep region. Broadly considered, however, the benthos is also influenced by factors related to depth as a vertical dimension. Thus, benthic communities may often be influencend by two dimensions, vertical and horizontal. This is not always the case, however. We have seen, for example, that in the shallow waters of lakes, community composition is often related to the sedimentary nature of the substrate which, in turn, may be independent of depth.

Longitudinal succession of communities in streams is now appreciated and accepted as principle. In this impressive linear arrangement of plant and animal assemblages, we observe horizontal zonation clearly structured by environmental factors attributable to flowing water and by evolutionary adaptations of the organisms. Although we did not consider the nature of estuarine communities in our previous discussions, the principles of zonation operate in this major community. Horizontal zonation of plankton, nekton, and benthos from the freshwater upper reaches to the more saline mouth of the estuary is usually clearly demonstrable.

PERIODISM IN AQUATIC COMMUNITIES

Periodism in communities refers to the recurrence of events, activities, and changes in the component species populations which, in turn, are ultimately felt in the nature of the community. A given recurrence may be sporadic and irregular, or it may be cyclic and regular. Cyclic phenomena are commonplace in nature and, indeed, are obvious and important in aquatic communities. The most conspicuous cycles are those associated with the earth's daily rotation, the 28-day movement of the moon around the earth, and the yearly orbiting of the earth about the sun. These periodisms are best recognized in terms of activities and events related to diel (daily), lunar (tidal), and annual phenomena. Endogenous and long-term cycles may also play a part.

Daily Cycles

Although relatively short-termed, diel (24-hr) cycles in aquatic communities can often exert considerable influence on community metabolism and structure. The response of plants to light manifested in photosyn-

Table 14·4. Vertical Movements of Zooplankton in Lake Lucerne

(Data from Allee, *et al.*, 1949. "Principles of Animal Ecology," W. B. Saunders Co., Philadelphia, Pa., after Worthington.)

Species	Extent of Vertical Movement (m)	Time of 1 m Descent (min)	Time of 1 m Ascent (min)	
			Noon to Dusk	Dusk and Later
Daphnia longispina, adults	10 to 40	5	25	6.3
D. longispina, young	5 to 60	3.1	17.5	4.3
Bosmina coregoni, young	7 to 55	4.5	26	7.5
Diaptomus gracilis, adults	13 to 30	12	60	15
D. gracilis, young	20 to 37	20	60	12
Cyclops strenuus, young	12 to 46	7	60	5.5

thesis and primary production is one of the most important daily cycles in nature. In addition to the production of energy-containing substance, this process also is related to nocturnal-diurnal cycles in the chemical composition of natural waters. Recall that at night organismal respiration generally exceeds photosynthesis, while during the day the reverse occurs, resulting in shifts in a number of chemical relationships. Here is a cycle with widespread effects throughout the entire ecosystem.

Daily cycles of plankton activity are prominent in lakes and in the sea. Many species of zooplankton migrate toward the surface during the night and sink bottomward during the day. Rate and extent of movement vary among the species (see Table 14·4), resulting in considerable modification of vertical structure of the plankton community. This cyclic change in structure is further complicated by the fact that numerous plankton-feeding animals migrate with the zooplankton. Although light intensity (absolute, or variations in) is doubtless involved in these migrations, the complete explanation is not now known.

Lunar Cycles

Cycles of activity which are associated with 28-day phases of the moon are less well known in inland waters. In estuaries and the seas, however, many events are correlated with tides, and thus indirectly with moon-earth gravitational effects. An oft-cited example is that of the grunion (*Leuresthes tenuis*), a fish of the Pacific Coast. These fish appear at night in the surf along the California coast precisely at the time of the highest high tides of the lunar cycle. During the second through the fourth night of this series of "spring" tides the fish go onto the shore with the farthest-reaching wave. Here, in the sand, the eggs are deposited and fertilized. Being at this high point, the eggs are not disturbed by waves again until ready for hatching, about two weeks after spawning. At this time, the waves of another spring tide series wash the rapidly hatching eggs and young fish back into the sea. In this adaptation the entire spawning activity and development of eggs of the grunion are tuned to tidal cycles.

Annual Cycles

Annual cycles of plant and animal activity, usually associated with a given season, are familiar to all, and thus we will not dwell at length upon them here. The seasonal emergence of midges, craneflies, and mayflies, and the "blackfly season" attest to reproduction and development regulated by annual cycles. In coastal areas, "runs" of salmon or shad into freshwater streams occur with regularity each year. Winter quiescence and summer activity in lakes and streams are reflections of seasonal changes in metabolic activity of the organisms, these changes being related to annual temperature cycles of the environment.

Probably no phenomenon in fresh waters illustrates more vividly seasonal changes in community structure and metabolism than does the annual cycle of overturn and stagnation in temperate lakes. We have already considered the immensity and complexity of the dynamics in this process. This example serves to illustrate how seasonal variations add complications to the community, per se, and to the concept of communities.

Endogenous Cycles

One of the many complicating factors of community periodism rests on the fact that all cycles are not, as we might have concluded from the preceding discussion, regulated by environmental factors. It has been shown that a great number of population and community activities are regulated by cyclic behavior patterns possessed innately by the organisms. These patterns are termed *endogenous*, and, although synchronized with environmental events, they may function independently of external factors. In other words, endogenous periodisms continue to be manifested by the organisms even if the environmental factors are changed or made uniform. Color changes in the fiddler crab (*Uca*) are synchronized to diel or tidal cycles of a given area. When the cycles are changed, for example by transporting the crabs from the east coast of North America to the west coast, or the clues such as light or temperature are removed, the color-change rhythm persists. Studies of these activities suggest the presence of a "biological clock" within the organism. This mechanism has been called an "Endogenous Self-sustaining Oscillation," and abbreviated ES-SO.

Long-Term Cycles

One further kind of periodism should be included in our consideration. It involves the long-term cycles which constitute a major principle in geomorphology, and which we might term *geologic periodism*. During the earth's history, mountains have been thrown up and eroded in cyclic fashion, and glaciers have advanced and receded periodically. Such processes as these greatly influence the nature of lake basins, stream channels, and their waters and the communities developed in them. An example of the operation of geologic cycles is seen in a section of earth cut by the Caloosahatchee River in southern Florida. Within a vertical distance of a few meters, several alternating layers of freshwater material and marine fossils (mostly mollusks) record times of fluctuating sea level during the Pleistocene.

ECOLOGICAL SUCCESSION

Our preceding considerations have pointed vividly to the fact that communities change. We have seen change in respect to metabolism, stratification, and periodism. From a broader point of view a community has a beginning, and then undergoes a series of ecological changes until it reaches a relatively stable state. The orderly sequence of progressive change of communities over a given space is termed *ecological succession*. The sequence of changes has direction which can be described and a rate which can be measured. The entire sequence of community types, from the beginning, or *pioneer* stage, to the final, generally stable, stage, or *climax*, is called the *sere*. The changes which result in "new" communities at a given point are spoken of as seral changes, and the communities as *seral stages*.

The history of a lake offers a classic example of ecological succession and involves many concepts and principles which we have already learned. A typical "young" lake of the temperate zone is oligotrophic; it has little organic content and a poorly developed littoral zone. In time, erosion of the basin and inflow of nutrient substances contribute to eutrophication. Bottom sediments become enriched in organic debris, the littoral develops a zonation of higher plants, and the biota of the limnetic increases. Organic debris and inwashed sediments accumulate in the shore zone, resulting in the lakeward encroachment of the littoral. In time, the lake becomes more shallow; plants which formerly rimmed the shore extend farther and farther lakeward until they meet in the center. The earlier littoral zone is now a marsh, and the central part of the original lake is a pond. Continued filling eventually eradicates the water, and through successional stages involving different plant and animal communities a forest develops in the area formerly occupied by the lake.

As we learned in Chapter 2, the sediments of lakes usually contain a history of the events and conditions in the ecological succession leading up to the present. Such a history has been developed for Linsley Pond, Connecticut, and is shown in Figure 14·14. From a pioneer stage as an oligotrophic lake, eutrophication has proceeded, accompanied by deposition of gyttja to fill the bottom, and by ecological succession of community components, as shown in the figure. The early oligotrophic stage (14·14A) was characterized by relatively high transparency and a thin layer of bottom sediments of silty composition in which the dipteran *Calopsectra* (= *Tanytarsus*) was predominant; phytoplankton and zooplankton were increasing, and the littoral vegetation was sparse. The stage of maximum production (14·14B) was accompanied by the deposition of gyttja, decreased transparency, and a great abundance of plankton and littoral vegetation; plankton blooms were common. Through continued eutrophication, the present pond stage was reached (14·14C). Transparency is generally low, the bottom consists of a thick deposition of gyttja, littoral vegetation is dense but narrowly restricted vertically, and the dipterans *Tendipes* (= *Chironomus*) and *Chaoborus* predominate in the benthos. Community structure and metabolism are relatively stable at present. Although considered primarily with reference to sedimentation, Figure 2·13 in Chapter 2 serves as another example of ecological succession.

From broad-scaled sequences of changes, such as described above, ecological succession can be seen in a great variety of levels and in dif-

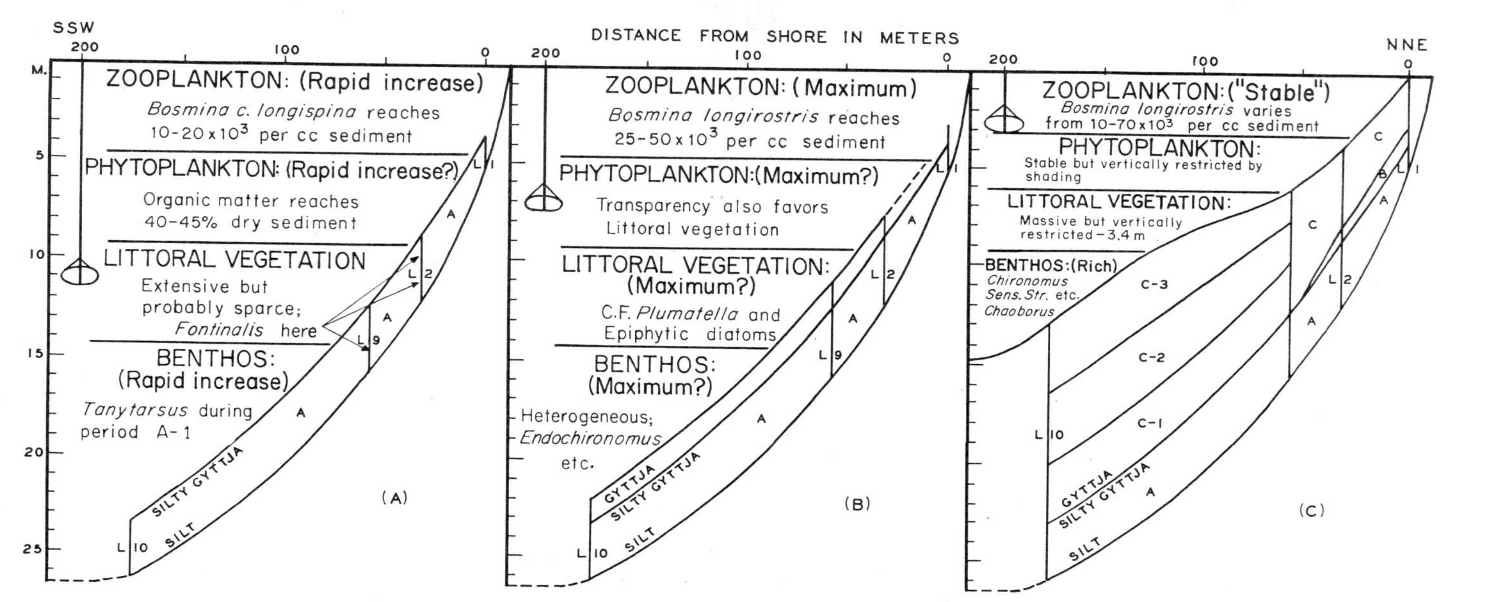

FIGURE 14·14. Schematic Representation of Bottom Profiles and Summary of Ecological Conditions in Linsley Pond, Connecticut. (A) during the times of spruce-fir dominance of the area, (B) at the time of maximum organic matter production, and (C) at present. Estimated Secchi disk transparencies are shown at upper left of each figure. Vertical dimension is shown to ten times the horizontal scale. (From Deevey, E. S., Jr., 1942. "Studies on Connecticut Lake Sediments. III. The Biostratonomy of Linsley Pond," *Am. J. Sci.*, Vol. 240, 313-324.)

fering degrees of magnitude. A simple laboratory hay infusion exhibits a marked succession of populations with age. Various population assemblages within a body of water may undergo seasonal succession in a somewhat cyclic fashion. Figure 14·4 illustrates succession in the microfauna of a small pond.

We indicated above that change is characteristic of communities and also, although somewhat paradoxically, that ecological succession leads to a relatively stable climax. Climax communities tend to be stable and in equilibrium with the environment as long as no intrinsic changes within the populations, or external disturbances, occur. But internal changes, such as fires, diseases, and the effects of man, and external disturbances, such as climate modification, erosion, and the intrusion of new species, do take place.

Community change is ultimately inevitable because the species populations comprising the communities are not static in themselves. Over long periods of time and many generations of organisms, evolution brings about changes which are bound to influence the composition of the community. Both the rate and direction that evolution takes may be strongly influenced by the physical, chemical, and biological features of the environment acting, on the one hand, to limit species development, or on the other, to provide new conditions in which organisms with novel modifications may flourish. It seems clear that to understand the momentary relationships and, further, the evolutionary aspects of aquatic communities, a thorough appreciation of the ecology of natural waters is mandatory. In the foregoing pages of this book we have attempted to set forth the principles of that ecology.

BIBLIOGRAPHY

Ahlstrom, E. H., 1943. "A Revision of the Rotatorian Genus *Keratella* with Descriptions of Three New Species and Five New Varieties," *Bull. Am. Museum Nat. Hist.*, Vol. 80, 411-457.

Alexander, W. B., Southgate, B. A., and Bassindale, R., 1932. "The Salinity of the Water Retained in the Muddy Foreshore of an Estuary," *J. Marine Biol. Assoc. United Kingdom*, N.S., Vol. 18, 297-298.

Alexander, W. B., Southgate, B. A., and Bassindale, R., 1935. "Survey of the River Tees. Part II. The Estuary–Chemical and Biological," *Water Pollution Research Tech. Paper*, No. 5, Dept. Sci. Ind. Research, London.

Allee, W. C., 1951. "Cooperation Among Animals," Henry Schuman, Inc., New York, N.Y.

Allee, W. C. and Bowen, E., 1932. "Studies in Animal Aggregations: Mass Protection Against Colloidal Silver Among Goldfishes," *J. Exptl. Zool.*, Vol. 61, 185-207.

Allee, W. C., Bowen, E., Welty, J., and Oesting, R., 1934. "The Effect of Homotypic Conditioning of Water on the Growth of Fishes, and Chemical Studies of the Factors Involved," *J. Exptl. Zool.*, Vol. 68, 183-213.

Allee, W. C., Emerson, A. E., Park, O., Park, T., and Schmidt, K. P., 1949. "Principles of Animal Ecology," W. B. Saunders Co., Philadelphia, Pa.

Allee, W. C., Frank, P., and Berman, M., 1946. "Homotypic and Heterotypic Conditioning in Relation to Survival and Growth of Certain Fishes," *Physiol. Zool.*, Vol. 19, 243-258.

Allee, W. C. and Schmidt, K. P., 1951. "Ecological Animal Geography" (2nd Ed.), John Wiley & Sons, Inc., New York, N.Y.

American Society of Civil Engineers Committee on Hydrology (no date). "Hydrology Handbook," published by the Society.

Anderson, G. C., 1958. "Some Limnological Features of a Shallow Saline Meromictic Lake," *Limnol. and Oceanog.*, Vol. 3, 259-269.

Andrewartha, H. G. and Birch, L. C., 1954. "The Distribution and Abundance of Animals," Univ. of Chicago Press, Chicago, Ill.

Antevs, E., 1922, "The Recession of the Last Ice Sheet in New England," *Am. Geog. Soc. Research Ser.*, Vol. 11, 1-10.

A.P.H.A. (American Public Health Association), 1955. "Standard Methods for the Examination of Water and Sewage" (10th Ed.), Am. Public Health Assoc., New York, N.Y.

Atwood, W. W., 1940. "The Physiographic Provinces of North America," Ginn and Co., Boston.

Bailey, R. M., Winn, H. E., and Smith, C. L., 1954. "Fishes from the Escambia River, Alabama and Florida, with Ecologic and Taxonomic Notes," *Proc. Acad. Nat. Sci. Philadelphia*, Vol. 106, 109-164.

Bamforth, S. S., 1958. "Ecological Studies on the Planktonic Protozoa of a Small Artificial Pond," *Limnol. and Oceanog.*, Vol. 3, 398-412.

Barlow, J. P., 1955. "Physical and Biological Processes Determining the Distribution of Zooplankton in a Tidal Estuary," *Biol. Bull.*, No. 109, 211-225.

Bassindale, R., 1938. "The Intertidal Fauna of the Mersey Estuary," *J. Marine Biol. Assoc. United Kingdom*, Vol. 23, 83-98.

Bassindale, R., 1942. "The Distribution of Amphipods in the Severn Estuary and Bristol Channel," *J. Animal Ecol.*, Vol. 2, 131-144.

Bassindale, R., 1943. "Studies on the Biology of the Bristol Channel. XI. The Physical Environment and the Intertidal Fauna of the Southern Shores of the Bristol Channel and the Severn Estuary," *J. Ecol.*, Vol. 31, 1-29.

Bassindale, R., 1943. "A Comparison of the Varying Salinity Conditions of the Tees and Severn Estuaries." *J. Animal Ecol.*, Vol. 12, 1-10.

Beadle, L. C., 1943. "Osmotic Regulation and the Faunas of Inland Waters," *Biol. Revs.*, Vol. 18, 172-183.

Berner, L., 1950. "The Mayflies of Florida," *Univ. Florida Studies, Biol. Sci. Ser.*, No. 4.

Berner, L., 1954. "The Occurrence of a Mayfly Nymph in Brackish Water," *Ecology*, Vol. 35, 98.

Berner, L. M., 1951. "Limnology of the Lower Missouri River." *Ecology*, Vol. 32, 1-12.

Bick, G. H., Hornuff, L. E., and Lambremont, E. N., 1953. "An Ecological Reconnaissance of a Naturally Acid Stream in Southern Louisiana," *J. Tenn. Acad. Sci.*, Vol. 28, 221-231.

Birge, E. A. and Juday, C., 1914. "A Limnological Study of the Finger Lakes of New York," *Bull. U. S. Bur. Fisheries*, No. 32, 525-609.

Birge, E. A. and Juday, C., 1921. "Further Limnological Observations on the Finger Lakes of New York," *Bull. U. S. Bur. Fisheries*, No. 37, 211-252.

Birge, E. A. and Juday, C., 1922. "The Inland Lakes of Wisconsin. The Plankton: 1. Its Quantity and Chemical Composition," *Bull. Wisconsin Geol. Nat. Hist. Survey*, No. 64, 1-222.

Black, A. P. and Brown, E., 1951. "Chemical Character of Florida's Water," *Florida Water Survey and Research Paper*, No. 6, Florida State Board Conserv., pp. 1-119.

Black, E. C., Fry, F. E. J., and Black, V. S., 1954. "The Influence of Carbon Dioxide on the Utilization of Oxygen by Some Fresh-water Fish," *Canadian J. Zool.*, Vol. 32, 408-420.

Blum, H. F., 1955. "Time's Arrow and Evolution," Princeton University Press, Princeton, N.J.

Blum, J. L., 1956. "The Ecology of River Algae," *Botan. Rev.*, Vol 22, 291-341.

Blum, J. L., 1957. "An Ecological Study of the Algae of the Saline River, Michigan," *Hydrobiologia*, Vol. 9, 361-408.

Borecky, G. W., 1956. "Population Density of the Limnetic Cladocera of Pymatuning Reservoir," *Ecology*, Vol. 37, 719-727.

Bourn, W. S. and Cottam, C., 1950. "Some Biological Effects of Ditching Tidewater Marshes," *U. S. Fish and Wildlife Serv. Research Rept.*, No. 19, 1-30.

Boyce, S. G., 1954. "The Salt Spray Community," *Ecol. Monographs*, Vol. 24, 29-67.

Boycott, A. E., 1936. "The Habits of Fresh Water Mollusca in Britain," *J. Animal Ecol.*, Vol. 5, 116-186.

Broekhuysen, G. J., 1935. "The Extremes in the Percentage of Dissolved Oxygen to Which the Fauna of a *Zostera* Field in the Tidal Zone at Nieuwdiep Can Be Exposed," *Arch. n'eerl. zool.*, Vol. 1, 339-346.

Brown, C. J. D., 1933. "A Limnological Study of Certain Freshwater Polyzoa with Special Reference to their Statoblasts," *Trans. Am. Microscop. Soc.*, Vol. 52, 271-316.

Brown, F. A., Jr., Fingerman, M., Sandeen, M. I., and Webb, H. M., 1953. "Persistent Diurnal and Tidal Rhythms of Color Change in the Fiddler Crab, *Uca pugnax*," *J. Exptl. Zool.*, Vol. 123, 29-60.

Brunson, R. B., 1950. "An Introduction to the Taxonomy of the Gastrotricha with a Study of Eighteen Species from Michigan," *Trans. Am. Microscop. Soc.*, Vol. 69, 325-352.

Bryson, R. A. and Stearns, C. R., 1959. "A Mechanism for the Mixing of the Waters of Lake Huron and South Bay, Manitoulin Island," *Limnol. and Oceanog.*, Vol. 4, 246-251.

Butcher, R. W., 1933. "Studies on the Ecology of Rivers. 1. On the Distribution of Macrophytic Vegetation in the Rivers of Britain," *J. Ecol.*, Vol. 21, 58-91.

Carriker, M. R., 1950. "A Preliminary List of the Literature on the Ecology of the Estuaries, with Emphasis on the Middle Atlantic Coast of the United States" (mimeographed), Rutgers University, New Brunswick, N.J.

Carriker, M. R., 1951. "Ecological Observations on the Distribution of Oyster Larvae in New Jersey Estuaries," *Ecol. Monographs*, Vol. 21, 19-38.

Chapman, V. J., 1940. "Studies in Salt-marsh Ecology, Sections VI and VII. Comparison with Marshes on the East Coast of North America," *J. Ecol.*, Vol. 28, 118-152.

Chu, S. P., 1943. "The Influence of the Mineral Composition of the Medium on the Growth of Planktonic Algae," *J. Ecol.*, Vol. 31, 109-148.

Chu, S. P., 1946. "The Utilization of Organic Phosphorus by Phytoplankton," *J. Marine Biol. Assoc. United Kingdom*, Vol. 26, 285-295.

Churchill, M. A., 1958. "Effects of Impoundments on Oxygen Resources," *in* "Oxygen Relationships in Streams," U. S. Public Health Service, pp. 107-129.

Clarke, F. W., 1924. "The Composition of River and Lake Waters of The United States," *U. S. Geol. Survey Profess. Papers*, No. 135.

Clarke, G. L., 1939. "The Utilization of Solar Energy by Aquatic Organisms," *in* Moulton (Ed.), "Problems in Lake Biology," *Publ. Am. Assoc. Advance. Sci.*, No. 10, pp. 27-38.

Clarke, G. L., 1954. "Elements of Ecology," John Wiley & Sons, Inc., New York, N.Y.

Clarke, G. L. and James, H. R., 1939. "Laboratory Analysis of the Selective Absorption of Light by Sea Water," *J. Opt. Soc. Am.*, Vol, 29, 43-55.

Coe, W. R., 1932. "Season of Attachment and Rate of Growth of Sedentary Marine Organisms at the Pier of the Scripps Institution of Oceanography, La Jolla, California," *Bull. Scripps. Inst. Oceanog., Tech Ser.*, No. 3, 37-86.

Coffin, C. C., Hayes, F. R., Jodrey, L. H., and Whiteway, S. G., 1949. "Exchange of Materials in a Lake as Studied by the Addition of Radioactive Phosphorus," *Canadian J. Research*, D., Vol. 27, 207-222.

Coker, R. E., 1954. "Streams, Lakes, Ponds," Univ. of North Carolina Press, Chapel Hill, N.C.

Collier, A. and Hedgpeth, J. W., 1950. "An Introduction to the Hydrography of Tidal Waters of Texas," *Publs. Inst. Marine Sci. Univ. Texas*, Vol. 1, 123-194.

Cooke, C. W., 1939. "Scenery of Florida," *Bull. Florida Geol. Survey*, No. 17.

Cooper, L. H. N. and Milne, A., 1938. "The Ecology of the Tamar Estuary. II. Underwater Illumination," *J. Marine Biol. Assoc. United Kingdom*, Vol. 22, 509-527.

Crombie, A. C., 1947. "Interspecific Competition," *J. Animal Ecol.*, Vol. 16, 44-73.

Curl, H., Jr., 1957. "A Source of Phosphorus for the Western Basin of Lake Erie," *Limnol. and Oceanog.*, Vol. 2, 315-320.

Davis, C. A., 1910. "Some Evidence of Recent Subsidence on the New England Coast," *Science*, N. S., Vol. 32, 63.

Davis, C. C., 1954. "A Preliminary Study of the Plankton of the Cleveland Harbor Area, Ohio. II. The Distribution and Quantity of the Phytoplankton," *Ecol. Monographs*, Vol. 24, 321-347.

Davis, C. C., 1955. "The Marine and Freshwater Plankton," Michigan State Univ. Press, East Lansing, Mich.

Davis, J. H., 1940. "The Ecology and Geologic Role of Mangroves in Florida," *Papers Tortugas Lab.*, Vol. 32, 302-412.

Dawson, C. E., 1955. "A Contribution to the Hydrography of Apalachicola Bay, Florida," *Publs. Inst. Marine Sci. Univ. Texas*, Vol. 4, 15-35.

Day, J. H., 1951. "The Ecology of South African Estuaries. Part 1. A Review of Estuarine Conditions in General," *Trans. Royal Soc. S. Africa*, Vol. 33, 53-91.

Deevey, E. S., Jr., 1940. "Limnological Studies in Connecticut. V. A Contribution to Regional Limnology," *Am. J. Sci.*, Vol. 238, 717-741.

Deevey, E. S., Jr., 1942. "Studies on Connecticut Lake Sediments. III. The Biostratonomy of Linsley Pond," *Am. J. Sci.*, Vol. 240, 313-324.

Deevey, E. S., Jr., 1950. "The Probability of Death," *Sci. American*, Vol. 182, 58-60.

Deevey, E. S., Jr., Gross, M. S., Hutchinson, G. E., and Kraybill, H. L., 1954. "The Natural C^{14} Contents of Materials from Hard-water Lakes," *Proc. Natl. Acad. Sci. Washington*, Vol. 40, 285-288.

Dendy, J. S., 1945. "Fish Distribution, Norris Reservoir, Tennessee, 1943. II. Depth Distribution of Fish in Relation to Environmental Factors, Norris Reservoir," *J. Tenn. Acad. Sci.*, Vol. 20, 114-135.

Dexter, R. W., 1947. "The Marine Communities of a Tidal Inlet at Cape Ann, Massachusetts: A Study in Bio-ecology," *Ecol. Monographs*, Vol. 17, 261-294.

Dice, L. R., 1952. "Natural Communities," Univ. Michigan Press, Ann Arbor, Mich.

Dineen, C. F., 1953. "An Ecological Study of a Minnesota Pond," *Am. Midland Naturalist*, Vol. 50, 349-376.

Domogalla, B. P., Juday, C., and Peterson, W. H., 1925. "The Forms of Nitrogen Found in Certain Lake Waters," *J. Biol. Chem.*, Vol. 63, 269-285.

Dorsey, N. E., 1940. "Properties of Ordinary Water-substance in All Its Phases: Water-Vapor, Water, and All the Ices," *Am. Chem. Soc. Monograph Ser.*, No. 81, Reinhold Publishing Corp., New York, N.Y.

Doty, M. S. and Newhouse, J., 1954. "The Distribution of Marine Algae in Estuarine Waters," *Am. J. Botany*, Vol. 41, 508-515.

Dugdale, R., Dugdale, V., Neess, J., and Goering, J., 1959. "Nitrogen Fixation in Lakes," *Science*, Vol. 130, 859-860.

Doudoroff, P. and Warren, C. E., 1957. "Biological Indices of Water Pollution, with Special Reference to Fish Populations," *in* "Biological Problems in Water Pollution," U. S. Public Health Service, pp. 144-163.

Eddy, S., 1925. "Fresh Water Algal Succession," *Trans. Am. Microsp. Soc.*, Vol. 44, 138-147.

Eddy, S., 1927. "The Plankton of Lake Michigan," *Illinois Nat. Hist. Survey Bull.*, No. 17, 203-232.

Eddy, S., 1934. "A Study of Fresh-water Plankton Communities," *Illinois Biol. Monographs*, Vol. 12, 1-93.

Edgren, R. A., Edgren, M. K., and Tiffany, L. H., 1953. "Some North American Turtles and Their Epizoophytic Algae," *Ecology*, Vol. 34, 733-740.

Edmondson, W. T., 1944. "Ecological Studies of Sessile Rotatoria. Part I. Factors Affecting Distribution," *Ecol. Monographs*, Vol. 14, 32-66.

Edmondson, W. T., 1956. "The Relation of Photosynthesis by Phytoplankton to Light in Lakes," *Ecology*, Vol. 37, 161-174.

✓Edmondson, W. T., 1957. "Trophic Relations of the Zooplankton," *Trans. Am. Microscop. Soc.*, Vol. 76, 225-245.

Edmondson, W. T. (Ed.), 1959. "Fresh-Water Biology," John Wiley & Sons, Inc., New York, N.Y.

Eggleton, F. E., 1931. "A Comparative Study of the Benthic Fauna of Four Northern Michigan Lakes," *Papers Michigan Acad. Sci.*, Vol. 20, 609-644.

Eggleton, F. E., 1956. "Limnology of a Meromictic, Interglacial, Plunge-basin Lake," *Trans. Am. Microscop. Soc.*, Vol. 75, 334-378.

Einsele, W., 1938. "Über Chemische und Kolloid-chemische Vorgänge in Eisenphosphat-systemen unter Limnochemischen und Limnogeologischen Gesichtspunkten," *Arch. Hydrobiol.*, Vol. 33, 361-387.

Ekman, S., 1953. "Zoogeography of the Sea," Sidgwick and Jackson, London.

Ellis, M. M., 1936. "Erosion Silt as a Factor in Aquatic Environments," *Ecology*, Vol. 17, 29-42.

Ellis, M. M., Westfall, B. A., and Ellis, M. D., 1946. "Determination of Water Quality," *U. S. Fish and Wildlife Serv. Research Rept.*, No. 9, 1-22.

✓Elton, C., 1946. "Competition and Structure of Ecological Communities," *J. Animal Ecol.*, Vol. 15, 54-68.

Emerson, R. and Green, L., 1938. "Effect of Hydrogen-ion Concentration on *Chlorella* Photosynthesis," *Plant Physiol.*, Vol. 13, 157-158.

Emery, K. O. and Stevenson, R. E., 1957. "Estuaries and Lagoons," *in* "Treatise on Marine Ecology and Paleoecology," *Geol. Soc. Am. Mem.*, No. 67, Vol. 1, pp. 673-750.

Eyster, C., 1958. "Bioassay of Water from a Concretion-forming Marl Lake," *Limnol. and Oceanog.*, Vol. 3, 455-458.

Fenneman, N. M., 1938. "Physiography of Eastern United States," McGraw-Hill Book Co., Inc., New York, N.Y.

Ferguson, G. E., Lingham, C. W., Love, S. K., and Vernon, R. O., 1947. "Springs of Florida," *Bull. Florida Geol. Survey*, No. 31, 1-197.

Flint, R. F., 1957. "Glacial and Pleistocene Geology," John Wiley and Sons, Inc., New York, N.Y.

Forbes, S. A., 1887. "The Lake as a Microcosm," reprint in: *Illinois Nat. Hist. Survey Bull.*, No. 15 (1925), 537-550.

Forrest, H., 1959. "Taxonomic Studies on the Hydras of North America. VII. Description of *Chlorohydra hadleyi*, New Species, with a Key to the North American Species of Hydras," *Am. Midland Naturalist*, Vol. 62, 440-448.

Frank, P. W., 1952. "A Laboratory Study of Intraspecies and Interspecies Competition in *Daphnia pulicaria* and *Simocephalus vetulus*," *Physiol. Zool.*, Vol. 25, 178-204.

Frank, P. W., 1957. "Coactions in Laboratory Populations of Two Species of *Daphnia*," *Ecology*, Vol. 38, 510-519.

Frey, D. G., 1950. "Carolina Bays in Relation to the North Carolina Coastal Plain," *J. Elisha Mitchell Sci. Soc.*, Vol. 66, 44-52.

Frey, D. G., 1955. "Langsee: A History of Meromixis," *Mem. ist. ital. idrobiol.*, Suppl. 8, 141-161.

Frey, D. G., 1955. "Distributional Ecology of the Cisco (*Coregonus artedii*) in Indiana," *Invest. Indiana Lakes and Streams*, Vol. 4, 177-208.

Fuller, M. L., 1912. "New Madrid Earthquake," *U. S. Geol. Survey Bull.*, No. 494.

/ Gaufin, A. R. and Tarzwell, C. M., 1955. "Environmental Changes in a Polluted Stream During Winter," *Am. Midland Naturalist*, Vol. 54, 78-88.

Gause, G. P., 1934. "The Struggle for Existence," Williams and Wilkins Co., Baltimore, Md.

√ Gerking, S. D., 1953. "Evidence for the Concepts of Home Range and Territory in Stream Fishes," *Ecology*, Vol. 34, 347-365.

Gilbert, G. K., 1890. "Lake Bonneville," *U. S. Geol. Survey Monographs*, Vol. 1.

Gorham, E., 1957. "Chemical Composition of Nova Scotian Waters," *Limnol. and Oceanog.*, Vol. 2, 12-21.

Gorham, E., 1957. "The Ionic Composition of Some Lowland Lake Waters from Cheshire, England," *Limnol. and Oceanog.*, Vol. 2, 22-27.

Gorham, E., 1958. "Observations on the Formation and Breakdown of the Oxidized Microzone at the Mud Surface in Lakes," *Limnol. and Oceanog.*, Vol. 3, 291-298.

Greenberg, B., 1947. "Some Relations Between Territory, Social Hierarchy and Leadership in the Green Sunfish (*Lepomis cyanellus*)," *Physiol. Zool.*, Vol. 20, 269-299.

Grinnell, J., 1917. "The Niche-Relationships of the California Thrasher," *Auk*, Vol. 34, 427-433.

√ Gunter, G., 1938. "Notes on Invasion of Fresh Water by Fishes of the Gulf of Mexico, with Special Reference to the Mississippi-Atchafalaya System," *Copeia*, Vol. 2, 69-72.

√ Gunter, G., 1942. "A List of the Fishes of the Mainland of North and Middle America Recorded from Both Fresh Water and Sea Water," *Am. Midland Naturalist*, Vol. 28, 305-326.

Gunter, G., 1945. "Studies on Marine Fishes of Texas," *Publs. Inst. Marine Sci. Univ. Texas*, Vol. 1, 1-190.

Gunter, G., 1947. "Paleoecological Import of Certain Relationships of Marine Animals to Salinity," *J. Paleontol.*, Vol. 21, 77-79.

Gunter, G., 1957. "Temperature," *in* "Treatise on Marine Ecology and Paleoecology," *Geol. Soc. Am. Mem.*, No. 67, Vol. 1, pp. 159-184.

Hansen, K., 1959. "The Terms Gyttja and Dy," *Hydrobiologia*, Vol. 13, 309-315.

Harrington, H. K. and Myers, F. J., 1926. "The Rotifer Fauna of Wisconsin. III. A Revision of the Genera *Lecane* and *Monostyla*," *Trans. Wisconsin Acad. Sci.*, Vol. 22, 315-423.

Harris, E., 1957. "Radiophosphorus Metabolism in Zooplankton and Microorganisms," *Canadian J. Zool.*, Vol. 35, 769-782.

Harshberger, J. W., 1909. "The Vegetation of the Salt Marshes and of the Salt and Fresh Water Ponds of Northern Coastal New Jersey," *Proc. Acad. Nat. Sci. Philadelphia*, Vol. 61, 373-400.

Hartley, P. H. T. and Spooner, G. M., 1938. "The Ecology of the Tamar Estuary. I. Introduction," *J. Marine Biol. Assoc. United Kingdom*, Vol. 27, 501-508.

Harvey, H. W., 1957. "The Chemistry and Fertility of Sea-Water," Cambridge Univ. Press, Cambridge.

Hasler, A. D., 1938, "Fish Biology and Limnology of Crater Lake," *J. Wildlife Mgmt.*, Vol. 2, 94-103.

Hasler, A. D., 1947. "Eutrophication of Lakes by Domestic Drainage," *Ecology*, Vol. 28, 383-395.

Hayes, F. R., 1955. "The Effect of Bacteria on the Exchange of Radiophosphorus at the Mud-Water Interface," *Verh. Int. Ver. Limnol.*, Vol. 12, 111-116.

Hayes, F. R., 1957. "On the Variation in Bottom Fauna and Fish Yield in Relation to Trophic Level and Lake Dimensions," *J. Fish. Research Board Canada*, Vol. 14, 1-32.

Hayes, F. R., McCarter, J. A., Cameron, M. L., and Livingstone, D. A., 1952. "On the Kinetics of Phosphorus Exchange in Lakes," *J. Ecol.*, Vol. 40, 202-216.

Hayes, F. R. and Phillips, J. E., 1958. "Lake Water and Sediment. IV. Radiophosphorus Equilibrium with Mud, Plants and Bacteria under Oxidized and Reduced Conditions," *Limnol. and Oceanog.*, Vol. 3, 459-475.

Hedgpeth, J. W., 1947. "The Laguna Madre of Texas," *Trans. 12th N. Am. Wildlife Conf.*, pp. 364-380.

Hedgpeth, J. W., 1951. "The Classification of Estuarine and Brackish Waters and the Hydrographic Climate," *Geol. Soc. Am., Rept. Comm. on Treatise on Marine Ecol. and Paleoecol.*, Vol. 1951, 49-56.

Hedgpeth, J. W., 1953. "An Introduction to the Zoogeography of the Northwestern Gulf of Mexico with Reference to the Invertebrate Fauna," *Publs. Inst. Marine Sci. Univ. Texas*, Vol. 3, 107-224.

Hedgpeth, J. W., 1957. "Classification of Marine Environments," *in* "Treatise on Marine Ecology and Paleoecology," *Geol. Soc. Am. Mem.*, No. 67, Vol. I, pp. 17-28.

Hedgpeth, J. W., 1957. "Concepts of Marine Ecology," *in* "Treatise on Marine Ecology and Paleoecology," *Geol. Soc. Am. Mem.*, No. 67, Vol. I, pp. 29-52.

Hela, I., Carpenter, C. A., and McNulty, J. K., 1957. "Hydrography of a Positive, Shallow, Tidal Bar-built Estuary (Report on the Hydrography of the Polluted Area of Biscayne Bay)," *Bull. Marine Sci. Gulf and Caribbean*, No. 7, 47-99.

Henderson, L. J., 1913. "The Fitness of the Environment," Macmillan Co., New York, N.Y.

Hendricks, S. B., 1955. "Necessary, Convenient, Commonplace," *in* "Yearbook of Agriculture," U.S.D.A., Washington, D.C., pp. 9-14.

Henson, E. B., 1959. "Evidence of Internal Wave Activity in Cayuga Lake, New York," *Limnol. and Oceanog.*, Vol. 4, 441-447.

Hesse, R., Allee, W. C., and Schmidt, K. P., 1951. "Ecological Animal Geography," John Wiley & Sons, Inc., New York, N.Y.

Hicks, S. D., 1959. "The Physical Oceanography of Narragansett Bay," *Limnol. and Oceanog.*, Vol. 4, 316-327.

Hjulstrom, F., 1935. "Studies of the Morphological Activity of Rivers as Illustrated by the River Fjris," *Bull. Univ. Upsala Geol. Inst.*, No. 25, 221-527.

Hooper, F. F., 1954. "Limnological Features of Weber Lake, Cheboygan County, Michigan," *Papers Mich. Acad. Sci.*, Vol. 39, 229-240.

Hopkins, D. M., 1949. "Thaw Lakes and Thaw Sinks in the Imruk Lake Area, Seward Peninsula, Alaska," *J. Geology*, Vol. 57, 119-131.

Hornuff, L. E., 1957. "A Survey of Four Oklahoma Streams with Reference to Production," *Oklahoma Fish. Research Lab. Rept.*, No. 63.

Hough, J. L., 1958. "Geology of the Great Lakes," Univ. of Illinois Press, Urbana, Ill.

Hrbáček, J., 1959. "Circulation of Water as a Main Factor Influencing the Development of Helmet in *Daphnia cucullata* Sars," *Hydrobiologia*, Vol. 13, 170-185.

Hulburt, E. M., 1956. "Distribution of Phosphorus in Great Pond, Massachusetts," *J. Marine Research*, Vol. 15, 181-192.

Hutchinson, G. E., 1938. "On the Relation Between the Oxygen Deficit and the Productivity and Typology of Lakes," *Intern. Rev. Hydrobiol.*, Vol. 36, 336-355.

Hutchinson, G. E., 1944. "Limnological Studies in Connecticut. VII. A Critical Examination of the Supposed Relationships Between Phytoplankton Periodicity and Chemical Changes in Lake Waters," *Ecology*, Vol. 25, 3-26.

Hutchinson, G. E., 1957. "A Treatise on Limnology. Vol. I. Geography, Physics, and Chemistry," John Wiley & Sons, Inc., New York, N.Y.

Hutchinson, G. E., Deevey, E. S., Jr., and Wollack, A., 1939. "The Oxidation-Reduction Potential of Lake Waters and Its Ecological Significance," *Proc. Natl. Acad. Sci. Washington*, Vol. 25, 87-90.

Hyman, L. H., 1929. "Taxonomic Studies on the Hydras of North America. I. General Remarks and Descriptions of *Hydra americana*," *Trans. Am. Microscop. Soc.*, Vol. 48, 242-255.

Hyman, L. H., 1930. "Taxonomic Studies on the Hydras of North America. II. The Characters of *Pelmatohydra oligactis* (Pallas)," *Trans. Am. Microscop. Soc.*, Vol. 49, 322-329.

Hyman, L. H., 1940-1959. "The Invertebrates," McGraw-Hill Book Co., Inc., New York, N.Y.

Irwin, W. H., 1945. "Methods of Precipitating Colloidal Soil Particles from Impounded Waters of Central Oklahoma," *Bull. Oklahoma A. & M. Coll.*, No. 42, 1-16.

Jackson, D. F. and Dence, W. A., 1958. "Primary Productivity in a Dichothermic Lake," *Am. Midland Naturalist*, Vol. 59, 511-517.

Johnson, D. W., 1919. "Shore Processes and Shoreline Development," John Wiley & Sons, Inc., New York, N.Y.

Johnson, D. W., 1925. "The New England-Acadian Shoreline," John Wiley & Sons, Inc., New York, N.Y.

Juday, C., 1922. "Quantitative Studies of the Bottom Fauna in the Deeper Waters of Lake Mendota," *Trans. Wisconsin Acad. Sci.*, Vol. 20, 461-493.

Juday, C., 1940. "The Annual Energy Budget of an Inland Lake," *Ecology*, Vol. 21, 438-450.

Juday, C., 1942. "The Summer Standing Crop of Plants and Animals in Four Wisconsin Lakes," *Trans. Wisconsin Acad. Sci.*, Vol. 29, 1-82.

Juday, C. and Birge, E. A., 1932. "Dissolved Oxygen and Oxygen Consumed in the Lake Waters of Northeastern Wisconsin," *Trans. Wisconsin Acad. Sci.*, Vol. 27, 415-486.

Juday, C., Birge, E. A., and Meloche, V. W., 1938. "Mineral Content of the Lake Waters of Northeastern Wisconsin," *Trans. Wisconsin Acad. Sci.*, Vol. 31, 223-276.

✓ Ketchum, B. H., 1947. "The Biochemical Relations Between Marine Organisms and Their Environment," *Ecol. Monographs*, Vol. 17, 309-315.

Ketchum, B. H., 1950. "The Exchanges of Fresh and Salt Waters in Tidal Estuaries," *Proc. Colloq. on Flushing of Estuaries*, U. S. Off. Naval Research.

✓ Ketchum, B. H., 1951. "The Flushing of Tidal Estuaries," *Sewage and Ind. Wastes*, Vol. 23, 198-209.

Ketchum, B. H., 1953. "Circulation in Estuaries," *Proc. 3rd Conf. Coastal Eng.*, Contribution No. 642 from Woods Hole Oceanographic Inst., pp. 65-76.

✓ Ketchum, B. H., 1954. "Relation Between Circulation and Planktonic Populations in Estuaries," *Ecology*, Vol. 35, 191-200.

Kofoid, C. A., 1903. "The Plankton of the Illinois River, 1894-1899, with Introductory Notes upon the Hydrography of the Illinois River and Its Basin. Part I. Quantitative Investigations and General Results," *Bull. Illinois State Lab. Nat. Hist.*, No. 6, 95-629.

Kofoid, C. A., 1908. "The Plankton of the Illinois River, 1894-1899, with Introductory Notes upon the Hydrography of the Illinois River and Its Basin. Part II. Constituent Organisms and their Seasonal Distribution." *Bull. Illinois State Lab. Nat. Hist.*, No. 8, 1-355.

Kolkwitz, R. and Marsson, M., 1909. "Ökologie der Fierischen Saprobien," *Intern. Rev. ges. Hydrobiol. Hydrog.*, Vol. 2, 126-152.

Krogh, A., 1939. "Osmotic Regulation in Aquatic Animals," Cambridge Univ. Press, Cambridge.

Kuenen, P. H., 1950. "Marine Geology," John Wiley & Sons, Inc., New York, N.Y.

Kuenen, P. H., 1955. "Realms of Water," John Wiley & Sons, Inc., New York, N.Y.

✓ Ladd, H. S., 1951. "Brackish-water and Marine Assemblages of the Texas Coast, with Special Reference to Mollusks," *Publs. Inst. Marine Sci. Univ. Texas*, Vol. 2, 125-164.

Langlois, T. H., 1954. "The Western End of Lake Erie and Its Ecology," J. W. Edwards Pub., Inc., Ann Arbor, Mich.

Langmuir, I., 1938. "Surface Motion of Water Induced by Wind," *Science*, Vol. 87, 119-123.

Layne, J. N., 1958. "Observation on Freshwater Dolphins in the Upper Amazon," *J. Mammalogy*, Vol. 39, 1-22.

Leet, L. D. and Judson, S., 1958. "Physical Geology" (2nd Ed.), Prentice-Hall, Inc., Englewood Cliffs, N.J.

Leopold, L. B. and Maddock, T., 1953. "The Hydraulic Geometry of Stream Channels and Some Physiographic Implications," *U. S. Geol. Survey Profess. Papers*, No. 252.

Leopold, L. B. and Miller, J. P., 1956. "Ephemeral Streams–Hydraulic Factors and Their Relation to the Drainage Net," *U. S. Geol. Survey Profess. Papers*, No. 282-A.

Liebig, J., 1840. "Chemistry in its Application to Agriculture and Physiology," Taylor and Walton, London.

Lindeman, R. L., 1942. "The Trophic-Dynamic Aspect of Ecology," *Ecology*, Vol. 23, 399-418.

Longwell, C. R., Knopf, A., and Flint, R. F., 1939. "A Textbook of Geology. Part I–Physical Geology" (2nd Ed.), John Wiley & Sons, Inc., New York, N.Y.

McAtee, W. L., 1941. "Wildlife of the Atlantic Coastal Marshes," *U. S. Fish and Wildlife Serv. Cir.*, No. 11.

McCombie, A. M., 1953. "Factors Influencing the Growth of Phytoplankton," *J. Fish. Research Bd. Canada*, Vol. 10, 253-282.

MacGinitie, G. E., 1935. "Ecological Aspects of a California Marine Estuary," *Am. Midland Naturalist*, Vol. 16, 629-765.

MacGinitie, G. E., 1939. "Littoral Marine Communities," *Am. Midland Naturalist*, Vol. 21, 28-55.

MacGinitie, G. E., 1939. "Some Effects of Fresh Water on the Fauna of a Marine Harbor," *Am. Midland Naturalist*, Vol. 21, 681-686.

MacGinitie, G. E. and MacGinitie, N., 1949. "Natural History of Marine Animals," McGraw-Hill Book Co., Inc., New York, N.Y.

Mann, K. H., 1958. "Annual Fluctuations in Sulphate and Bicarbonate Hardness in Ponds," *Limnol. and Oceanog.*, Vol. 3, 418-422.

Margalef, R., 1958. "Temporal Succession and Spatial Heterogeneity in Phytoplankton," *in* "Perspectives in Marine Biology," Univ. of California Press, Berkeley, Calif., pp. 323-349.

Marmer, H. A., 1926. "The Tide," Appleton, Inc., New York, N.Y.

Meinzer, O. E. (Ed.), 1942. "Physics of the Earth. IX. Hydrology," McGraw-Hill Book Co., Inc., New York, N.Y.

Metcalf, Z. P., 1930. "Salinity and Size," *Science*, Vol. 72, 526-527.

Millard, N. A. H. and Harrison, A. D., 1954. "The Ecology of South African Estuaries. Part V: Richard's Bay," *Trans. Roy. Soc. S. Africa*, Vol. 34, Pt. 1, 157-174.

Miller, D. E., 1936. "A Limnological Study of *Pelmatohydra* with Special Reference to Their Quantitative and Seasonal Distribution," *Trans. Am. Microscop. Soc.*, Vol. 55, 126-193.

Miller, W. R. and Egler, F. E., 1950. "Vegetation of the Wequetequock-Pawcatuck Tidal Marshes, Connecticut," *Ecol. Monographs*, Vol. 20, 141-172.

Milne, A., 1938. "The Ecology of the Tamar Estuary. III. Salinity and Temperature Conditions in the Lower Estuary," *J. Marine Biol. Assoc. United Kingdom*, N.S., Vol. 22, 529-542.

Milne, A., 1940. "The Ecology of the Tamar Estuary. IV. The Distribution of the Fauna and Flora on Buoys," *J. Marine Biol. Assoc. United Kingdom*, N.S., Vol. 24, 69-87.

Miner, R. W., 1950. "Fieldbook of Seashore Life," G. P. Putnam's Sons, New York, N.Y.

Möbius, K., 1883. "The Oyster and Oyster Culture," Transl. in: *Rept. U. S. Fish Comm.*, No. 1880, 683-751.

Moore, H. B., 1958. "Marine Ecology," John Wiley & Sons, Inc., New York, N.Y.

Moore, W. G., 1950. "Limnological Studies of Louisiana Lakes. I. Lake Providence," *Ecology*, Vol. 31, 86-99.

Mortimer, C. H., 1941. "The Exchange of Dissolved Substances Between Mud and Water in Lakes," *J. Ecol.*, Vol. 29, 280-329.

Mortimer, C. H., 1942. "The Exchange of Dissolved Substances Between Mud and Water in Lakes," *J. Ecol.*, Vol. 30, 147-201.

Mortimer, C. H., 1952. "Water Movements in Lakes During Summer Stratification; Evidence from the Distribution of Temperature in Windermere," *Phil. Trans. Roy Soc. London*, Ser. B, Vol. 236, 355-404.

Mortimer, C. H., 1953. "The Resonant Response of Stratified Lakes to Wind," *Schweiz. Z. Hydrol.*, Vol. 15, 94-151.

Mortimer, C. H., 1954. "Models of the Flow Pattern in Lakes," *Weather*, Vol. 9, 177-184.

Mortimer, C. H., 1956. "An Explorer of Lakes," *in* Sellery, G. C., "E. A. Birge," Univ. of Wisconsin Press, Madison, Wis., pp. 165-211.

Mortimer, C. H., 1959. Rev. of: Hutchinson, G. E., "A Treatise on Limnology," *Limnol. and Oceanog.*, Vol. 4, 108-113.

Moyle, J. B., 1949. "Some Indices of Lake Productivity," *Trans. Am. Fish. Soc.*, Vol. 76 (1946), 322-334.

Myers, G. S., 1949. "Usage of Anadromous, Catadromous and Allied Terms for Migratory Fishes," *Copeia*, Vol. 1949, 89-97.

Nash, C. B., 1947. "Environmental Characteristics of a River Estuary," *J. Marine Research*, Vol. 6, 147-174.

Needham, J., 1930. "On the Penetration of Marine Organisms into Fresh Water," *Biol. Zentr.*, Vol. 50, 504-509.

Needham, J. G. and Westfall, M. J., 1955. "A Manual of Dragonflies of North America," Univ. of California Press, Berkeley, Calif.

Neil, J. H., 1957. "Investigations and Problems in Ontario," *in* "Biological Problems in Water Pollution," U. S. Public Health Service, pp. 184-187.

Nelson, T., 1947. "Some Contributions from the Land in Determining Conditions of Life in the Sea," *Ecol. Monographs*, Vol. 17, 337-346.

Newcombe, C. E., Horne, W. A., and Shepherd, B. B., 1939. "Studies on the Physics and Chemistry of Estuarine Waters in Chesapeake Bay," *J. Marine Research*, Vol. 2, 87-116.

Nicol, E. A. T., 1935. "The Ecology of a Salt-marsh," *J. Marine Biol. Assoc. United Kingdom*, Vol. 20, 203-261.

Noel, M. S., 1954. "Animal Ecology of a New Mexico Springbrook," *Hydrobiologia*, Vol. 6, 120-135.

Norris, R. M., 1953. "Buried Oyster Reefs in Some Texas Bays," *J. Paleontol.*, Vol. 27, 569-576.

Odum, E. P., 1959. "Fundamentals of Ecology," W. B. Saunders Co., Philadelphia, Pa.

Odum, H. T., 1953. "Dissolved Phosphorus in Florida Waters," *Rept. Florida Geol. Survey*, Vol. 9, 1-40.

Odum, H. T., 1956. "Primary Production in Flowing Waters," *Limnol. and Oceanog.*, Vol. 1, 102-117.

Odum, H. T., 1957. "Trophic Structure and Productivity of Silver Springs, Florida," *Ecol. Monographs*, Vol. 27, 55-112.

Odum, H. T. and Caldwell, D. K., 1955. "Fish Respiration in the Natural Oxygen Gradient of an Anaerobic Spring in Florida," *Copeia*, Vol. 1955, 104-106.

Ohle, W., 1934. "Chemische und Physikalische Untersuchungen Norddeutscher Seen," *Arch. Hydrobiol.*, Vol. 26, 386-464.

Ortmann, A. E., 1902. "The Geographical Distribution of Freshwater Decapods and Its Bearing upon Ancient Geography," *Proc. Am. Phil. Soc.*, Vol. 41, 267-400.

Patrick, R., 1949. "A Proposed Biological Measure of Stream Conditions, Based on a Survey of the Conestoga Basin, Lancaster County, Pennsylvania," *Proc. Acad. Nat. Sci. Philadelphia*, Vol. 101, 277-341.

Pauling, L. C., 1960. "Nature of the Chemical Bond and the Structure of Molecules and Crystals: an Introduction to Modern Structural Chemistry," Cornell Univ. Press, Ithaca, N.Y.

Pearse, A. S., 1950. "The Emigrations of Animals from the Sea," Sherwood Press, Dryden, N.Y.

Pearse, A. S. and Gunter, G., 1957. "Salinity," *in* "Treatise on Marine Ecology and Paleoecology," *Geol. Soc. Am. Mem.*, No. 67, Vol. I, pp. 129-158.

Penfound, W. T., 1956. "Primary Production of Vascular Plants," *Limnol. and Oceanog.*, Vol. 1, 92-101.

Penfound, W. T. and Hathaway, E. S., 1938. "Plant Communities in the Marshlands of Southeastern Louisiana," *Ecol. Monographs*, Vol. 8, 1-56.

Pennak, R. W., 1944. "Diurnal Movements of Zooplankton Organisms," *Ecology*, Vol. 25, 387-403.

Pennak, R. W., 1949. "Annual Limnological Cycles in Some Colorado Lakes," *Ecol. Monographs*, Vol. 19, 233-267.

Pennak, R. W., 1953. "Freshwater Invertebrates of the United States," Ronald Press Co., New York, N.Y.

Pennak, R. W., 1955. "Comparative Limnology of Eight Colorado Mountain Lakes," *Univ. Colorado Ser. in Biol.*, No. 2, 1-75.

Pennak, R. W., 1955. "Persistent Changes in the Dominant Species Composition of Limnetic Entomostracan Populations in a Colorado Mountain Lake," *Trans. Am. Microscop. Soc.*, Vol. 2, 116-118.

Pennak, R. W., 1957. "Species Composition of Limnetic Zooplankton Communities," *Limnol. and Oceanog.*, Vol. 2, 222-232.

Percival, E., 1929. "A Report on the Fauna of the Estuaries of the River Tamar and the River Lynker," *J. Marine Biol. Assoc. United Kingdom*, Vol. 16, 81-108.

Perraton, C., 1953. "Salt Marshes of the Hampshire-Sussex Border," *J. Ecol.*, Vol. 41, 240-247.

Phleger, F. B. and Walton, W. R., 1950. "Ecology of Marsh and Bay Foraminifera, Barnstable, Mass.," *Am. J. Sci.*, Vol. 248, 274-295.

Picken, L. R., 1937. "The Structure of Some Protozoan Communities," *J. Ecol.*, Vol. 25, 368-384.

Pierce, E. L., 1947. "An Annual Cycle of the Plankton and Chemistry of Four Aquatic Habitats in Northern Florida," *Univ. Florida Studies, Biol. Sci. Ser.*, Vol. IV, No. 3.

Pierce, E. L., 1958. "The Chaetognatha of the Inshore Waters of North Carolina," *Limnol. and Oceanog.*, Vol. 3, 166-170.

Pittendrigh, Colin S., 1958. "Perspectives in the Study of Biological Clocks," *in* "Perspectives in Marine Biology," Univ. of California Press, Berkeley, Calif., pp. 239-268.

Potzger, J. E., 1956. "Pollen Profiles as Indicators in the History of Lake Filling and Bog Formation," *Ecology*, Vol. 37, 476-483.

Powell, A. W. B., 1939. "Notes on the Importance of Recent Animal Ecology as a Basis of Paleoecology," *Proc. 6th Pac. Sci. Cong.*, pp. 607-617.

Prescott, G. W., 1951. "Algae of the Western Great Lakes," *Cranbrook Institute of Science Bull.*, No. 31, Bloomfield Hills, Mich.

Prescott, G. W., 1954. "How to Know the Fresh Water Algae," Wm. C. Brown Co., Dubuque, Iowa.

Pritchard, D. W., 1951. "The Physical Hydrography of Estuaries and Some Applications to Biological Problems," *Trans. N. Am. Wildlife Conf.*, Vol. 16, 368-376.

Pritchard, D. W., 1952. "Salinity Distribution and Circulation in the Chesapeake Bay Estuarine System," *J. Marine Research*, Vol. 11, 106-123.

Pritchard, D. W., 1953. "Distribution of Oyster Larvae in Relation to Hydrographic Conditions," *Proc. Gulf and Carib. Fisheries Inst.*, 5th Session, Vol. 1952, 123-132.

Pritchard, D. W., 1956. "The Structure of a Coastal Plain Estuary," *J. Marine Research*, Vol. 15, 33-42.

Provasoli, L., 1958. "Nutrition and Ecology of Protozoa and Algae," *Ann. Rev. Microbiol.*, Vol. 12, 279-308.

Ragotzkie, R. A. and Bryson, R. A., 1953. "Correlation of Currents with the Distribution of Adult *Daphnia* in Lake Mendota," *J. Marine Research*, Vol. 12, 157-172.

Raney, E. C. and Massman, W. H., 1953. "The Fishes of the Pamunkey River, Virginia," *J. Wash. Acad. Sci.*, Vol. 43, 424-432.

Rawson, D. S., 1930. "The Bottom Fauna of Lake Simcoe and Its Role in the Ecology of the Lake," *Univ. of Toronto Studies, Publs. Ont. Fish. Research Lab.*, No. 40.

Rawson, D. S., 1939. "Some Physical and Chemical Factors in the Metabolism of Lakes," *Problems in Lake Biology* (A.A.A.S. Publ. No. 10), pp. 9-26.

Rawson, D. S., 1944. "The Calculation of Oxygen Saturation Values and Their Correction for Altitude," *Limnol. Soc. Am. Spec. Publ.*, No. 15.

Rawson, D. S., 1951. "The Total Mineral Content of Lake Waters," *Ecology*, Vol. 34, 669-672.

Rawson, D. S., 1956. "Algal Indicators of Lake Types," *Limnol. and Oceanog.*, Vol. 1, 18-25.

Reid, D. M., 1930. "Salinity Interchange Between Sea Water in Sand and Overflowing Fresh Water at Low Tide, I," *J. Marine Biol. Assoc. United Kingdom*, Vol. 16, 609-614.

Reid, D. M., 1932. "Salinity Interchange Between Salt Water in Sand and Overflowing Fresh Water at Low Tide, II," *J. Mar. Biol. Assoc. United Kingdom*, Vol. 18, 299-306.

Reid, G. K., 1950. "Food of the Black Crappie, *Pomoxis nigro-maculatus* (Lesueur), in Orange Lake, Florida," *Trans. Am. Fish. Soc.*, Vol. 79, 145-154.

Reid, G. K., 1952. "Some Considerations and Problems in the Ecology of Floating Islands," *Quart. J. Florida Acad. Sci.*, Vol. 15, 63-66.

Reid, G. K., 1954. "An Ecological Study of the Gulf of Mexico Fishes, in the Vicinity of Cedar Key, Florida," *Bull. Marine Sci. Gulf and Caribbean*, No. 4, 1-94.

Reid, G. K., 1955. "A Summer Study of the Biology and Ecology of East Bay, Texas," *Texas J. Sci.*, Vol. 7, 316-343.

Reid, G. K., 1957. "Biologic and Hydrographic Adjustment in a Disturbed Gulf Coast Estuary," *Limnol. and Oceanog.*, Vol. 2, 198-212.

Reimers, N., 1958. "Conditions of Existence, Growth, and Longevity of Brook

Trout in a Small, High-altitude Lake of the Eastern Sierra Nevada," *Calif. Fish and Game*, Vol. 44, 319-333.

Reimers, N., Maciolek, J. A., and Pister, E. P., 1955. "Limnological Study of the Lakes in Convict Creek Basin, Mono County, California," *U. S. Fish and Wildlife Serv. Fish. Bull.*, No. 103.

Richards, F. A. and Corwin, N., 1956. "Some Oceanographic Applications of Recent Determinations in the Solubility of Oxygen in Sea Water," *Limnol. and Oceanog.*, Vol. 1, 263-267.

Richman, S., 1958. "The Transformation of Energy by *Daphnia pulex*," *Ecol. Monographs*, Vol. 18, 273-291.

Ricker, W. E., 1934. "An Ecological Classification of Certain Ontario Streams," *Publs. Ontario Fish. Research Board, Biol. Ser.*, Vol. 37, 1-114.

Ricker, W. E., 1937. "Physical and Chemical Characteristics of Cultus Lake, British Columbia," *J. Biol. Board Canada*, Vol. 3, 363-402.

Ricker, W. E., 1952. "The Benthos of Cultus Lake," *J. Fish. Research Board Canada*, Vol. 9, 204-212.

Ries, H. and Watson, T. L., "Elements of Engineering Geology" (2nd Ed., Rev.), John Wiley & Sons, Inc., New York, N.Y.

Rigler, F. H., 1956. "A Tracer Study of the Phosphorus Cycle in Lake Water," *Ecology*, Vol. 37, 550-562.

Riley, G. A., 1937. "The Significance of the Mississippi River Drainage for Biological Conditions in the Northern Gulf of Mexico," *J. Marine Research*, Vol. 1, 60-74.

Riley, G. A., 1946. "Factors Controlling Phytoplankton Populations on Georges Bank," *J. Marine Research*, Vol. 6, 54-71.

Riley, G. A., 1957. "Phytoplankton of the North Central Sargasso Sea, 1950-52," *Limnol. and Oceanog.*, Vol. 2, 252-270.

Riley, G. A., Strommel, H., and Bumpus, D. F., 1949. "Quantitative Ecology of the Plankton of the Western Atlantic," *Bull. Bingham Oceanog. Coll.*, No. 12, 1-169.

Robeck, G. C., Henderson, C., and Palange, R. C., 1954. "Water Quality Studies on the Columbia River," U. S. Public Health Service.

Rodeheffer, T., 1941. "The Movements of Marked Fish in Douglas Lake, Michigan," *Papers Mich. Acad. Sci.*, Vol. 26, 265-280.

Rodhe, W., 1948. "Environmental Requirements of Fresh Water Plankton Algae," *Symbolae Botan. Upsaliensis*, No. 10.

Rogick, M. D., 1935. "Studies on Freshwater Bryozoa. II. The Bryozoa of Lake Erie," *Trans. Am. Microsp. Soc.*, Vol. 54, 245-263.

Rubey, W. W., 1951. "Geologic History of Sea Water. An Attempt to State the Problem," *Bull. Geol. Soc. Am.*, No. 62, 1111-1148.

Russell, I. C., 1895. "Lakes of North America," Ginn & Co., Boston, Mass.

Ruttner, F., 1930. "Das Plankton des Lunzer Untersees," *Intern. Rev. Hydrobiol.*, Vol. 23, 1-287.

Ruttner, F., 1953. "Fundamentals of Limnology" (transl. by D. G. Frey and F. E. J. Fry), Univ. of Toronto Press, Toronto, Canada.

Ryther, J. H., 1956. "The Measurement of Primary Production," *Limnol. and Oceanog.*, Vol. 1, 72-84.

Saur, J. F. T. and Anderson, E. R., 1956. "The Heat Budget of a Body of Water of Varying Volume," *Limnol. and Oceanog.*, Vol. 1, 247-251.

Schneller, M. V., 1955. "Oxygen Depletion in Salt Creek, Indiana," *Invest. Indiana Lakes and Streams*, Vol. 4, 163-175.

Segerstråle, S. G., 1957. "Baltic Sea," *in* "Treatise on Marine Ecology and Paleoecology," *Geol. Soc. Am. Mem.*, No. 67, Vol. I, pp. 751-800.

Shapiro, J., 1957. "Chemical and Biological Studies on the Yellow Organic Acids of Lake Water," *Limnol. and Oceanog.*, Vol. 2, 161-179.

Shelford, V. E., 1913. "Animal Communities in Temperate America," Univ. of Chicago Press, Chicago, Ill.

Shelford, V. E., Weese, A. O., Rice, L. A., Rasmussen, D. I., and Maclean, A., 1935. "Some Marine Biotic Communities of the Pacific Coast of North America: Part I, General Survey of the Communities," *Ecol. Monographs*, Vol. 5, 249-293.

Shepard, F. P., 1937. "Revised Classification of Marine Shorelines," *J. Geol.*, Vol. 45, 602-624.

Silliman, R. P. and Gutsell, J. S., 1958. "Experimental Exploitation of Fish Populations," *U. S. Fish and Wildlife Serv. Fish. Bull.*, No. 133, 215-252.

Simpson, G. G., 1944. "Tempo and Mode in Evolution," Columbia Univ. Press, New York, N.Y.

Slack, K. V., 1955. "A Study of the Factors Affecting Stream Productivity by the Comparative Method," *Invest. Indiana Lakes and Streams,* Vol. 4, 3-47.

Sloan, W. C., 1956. "The Distribution of Aquatic Insects in Two Florida Springs," *Ecology,* Vol. 37, 81-98.

Slobodkin, L. B., 1954. "Population Dynamics in *Daphnia obtusa* Kurz," *Ecol. Monographs,* Vol. 24, 69-88.

Smith, R. I., 1953. "The Distribution of the Polychaete, *Neanthes lighti,* in the Salinas River Estuary, California," *Biol. Bull.,* No. 105, 335-347.

Southwick, C. H., 1956. "The Logistic Theory of Population Growth–Past and Present Attitudes," *Turtox News,* Vol. 37, 131-133.

Spencer, R. S., 1956. "Studies in Australian Estuarine Hydrology. II. The Swan River," *Australian J. Marine and Freshwater Research,* Vol. 7, 193-253.

Spooner, G. M., 1947. "The Distribution of *Gammarus* Species in Estuaries," *J. Marine Biol. Assoc. United Kingdom,* Vol. 27, 1-52.

Spooner, G. M. and Moore, H. B., 1940. "The Ecology of the Tamar Estuary. IV. An Account of the Macrofauna of the Intertidal Muds," *J. Marine Biol. Assoc. United Kingdom,* Vol. 24, 283-330.

Steen, H., 1958. "Determinations of the Solubility of Oxygen in Pure Water," *Limnol. and Oceanog.*, Vol. 3, 423-426.

Stickney, A. P., 1959. "Ecology of the Sheepscot River Estuary," *U. S. Fish and Wildlife Serv. Special Sci. Rept.*, No. 309, 1-21.

Streeter, H. W. and Phelps, E. B., 1958. "A Study of the Pollution and Natural Purification of the Ohio River. III. Factors Concerned in the Phenomena of Oxidation and Reaeration," *U. S. Public Health Serv. Public Health Bull.*, No. 146.

Sublette, J. E., 1955. "The Physico-Chemical and Biological Features of Lake Texoma (Denison Reservoir), Oklahoma and Texas: A Preliminary Study," *Texas J. Sci.*, Vol. 7, 164-182.

Sublette, J. E., 1957. "The Ecology of the Macroscopic Bottom Fauna in Lake Texoma (Denison Reservoir), Oklahoma and Texas," *Am. Midland Naturalist,* Vol. 57, 371-402.

Sverdrup, H. U., Johnson, M. W., and Fleming, R. H., 1942. "The Oceans, Their Physics, Chemistry, and General Biology," Prentice-Hall, Inc., New York, N.Y.

Sylvester, R. O., 1958. "Water Quality Studies in the Columbia River Basin," *U. S. Fish and Wildlife Serv. Spec. Sci. Rept.*, No. 239, 1-134.

Tansley, A. G., 1935. "The Use and Abuse of Vegetational Concepts and Terms," *Ecology*, Vol. 16, 284-307.

Tansley, A. G., 1939. "The British Islands and Their Vegetation," Cambridge Univ. Press, Cambridge.

Taylor, W. R., 1939. "Marine Algae of the Northeastern Coast of North America." Univ. of Michigan Press, Ann Arbor, Mich.

Teal, J. M., 1957. "Community Metabolism in a Temperate Cold Spring," *Ecol. Monographs*, Vol. 27, 283-302.

Thompson, D. H. and Hunt, F. D., 1930. "The Fishes of Champaign County: A Study of the Distribution and Abundance of Fishes in Small Streams," *Illinois Nat. Hist. Survey Bull.*, No. 19, 1-101.

Thorson, G., 1957. "Bottom Communities (Sublittoral or Shallow Shelf)," *in* "Treatise on Marine Ecology and Paleoecology," *Geol. Soc. Am. Mem.*, No. 67, Vol. 1, pp. 461-534.

Thwaites, F. T., 1956. "Outline of Glacial Geology," published by the author, Madison, Wis.

Tiffany, L. H., 1951. "The Ecology of Freshwater Algae," *in* Smith, G. M., "Manual of Phycology," Chronica Botanica Co., Waltham, Mass.

Tiffany, L. H. and Britton, M. E., 1952. "The Algae of Illinois," Univ. of Chicago Press, Chicago, Ill.

Tonolli, V., 1949. "Ripartizione Spaziale e Migrazioni Verticali dello Zooplancton—Ricerche e Considerazioni," *Mem. ist. ital. idrobiol.*, Vol. 5, 211-228.

Tressler, W. L., Tiffany, L. H., and Spencer, W. P., 1940. "Limnological Studies of Buckeye Lake, Ohio," *Ohio J. Sci.*, Vol. 40, 261-290.

Tucker, A., 1957. "The Relation of Phytoplankton Periodicity to the Nature of the Physico-chemical Environment with Special Reference to Phosphorus," *Am. Midland Naturalist*, Vol. 57, 300-370.

Twenhofel, W. H., 1939. "Principles of Sedimentation," McGraw-Hill Book Co., Inc., New York, N.Y.

Twitty, V. C., 1959. "Migration and Speciation in Newts," *Science*, Vol. 130, 1735-1743.

Vallentyne, J. R., 1957. "Principles of Modern Limnology," *Am. Scientist*, Vol. 45, 218-244.

Vallentyne, J. R., 1957a. "The Molecular Nature of Organic Matter in Lakes and Oceans with Lesser Reference to Sewage and Terrestrial Soils," *J. Fish. Research Board Canada*, Vol. 14, 33-82.

Verduin, J., 1954. "Phytoplankton and Turbidity in Western Lake Erie," *Ecology*, Vol. 35, 550-561.

Verduin, J., 1956. "Energy Fixation and Utilization by Natural Communities in Western Lake Erie," *Ecology*, Vol. 37, 40-50.

Verduin, J., 1956. "Primary Production in Lakes," *Limnol. and Oceanog.*, Vol. 1, 85-91.

Verduin, J., Whitmer, E. E., and Cowell, B. C., 1959. "Maximal Photosynthetic Rates in Nature," *Science*, Vol. 130, 268-269.

Walford, L. A., 1958. "Living Resources of the Sea," Ronald Press Co., New York, N.Y.

Welch, P. S., 1948. "Limnological Methods," Blakiston Co., New York, N.Y.

Welch, P. S., 1952. "Limnology," McGraw-Hill Book Co., Inc., New York, N.Y.

Wesenberg-Lund, C., 1905. "A Comparative Study of the Lakes of Scotland and Denmark," *Proc. Roy. Soc. Edinburgh*, Vol. 25, 401-448.
Whipple, G. C., 1927. "The Microscopy of Drinking Water" (4th Ed.), John Wiley & Sons, Inc., New York, N.Y.
Whitford, L. A., 1956. "Communities of Algae in Springs and Spring Streams of Florida," *Ecology*, Vol. 37, 433-442.
Whittaker, J. R. and Vallentyne, J. R., 1957. "On the Occurrence of Free Sugars in Lake Sediment Extracts," *Limnol. and Oceanog.*, Vol. 2, 98-110.
Wilson, J. N., 1957. "Effects of Turbidity and Silt on Aquatic Life," *in* "Biological Problems in Water Pollution," U. S. Public Health Serv., pp. 235-239.
Wisler, C. O. and Brater, E. F., 1949. "Hydrology," John Wiley & Sons, Inc., New York, N.Y.
Wistendahl, W. A., 1958. "The Flood Plain of the Raritan River, New Jersey," *Ecol. Monographs.*, Vol. 28, 129-153.
Wolfe, P. E., 1953. "Periglacial Frost-thaw Basins in New Jersey," *J. Geol.*, Vol. 61, 133-141.
Wolle, F. A., 1887. "Freshwater Algae of the United States," Bethlehem, Pa.
Wolman, M. G. and Leopold, L. B., 1957. "River Flood Plains: Some Observations on Their Formation," *U. S. Geol. Survey Profess. Paper*, No. 282-C.
Woltereck, R., 1932. "Races, Associations and Stratification of Pelagic Daphnids in Some Lakes of Wisconsin and Other Regions of the United States and Canada," *Trans. Wisconsin Acad. Sci.*, Vol. 27, 487-522.
Wood, K. G., 1953. "The Bottom Fauna of Louisa and Redrock Lakes, Algonquin Park, Ontario," *Trans. Am. Fish. Soc.*, Vol. 82 (1952), 203-212.
Wood, R. D., 1959. "A Naturally Occurring Visible Thermocline," *Ecology*, Vol. 40, 153-154.
Woods, W. J., 1960. "An Ecological Study of Stony Brook, New Jersey" (unpublished dissertation), Rutgers University, New Brunswick, N.J.
Woods Hole Oceanographic Institute, 1951. "Report on a Survey of the Hydrography of Great South Bay Made During the Summer of 1950 for the Town of Islip, N.Y." (mimeographed).
Wright, J. C., 1954. "The Hydrology of Atwood Lake, a Flood-control Reservoir," *Ecology*, Vol. 35, 305-316.
Yonge, C. M., 1949. "The Sea Shore," Wm. Collins Sons & Co. Ltd., London.
Yoshimura, S., 1931. "Contributions to the Knowledge of the Stratification of Iron and Manganese in Lake Waters of Japan," *Japanese J. Geol. and Geog.*, Vol. 9, 61-69.
Yoshimura, S. and Masuko, K., 1935. "Kato- and Dichothermy During the Autumnal Circulation Period in Small Ponds of Miyagi Prefecture, Japan," *Proc. Imp. Acad. Tokyo*, Vol. 11, 146-148.
Young, F. N. and Zimmerman, J. R., 1956. "Variations in the Temperature in Small Aquatic Situations," *Ecology*, Vol. 37, 609-611.
Zernitz, E. R., 1932. "Drainage Patterns and Their Significance." *J. Geol.*, Vol. 40, 498-521.
ZoBell, C. E., 1946. "Marine Microbiology," Chronica Botanica Co., Waltham, Mass.
ZoBell, C. E. and Feltham, C. B., 1942. "The Bacterial Flora of a Marine Mud Flat as an Ecological Factor," *Ecology*, Vol. 23, 69-78.

INDEX